The Confluence of Cryptography, Blockchain and Artificial Intelligence

Editors

Ankita, Nayancy and Rajat Verma

Assistant Professor, Department of Computer Science and Engineering

Pranveer Singh Institute of Technology

Kanpur, India

CRC Press

Taylor & Francis Group

Boca Raton London New York

CRC Press is an imprint of the

Taylor & Francis Group, an **informa** business

A SCIENCE PUBLISHERS BOOK

First edition published 2025
by CRC Press
2385 NW Executive Center Drive, Suite 320, Boca Raton FL 33431

and by CRC Press
4 Park Square, Milton Park, Abingdon, Oxon, OX14 4RN

CRC Press is an imprint of Taylor & Francis Group, LLC

Library of Congress Cataloging-in-Publication Data (applied for)

ISBN: 978-1-032-69317-0 (hbk)
ISBN: 978-1-032-71152-2 (pbk)
ISBN: 978-1-032-71151-5 (ebk)

DOI: 10.1201/9781032711515

Typeset in Times New Roman
by Prime Publishing Services

Preface

The landscape of technology is constantly evolving, and at the forefront of this evolution lies the intricate convergence of three powerful forces: Cryptography, Blockchain, and Artificial Intelligence (AI). These fields, each with their own rich history and unique potential, are now beginning to interweave, creating a tapestry of possibilities that promise to revolutionize numerous aspects of our world.

This edited book, *The Confluence of Cryptography, Blockchain and Artificial Intelligence: Advances*, delves into this exciting new frontier, exploring the ways in which these three domains are coming together to shape the future. It brings together the insights of leading researchers and practitioners who are actively exploring the potential of this confluence.

The book is divided into several sections, each focusing on a specific aspect of this convergence. The first section lays the groundwork by providing a comprehensive overview of Cryptography, Blockchain, and AI, individually. Subsequent sections delve deeper into the specific areas where these fields intersect, exploring themes such as:

- **Secure and Scalable AI:** How cryptography and blockchain can be leveraged to ensure the security and scalability of AI systems.
- **Decentralized AI:** Exploring the potential of blockchain technology to enable the development and deployment of decentralized AI systems.
- **Explainable AI:** How AI can be made more transparent and interpretable through the use of cryptographic techniques.
- **AI-powered Security:** Leveraging AI to enhance the security of cryptographic systems and blockchain networks.

The Confluence of Cryptography, Blockchain and Artificial Intelligence: Advances is a timely and essential resource for anyone interested in understanding the potential of these converging technologies. It provides a comprehensive overview of the current state of the art, while also exploring the exciting possibilities that lie ahead. We believe that this book will be a valuable resource for researchers, practitioners, and policymakers alike, as they work together to harness the power of this confluence to build a more secure, efficient, and intelligent future.

Contents

CHAPTER 1

The Introduction of Cryptography, Blockchain, and Artificial Intelligence in the Modern World of Computing

Abhishek Rawat, Rajat Verma* and *Raghuraj Singh Suryavanshi*

1. Introduction

Innovations in the ever changing field of computing are changing how we connect with information and do business [1]. Among these groundbreaking advancements, three remarkable concepts have emerged as true game-changers: Cryptography, Blockchain, and Artificial Intelligence. These state-of-the-art technologies have completely changed how choices are made, transactions are carried out, and data is secured. The science of data encryption and decryption has a very long history dating back thousands of years and is known as cryptography [2]. However, in the modern world of computing, cryptography has taken on a whole new level of importance. With cyber threats becoming increasingly sophisticated, the need to protect sensitive data has never been more critical [3]. A strong framework for digital communication security is offered by cryptography.

Blockchain has become a game-changing technology in the field of transparent and decentralized transactions. At first, Blockchain was touted as the technology behind cryptocurrencies, but it has quickly gained recognition for its ability to change

Department of Computer Science & Engineering, Pranveer Singh Institute of Technology, Kanpur, Uttar Pradesh, India.

* Coresponding author

a variety of industries [4]. Blockchain makes transactions safe and transparent by generating an immutable, decentralized ledger that eliminates the need for middlemen. Its impact extends beyond finance, reaching supply chain management, healthcare, and even governance. Artificial intelligence is the outcome of decades of research and progress in computer science (AI) [5]. Since artificial intelligence (AI) can replicate the intellect of humans and learn from massive volumes of data. Artificial Intelligence is revolutionizing how we work, communicate, and make decisions [6]. Examples include predictive analytics, personalized recommendations, driverless vehicles, and natural language processing.

Together, Cryptography, Blockchain, and Artificial Intelligence are reshaping the modern world of computing in ways we never thought possible [7]. They are breaking down barriers, enhancing security, and unlocking new possibilities for innovation. We must negotiate the ethical, legal, and sociological ramifications of these revolutionary technologies as we learn more about them to maximize their potential for the benefit of society [8]. This chapter explores the introduction of Cryptography, Blockchain, and Artificial Intelligence in the modern world of computing. It will explore their history, explore their uses, and discuss the difficulties and possibilities that lay ahead. Come along on this fascinating journey with us as we explore the potential and power of these cutting-edge technologies to influence computing in the future [9].

2. Cryptography—Safeguarding Information

Current information security systems must include cryptography as a fundamental component to secure sensitive data and prevent illegal access. With the rising prevalence of data breaches, the need for strong cryptographic controls has become more important than ever [10]. Through the use of cipher text, an unintelligible format, cryptography ensures that even if a third party obtains access to the information they are unable to decipher the data. This is made possible by several encryption methods, including Data Encryption Standard such as Advanced Encryption Standard, which modifies data so that it is computationally challenging to decipher without the right key [11].

Another essential component of encryption is data integrity. It guarantees that data is transmitted or stored with integrity and without modification. One kind of cryptographic method that is essential to data integrity is hash functions. These take any size of incoming data and output fixed-size hash results. By contrasting the value of the hash that was determined at the point of reception to the original hash value, data integrity may be verified [12]. Another important component of cryptography is authentication. It entails confirming the legitimacy of people or systems that are a part of a communication process. Asymmetric key cryptography is employed by digital signatures, which are frequently used for authentication. Using their private key, the sender generates a unique digital signature that anyone can validate with the public key. This guarantees that the document has been transmitted by the intended recipient and has not been tampered with [13].

2.1 The Origins of Cryptography

The study of safe communication, known as cryptography, has its roots in prehistoric times. Cryptography has been essential for safeguarding sensitive data throughout history [14]. The ancient Egyptians, for example, used hieroglyphic symbols to encrypt their messages, ensuring that only those with knowledge of the symbols could understand the content. Caesar cipher shifts every character in the English alphabet by a fixed number of points, increasing the difficulty with which opponents can interpret the message. Basic encryption was offered by the Caesar cipher, which had shortcomings and was susceptible to known plaintext and brute force attacks.

The German-created Enigma machine emerged as a crucial cryptography instrument during World War II. Enigma employed revolving rotor disks for encryption and decryption, rendering communications exceedingly challenging for attackers to decode. Polish cryptographers eventually cracked the machine's cipher, and British cryptographers took over the technology to devise a way to extract the daily key [15]. The deciphering of the Enigma code was essential to the Allies' victory in the conflict. Over time, cryptography has changed, and new encryption algorithms have been created to overcome the shortcomings of older ones. Today, cryptography is essential for secure communication, corporate cyber security, and various other applications. It permits authentication, preserves data integrity, and guarantees data confidentiality. Data is encoded and decoded using contemporary cryptographic techniques, like the AES, offering a high level of security.

2.2 Cryptographic Algorithms

Cryptography is the cornerstone of secure communication. There are two primary types of cryptography: symmetric and asymmetric.

- **Symmetric Cryptography:** A single key is used in symmetric cryptography for both encoding and decryption. Although this method is quicker and more effective, safely transferring the key between the persons involved in communication presents a challenge. Two instances of symmetric encryption techniques are the AES, and DES [16]. These algorithms jumble the data using mathematical operations such that only the right key can decode it.
- **Asymmetric Cryptography:** Encryption uses a public key, while decryption uses a private key. This approach ensures secure contact without necessitating a key exchange, despite being more labor-intensive. The most popular asymmetric encryption method, The RSA algorithm relies on the challenge of big prime numbers as a security measure [17].

Cryptographic hash functions are also essential for guaranteeing authentication and data integrity. Because hash functions produce an output with a fixed size (hash) from any input, attempting to analyze the original data is not computationally viable. The Secure Hash Algorithm (SHA) family and the message-based Digest Algorithm (MD5) are two popular hash algorithms [18]. Cryptographic algorithms form the backbone of secure communication, providing confidentiality, integrity, and authentication. Digital signatures, safe key exchange, data integrity checks, and secure file encryption are just a few of the uses for them [19]. Organizations can guarantee secure connections and safeguard sensitive data by utilizing these algorithms. The necessity to keep ahead of emerging threats and guarantee the security of digital systems drives the ongoing development of cryptography algorithms and procedures. Even as technology advances, cryptography is crucial for safeguarding confidential information and maintaining the authenticity of communication channels.

2.3 Applications of Cryptography

Cryptography has several uses in protecting data and facilitating safe communication in different fields. These are a few important uses:

- **Secure Communication:** Cryptography ensures confidentiality and integrity in email exchanges, messaging platforms, and online transactions. Data encryption limits access to only those who are allowed and have the right decryption key. For instance, to employ cryptographic methods to safeguard online communication, including HTTPS connections [20].
- **Digital Signatures:** Asymmetric cryptography is used by digital signatures to confirm the authenticity and reliability of digital documents and transactions [21]. A sender can demonstrate that a document came from them and hasn't been altered by signing it with their private key. This is widely used in areas such as digital contracts, electronic voting, and financial transactions.
- **Password Storage:** Cryptography is used to protect user credentials by securely storing passwords [22]. Techniques like salting (adding random data to passwords) and key stretching (iterating cryptographic functions to slow down password cracking) make it more challenging for attackers to retrieve passwords from compromised databases.
- **Secure Web Browsing:** A key component of internet browsing security is cryptography. Users are protected from man-in-the-middle and eavesdropping attacks by it. Technologies like SSL and TLS encrypt data exchanged between internet servers and clients using public key cryptography to create an encrypted link for communication [23].
- **Digital Currencies:** The security of virtual currencies like Bitcoin depends on cryptography. Transactions are secured by complex algorithms and cryptographic keys, which make it difficult to tamper with or falsify them [24]. Digital signatures are generated and verified using a public key, which guarantees the integrity and validity of transactions.
- **Data Integrity and Authentication:** Data integrity and authenticity are vitally dependent on cryptographic hash functions. Reverse-engineering the original data becomes computationally impractical due to the fixed-size hash values that hash functions create from incoming data [25]. You can use these hash codes to verify that the information being transmitted is correct and hasn't been tampered with.

A key component of information security, cryptography ensures the integrity of digital systems, permits secure communication and safeguards sensitive data. It has applications in a wide range of sectors, such as e-commerce, government, and healthcare. Organizations can preserve user privacy, protect their information, and uphold their customers' trust by utilizing cryptography approaches.

3. Blockchain—Empowering Trust and Transparency

Blockchain technology is an effective instrument for promoting transparency and trust across a range of industries. By leveraging cryptographic techniques and decentralized networks, blockchain enables secure and transparent transactions, data sharing, and information management [26]. Here are some key aspects of how blockchain empowers trust and transparency:

- **Transparency and Traceability:** SCM and the provenance of consumer items are made more transparent with Blockchain. It offers an immutable, decentralized ledger that logs each transaction and occurrence, producing an open, auditable trail [27]. Because of this transparency, customers can confirm the legitimacy and integrity of items, guaranteeing fair trade, ethical sourcing, and quality control. Additionally, Blockchain lowers the risk of fraud by increasing asset tracking precision throughout the supply chain.
- **Data Security and Privacy:** Blockchain uses cryptographic algorithms to protect data and maintain anonymity. Public-key cryptography makes sure the data can only be accessed by a verified person. Data breaches and unauthorized tampering are reduced by blockchain technology, which encrypts data and stores it across a distributed network [28]. Furthermore, blockchain gives consumers autonomy over their data, enabling them to share it however and with whomever they choose.
- **Trust and Decentralization:** Blockchain eliminates the need for intermediaries and centralized agencies, enabling trustless transactions. Since no one individual has authority over the entire network, manipulation, and fraud are unlikely to occur on the blockchain. Consensus methods make sure that multiple parties validate transactions, which improves security and trust [29,68–69].
- **Digital Signatures and Smart Contracts:** Smart contracts and digital signatures are made possible by blockchain technology, which increases transparency and trust. Digital signatures based on asymmetric cryptography provide a means of verifying the consistency and validity of electronic transactions and documents. Self-executing contracts, or smart contracts, are stored on the blockchain. They eliminate the demand for middlemen by automating and enforcing contract terms, which ensures transparency [30].
- **Empowering Individuals:** Blockchain allows for peer-to-peer transactions and gives users authority over their data, which empowers people. Since blockchain offers a platform users can rely on to secure their data and distribute it whenever they see fit, users might feel secure about the security of information [31].

Despite being in its infancy, blockchain technology is already being used in several industries, including healthcare and SCM. Its capacity to bring about stability, transparency, and trust has the power to overthrow current structures and create fresh chances for creativity and cooperation.

3.1 Types of Blockchain

Blockchain networks come in four varieties, each with unique properties and applications. The four types of blockchain that are most frequently cited are as follows:

- **Public Blockchain:** Anyone can join and take part in a permissionless network, also known as a public blockchain, without any restrictions. It allows anyone with internet access to explore the blockchain, conduct transactions, and maintain a copy of the ledger [32]. One instance of a public blockchain is Bitcoin. Because every transaction on a public blockchain is transparent and decentralized, everyone on the network can see them.
- **Private Blockchain:** Private blockchains, as opposed to public blockchains, are permissioned networks in which participation and access are limited to a particular member group [33]. Organizations frequently utilize private blockchains to preserve network control and guarantee confidentiality and privacy. In a private blockchain, users are usually verified as are well-known organizations and the network is not accessible to the general public.
- **Consortium Blockchain:** A blockchain consortium is a cross between a public and private blockchain. A consortium or collection of organizations that concur on the guidelines and regulations governing the blockchain network is in charge of it [34]. Blockchains operated by consortiums provide a compromise between control and decentralization. They are frequently employed in situations like supply chain management or consortia that are industry-specific, where numerous firms must cooperate and safely share data.
- **Hybrid Blockchain:** The characteristics of public and private blockchains are combined in hybrid blockchains. They permit private ownership and public involvement in specific network functions. Hybrid blockchains are frequently employed in situations where privacy and transparency must coexist [35]. For instance, in supply chain management, private information may be kept secret while public information may be available.

3.2 Consensus Mechanism

A consensus system enables a distributed system of users to agree on Bitcoin's current state and accept transactions. Blockchain ensures that all miners reach a consensus about the validity and order of transactions to maintain their integrity and security [36]. In blockchain networks, consensus procedures are essential for establishing transparency and confidence. Consensus methods come in different varieties, each with unique features and advantages. A few common techniques consist of:

- **Proof of Work (PoW):** Miners, who are the users of Bitcoin along with other cryptocurrencies, must solve difficult mathematical problems to verify transactions and mine new blocks. The initial miner to figure out the puzzle and add a block to the existing blockchain wins rewards [37]. Although PoW is widely recognized for its security, it is computationally and energy-intensive.

- **Proof of Stake (PoS):** The likelihood of validating a new block in proof-of-work (PoS) is based on the ownership or stake of the cryptocurrency held by the participants [38]. A higher stakeholder's probability of being selected to validate transactions is higher. Even though PoS is believed to be more environmentally friendly than PoW, if a few numbers of users hold most of the money, centralization concerns might surface.
- **Delegated Proof of Stake (DPoS):** DPoS a variation on PoS, enables users to choose a limited group of delegates to approve transactions and mine new blocks [39]. This approach seeks to decrease the number of parties participating in block validation to achieve faster transaction processing speeds and scalability.
- **Proof of Authority (PoA):** Since pre-approved validators have the power to approve transactions and add new blocks, they are crucial to PoA [40]. Validators are typically reputable organizations or consortia members who are well-known and trustworthy. PoA is appropriate for private or consortium blockchains and provides high transaction throughput.
- **Practical Byzantine Fault Tolerance (PBFT):** To decide on the sequence and validity of transactions, a predefined group of validators must agree. Enterprise blockchain applications frequently use PBFT because it offers quick transaction finality [41].

These are only a few instances of consensus mechanisms: several hybrid models and variants are either in use or being developed. The blockchain network's unique needs, including security, scalability, energy efficiency, and the intended degree of decentralization, determine the consensus mechanism to be used. Blockchain networks' dependability and integrity depend heavily on consensus processes.. They foster trust and transparency by empowering users to collectively determine the state of the blockchain to approve transactions without relying on a central authority. Consensus procedures will become increasingly important as blockchain technology develops and will influence how decentralized networks operate in the future.

3.3 Cryptocurrency and Smart Contracts:

In recent years, cryptocurrencies like Bitcoin and Ethereum have attracted a lot of attention. Powered by blockchain technology, these digital currencies offer unique features and use cases that contribute to trust and transparency in a variety of ways [42].

- **Bitcoin and Beyond:** The idea of a digital currency was first made possible by Bitcoin, allowing peer-to-peer transactions to take place without the use of middlemen like banks. The blockchain, a decentralized, open ledger with immutability and security, records transactions conducted with Bitcoin [43]. Since the release of Bitcoin, a large number of other cryptocurrencies have appeared, each with special characteristics and applications. For instance, whereas smart contract networks like Ethereum facilitate the execution of programmed contracts and decentralized apps (DApps), privacy-focused cryptocurrencies like Monero seek to improve anonymity.

- **Smart Contracts:** Smart contracts are self-executing programs that have predefined criteria that are written in code. These agreements automatically come into effect and enforce the agreed terms once the prerequisites are met. Financial transactions, voting procedures, and SCM are just tasks that smart contracts can automate and decentralize. Ethereum, one of the most famous blockchain platforms, made smart contracts popular and provides a stable environment in which they may be developed and implemented [44]. Because smart contracts increase efficiency and decrease the need for middlemen, they have the potential to optimize and streamline several industries. Smart contracts can be used for a variety of purposes and are not just for financial transactions. For example, they can facilitate the automatic execution of insurance claims, enable decentralized governance in organizations, and even power decentralized autonomous organizations (DAOs). The conditions and performance of smart contract agreements are verifiable and tamper-proof thanks to the blockchain's transparency and immutability.

Although smart contracts and cryptocurrencies have many advantages, it's vital to remember that they also have drawbacks and issues. Cryptocurrencies have been subject to volatility and regulatory scrutiny, and smart contracts require careful coding and auditing to ensure their security and accuracy. Additionally, scalability and energy consumption are ongoing concerns for Blockchain Networks. Cryptocurrencies and Smart Contracts, powered by Blockchain, have revolutionized the way we think about trust and transparency in various domains [45]. They enable peer-to-peer transactions, automate processes, and provide verifiable and tamper-proof records. Technology can change sectors, improve productivity, and give people new and exciting opportunities as it develops.

3.4 Application of Blockchain

Blockchain has found applications beyond currencies, revolutionizing various industries. Here are some additional details on how blockchain is transforming specific sectors:

- **Supply Chain Management:** Blockchain ensures transparency and authenticity by enabling end-to-end product tracking and verification [46]. It provides an unbreakable record of each step in the manufacturing chain. Because customers can confirm the origin and legitimacy of the things they buy, this increases trust and lowers fraud.
- **Healthcare:** Blockchain technology makes it possible to safely store medical records, facilitating communication between various healthcare professionals [47]. Patients can oversee their medical records, allowing medical professionals access as necessary. By decreasing fraud and automating the verification process, blockchain also simplifies insurance claims.
- **Voting Systems:** Voting systems can benefit from increased security, immutability, and transparency thanks to blockchain [48]. It enables a more reliable and auditable electoral process by offering a decentralized record of

votes. Voter fraud, manipulation, and conflicts can be addressed via blockchain-based voting systems, protecting democratic processes.

- **Product Authentication:** Blockchain innovation can confirm the authenticity of products and put an end to counterfeiting [49]. It enables safe and unalterable documentation of the product's source and its passage through the supply chain. This aids in guaranteeing the caliber and reliability of goods.
- **Traceability and Quality Control:** Blockchain enables visibility and tracking of products and commodities along the whole supply chain [50]. It makes traceability and quality control possible by enabling the construction of a strong audit trail. By connecting supply chain networks through a decentralized system, blockchain enables smooth communication between suppliers and manufacturers.
- **Fraud Prevention:** Supply chain fraud can be identified and stopped with the use of blockchain. It offers a transparent and safe space for information tracking and sharing [51]. Pharmaceutical companies, for example, are using blockchain to verify the authenticity of products, identify and trace potential frauds, and improve quality and trust.

These are just a few industries utilizing blockchain technology. Blockchain technology's efficiency, security, and transparency have the power to drastically alter how things are done now and open up new avenues for openness and confidence.

4. Artificial Intelligence—Unleashing Intelligent Computing

In the field of computer science, AI is used to make intelligent machines that are capable of tasks that normally need human intelligence. It entails creating models and algorithms that let computers solve complicated problems, acquire from and modify data, and make decisions. Supervised AI, computer vision, and robotics are just a few of the methods and strategies that are included in the umbrella term AI. With the help of these methods, machines can now observe, comprehend, reason, and interact with others just like humans. An essential part of artificial intelligence is machine learning, which trains computers on massive datasets to identify patterns and offer predictions or opinions without the need for explicit programming [52].

Artificial Intelligence has been applied across multiple industries and areas, transforming our daily lives and workplaces. It has revolutionized healthcare by making precise diagnoses, individualized treatment regimens, and effective management of medical records possible. In the finance industry, Algorithmic trading, risk assessment, and fraud detection are applications of artificial intelligence. Virtual assistants powered by AI are proliferating; they provide information retrieval, customer support, and personalized recommendations. Additionally, AI has made important advancements in several domains, such as autonomous cars, and computer vision. AI algorithms are utilized by self-driving cars to sense & navigate their environment [53]. Computer vision allows computers to perform tasks like object detection and facial recognition by interpreting and analyzing visual data. Robotics creates intelligent machines that can interact with the real world by fusing AI with physical systems. The creation and application of AI bring up significant issues and difficulties. Some of the aspects that need careful study are employment effects, privacy issues, and ethical concerns. As artificial intelligence (AI) develops, it is imperative to guarantee appropriate development and application to optimize its advantages while mitigating any hazards [54]. In conclusion, the rapidly developing field of AI goals to create intelligent machines that are carrying out tasks that have historically required human intelligence. Its applications are widely ranging, transforming our everyday lives, work environments, and relationships with technology. AI has enormous potential to spur innovation, increase productivity, and influence societal trends as it develops.

4.1 Key Concepts in AI

To understand AI deeply, it is important to grasp its core concepts. Some key elements in AI include:

- **Machine Learning:** The creation of models and methods that enable computers to learn from data and grow more efficiently over time is the aim of the machine learning (ML) branch of AI [55]. Algorithms for machine learning, or ML, are taught on past data to find links and patterns to provide forecasts, categorize data, aggregate data points, decrease dimensions, and create new content. This is accomplished through a procedure known as training, in which the algorithm optimizes its performance by modifying its parameters considering the supplied data. The problem to be addressed along with the sort of data used dictates the type of machine learning approach chosen. While recurrent neural networks, or RNNs, are commonly used in tasks involving natural language processing, convolutional neural networks, also known as CNNs, are often used in image identification tasks.

  ML is best suited for situations with large amounts of data, such as recordings from previous customer conversations, sensor logs, or ATM transactions [56]. More data makes it easier for machine learning algorithms to learn and produce reliable predictions or judgments. There are numerous uses for machine learning across numerous sectors. In the financial domain, it can be applied to credit rating. In the healthcare sector ML algorithms can help in illness diagnosis and treatment planning. Recommendation systems for tailored content suggestions on stream services and product suggestions in e-commerce platforms also use machine learning. The capacity of machine learning to continuously enhance performance over time is one of its main features. The learning process in ML is automated and improved based on the experiences of the machines throughout the process. The algorithms can adjust to new data and improve their predictions or judgments thanks to this repeated feedback loop. Machine learning is an essential component in AI systems development. With the use of ML, computers can improve their performance and gain the same understanding and capacity for data-driven decision-making as people.

- **Neural networks:** These networks are one type of artificial intelligence model that takes its cues from the structure and functions of the human brain. They consist of networked nodes in sometimes-called neurons that process and exchange information to produce predictions or judgments. Artificial neural networks (also known as are composed of layers made up of nodes: an input layer, and several layers that are concealed. Each node is connected to other nodes and has a certain amount of weight and threshold associated with it [57]. When the network is in operation, each link transmits data to the node, which processes it using a nonlinear function, multiplies it by the weight allocated to it, and produces an output. The connections between nodes, known as edges, can transmit signals to other neurons. The signal strength at a connection is determined by the weight connected to that connection. The weights are adjusted by the network as it learns from the data it receives, allowing it to improve and develop over time.

Neural networks can learn and simulate intricate correlations between data input and output. Based on the patterns they discover from the training data, they can conclude, generalize, and forecast. Because of this, they are appropriate for applications including predictive modeling, natural language processing, and image recognition. Neural networks are modeled after the structure and functions of the human brain, yet it is still debatable to what extent artificial neural networks replicate brain function. Neural networks, on the other hand, have proven to be useful tools in the field of artificial intelligence, enabling machines to process data that is comparable to human cognitive functions. In conclusion, neural networks are a crucial component of ML and AI. They let computers learn from data and make inferences or predictions by mimicking the structure and functions of the human brain. These networks have become an essential tool in many applications, from image identification to natural language processing, because of their capacity to represent complicated interactions.

- **Natural Language Processing:** It facilitates the understanding, translation, and production of human language by robots, which eases the work of tasks like sentiment evaluation, translation of languages, and chatbot conversations. NLP uses a range of techniques and algorithms to analyze and interpret natural language input. Named object recognition, syntactic parsing, discourse analysis, semantic assessment, and part-of-speech tagging are some of the tasks it involves [58]. Through the use of these techniques, machines can extract sentiment, context, and meaning from text or speech data. Sentiment analysis is one of the main applications of NLP, where data scientists evaluate remarks on social media or customer service notes to determine how well a company's brand is doing or what needs to be improved. NLP also enables the development of personal assistants like Alexa, chatbots for customer support, and voice-operated GPS systems.

  NLP combines machine learning, deep learning models, statistics, and computational linguistics. With the use of these technologies, computers are now able to process natural language, comprehend its entire meaning, and draw conclusions and information from speech or text data. There are a lot of advantages to NLP. It enhances communication between people and computers, making human-computer interactions more natural. Additionally, it facilitates the production of accessible summaries of difficult materials, improves the productivity and quality of documentation, and supports several applications, including customer assistance and language translation. In conclusion, the crucial field of AI known as NLP makes it possible for robots to understand, translate, and create human language. It is highly versatile, contributing significantly to the enhancement of human-computer interactions and the ease of language processing jobs.

- **Computer vision:** The Artificial Intelligence, or AI, subfield of computer vision aims to enable machines to interpret and evaluate visual information included in images or videos. It includes things like segmenting images, classifying images, and recognizing objects. Computer vision aims to imitate the way humans see things by collecting information from digital photographs or videos and analyzing them to determine attributes. It attempts to automate tasks performed by the visual system of humans, such as deriving, interpreting, and extracting information from images. One of the main objectives of computer vision is to teach robots how to extract data from pixels and translate digital visual stuff into explicit descriptions [59]. Subsequently, this data is converted into a machine-readable language to aid in decision-making.

  Computer vision algorithms are necessary for a wide range of applications. Object recognition technology allows machines to identify and classify objects in images or movies. Assigning a label or category to an image depends on its content. This is known as image categorization. Image segmentation aims to partition images into multiple sets of views, while object detection determines the location of objects in an image. Applications like augmented reality, surveillance systems, and driverless cars depend on these tasks. RNN and deep learning are two technologies used in computer vision to handle visual input. One kind of deep learning technique is a CNN, which can analyze images and understand complex features; a random neural network (RNN) can analyze movies and understand the relationships between images. The increasing presence of image data has led to considerable developments in the era of computer vision. Applications for it can be found in many fields, such as robotics, security, healthcare, and entertainment. In summary, computer vision is the branch of computational intelligence that focuses on enabling robots to perceive, understand, and analyze visual information. It covers tasks including object recognition, image classification, and segmentation [60]. Many applications rely on computer vision algorithms, which use neural networks and deep learning to handle visual data.

4.2 Applications of AI

Artificial Intelligence has transformed multiple sectors and has become a crucial aspect of our daily existence. Several noteworthy uses of AI include:

- **Voice Assistants:** Voice assistants driven by AI, such as Siri, Alexa, and Google Assistant, are already ubiquitous in homes, enabling users to communicate with their gadgets using natural language instructions. These sophisticated virtual assistants are capable of carrying out a wide range of duties, frequently on several platforms and devices. Through voice-based engagements, they commonly engage in natural language conversations with users, facilitating quick and simple information access, control over smart home appliances, and calendar management through integration with a variety of services and apps. Voice assistants use artificially intelligent (AI) technology that processes natural language to comprehend human speech and respond in a human-sounding manner [61]. They depend on natural language processing and synthesis as well as machine learning to work well and improve over time. Machine learning, a branch of artificial intelligence, allows voice assistants, such as Alexa and Siri, to learn from information and past experiences (AI). This allows them to respond more intelligently and complete tasks with greater accuracy. Most of these voice assistants are cloud-based software that depends on internet-connected gadgets and apps to function. They are present in gadgets including automobiles, smart speakers, headphones, cellphones, and household appliances. Their reach is extensive, as they are compatible with thousands of devices from different brands.

Voice assistants have completely changed how people use digital gadgets and obtain information. Digital marketers are optimizing their content to align with these conversational queries, using long-tail keywords and natural language optimization to capture voice search traffic. Software giants are still vying to be the virtual assistant preference for smart devices worldwide, despite the growing appeal and extensive use of voice-controlled assistants such as Siri, Alexa, the Google Assistant. The battle for ownership of the smart speaker space is intense, with companies striving to offer the best user experience and expand their market share. In conclusion, voice assistants driven by artificial intelligence have become a common fixture in many homes, enabling users to communicate with their gadgets using natural language instructions.

- **Recommendation Engines:** Artificial intelligence (AI) systems known as recommendation engines examine user preferences and behavior to generate tailored recommendations for applications like social media, streaming media, and e-commerce. These engines interpret user data and produce highly precise and customized content or product recommendations by utilizing machine learning and deep learning techniques. Recommendation systems have seen a substantial change from conventional techniques with the introduction of AI. From there, they may learn from user habits, choices, and interactions to make recommendations that are more individualized for each user [62]. Their ability to identify minor differences and patterns that older approaches might miss leads to more accurate and individualized recommendations. To provide personalized product recommendations, AI-driven engines for recommendations in e-commerce analyze customer behavior and preferences. These engines increase consumer engagement and maintain users' interest in the platform by learning about the interests and preferences of their users. By classifying customers and making product recommendations based on their preferences, they enhance the overall shopping experience.

  Streaming services such as Netflix and Spotify also employ recommendation engines to keep users engaged and happy. These search engines examine what people have watched or listened to in the past and suggest films, albums, or videos based on those selections. Recommendation engines are used by social networking sites like Facebook and Instagram to recommend friends to establish relationships with, posts to engage with, and groups to join. By creating individualized feeds based on user behaviors and interests, these engines increase user engagement and platform time. The impact of AI-powered recommendation systems is not limited to e-commerce and entertainment industries. These days, they are applied in many different fields, such as healthcare, where AI systems may evaluate patient data and provide individualized treatment regimens based on symptoms and medical history. In summary, recommendation engines driven by AI are essential for giving users customized recommendations across a range of industries. Using machine learning as well as deep learning techniques, they analyze user behavior and preferences to provide accurate and personalized recommendations. These engines improve user satisfaction, user experience, and customer engagement in social media, video streaming services, and e-commerce.

- **Autonomous Vehicles:** The creation of self-driving cars depends on artificial intelligence (AI. Self-driving cars use machine learning, deep learning, and artificial intelligence algorithms to evaluate input from multiple sensors and make decisions in real time. AI including ML and NLP are used in self-driving cars. Thanks to the use of ML, the cars might eventually improve what they do by learning from data [63]. Computer vision analyzes visual input from cameras and sensors to give the cars the ability to see and comprehend their surroundings. Natural language processing can be used for voice commands and interactions with passengers. For self-driving automobiles, there are generally five automation levels: level 0 (no automation) to level 5 (full autonomy). As automation levels rise, artificial intelligence (AI) algorithms become increasingly important in managing the vehicle's functions, hence decreasing the necessity for human involvement.

  Companies like Aurora, Tesla, and traditional automakers such as Audi, BMW, and Ford are actively involved in the development and testing of self-driving cars. These businesses are using AI technologies to raise the standard of autonomous vehicles' ease, effectiveness, and safety. In conclusion, artificial intelligence is critical to the development of cars that can drive themselves. These vehicles use it to view their surroundings, make decisions, and travel securely. Among other AI technologies, sensor information analysis, machine learning, machine vision, and processing of natural languages improve the performance of self-driving cars. Many companies and researchers are working to improve the overall effectiveness and security of these vehicles, as well as boost the automation levels, in the ongoing process of building self-driving automobiles.

- **Healthcare:** Indeed, artificial intelligence (AI) is being applied to several healthcare domains, such as medication development, customized medicine, and medical diagnostics. These uses of AI in healthcare seek to raise general standards of care, expedite procedures, and improve patient outcomes. To assist medical professionals in making timely and accurate patient diagnoses, AI systems can analyze enormous volumes of patient data, including results of tests, imaging scans, or medical records. Algorithms using machine learning can improve diagnosis accuracy and aid in the early detection of diseases by identifying connections and patterns in the data. Indeed, artificial intelligence (AI) is being used in many healthcare domains, such as pharmaceutical development, customized medicine, and medical diagnostics. These uses of AI within medical settings are intended to improve overall standards of care, expedite procedures, and improve patient outcomes. AI systems can evaluate vast amounts of patient data, such as outcomes of tests, imaging scans, & medical records, to help medical personnel diagnose patients quickly and accurately [64]. By finding connections and trends in the information, learning algorithms can increase the accuracy of diagnoses and help in the earlier identification of diseases.

The application of machine learning (AI) to health care has an opportunity to improve patient outcomes, increase efficiency, and reduce the cost of providing treatment. It can help medical practitioners make better decisions, boost the timeliness and accuracy of diagnoses, and enhance customer service and satisfaction. It's important to keep in mind that using AI in healthcare is not without challenges. These include resolving privacy and legal issues, guaranteeing data availability and quality, and incorporating AI technologies into current healthcare systems. In conclusion, AI is being applied to personalized medicine, medication development, and medical diagnostics to enhance patient outcomes and expedite medical procedures. Large data sets can be analyzed by AI algorithms to help with precise diagnosis, find promising medication candidates, and create individualized treatment regimens. AI in healthcare has enormous potential benefits, even though there are obstacles to be solved.

- **Intelligent Surveillance Systems:** Artificial intelligence (AI)-driven surveillance systems are becoming more commonplace in a variety of contexts, such as public areas, transit hubs, and private businesses. These systems examine video feeds in real time using AI algorithms, allowing them to identify and notify authorities of possible security risks. Surveillance systems with AI capabilities can identify and track objects, identify faces, and identify suspicious activities through the use of computer vision and machine learning algorithms. By analyzing patterns, motions, and abnormalities in video data, these systems enable security professionals to react to possible threats promptly and effectively. The advantages of using AI in surveillance systems are numerous. Large volumes of video footage may be processed in real-time by AI algorithms, allowing them to quickly identify and address possible threats [65]. They can distinguish between people and objects, spot strange activity, and recognize faces. This feature increases the precision and efficacy of surveillance systems, lowering false alarms and enhancing security in general.

  Intelligent surveillance systems also offer benefits such as improved situational awareness, reduced reliance on human operators, and the ability to analyze video feeds continuously without fatigue. AI algorithms can work day and night, process a large number of video feeds simultaneously, and implement advanced analytics in real time. Security staff may now access better, faster, and more accurate data, empowering them to decide wisely and respond appropriately. But it's crucial to think about the privacy and ethical ramifications of AI-powered surveillance technologies. The widespread application of facial recognition software and the potential for mass surveillance raises concerns about privacy, civil liberties, and potential technological abuse. In deploying these technologies, finding a balance between security and privacy is essential. In conclusion, real-time video stream analysis, threat detection, and notification of authorities are all possible with AI-powered surveillance systems. These systems use machine learning and computer vision techniques to improve the precision and efficiency of monitoring [66]. Even while they have many advantages in terms of increased efficiency and security, their implementation should carefully take privacy and ethical issues into account.

5. Conclusion

The advent of blockchain, artificial intelligence, and cryptography has significantly changed the modern computer landscape. With its capacity to safeguard confidential data and maintain data integrity, cryptography has emerged as a crucial instrument for guaranteeing the confidentiality and security of digital communications. Its uses, which include data encryption and safe online transactions, have completely changed how people and businesses communicate online. By offering transparent and decentralized environments for a variety of applications, such as supply chain management, cryptocurrencies, and smart contracts, blockchain technology has completely upended conventional procedures. Because of its unchangeable and tamper-proof characteristics, trust has been established and the need for middlemen has been removed from numerous processes, opening the door to more reliable and efficient transactions. Artificial intelligence, powered by advanced algorithms and machine learning, has revolutionized the way we process and analyze vast amounts of data. From medical diagnostics to autonomous vehicles, AI has enabled us to make more accurate predictions, automate complex tasks, and uncover insights that were previously inaccessible. Its impact on industries such as healthcare, finance, and manufacturing has been profound, enhancing productivity, decision-making, and overall efficiency.

The future of computing could be shaped by the convergence of blockchain, artificial intelligence, and cryptography. We may anticipate even more developments in fields like cyber-security, data privacy, decentralized systems, and intelligent automation as these technologies continue to advance and blend. There are obstacles and things to take into account, though, just like with any innovative technology. To guarantee accountable and helpful incorporation of cutting-edge technology into daily lives, it is essential to take into account issues such as issues related to privacy, legal consequences, and ethical consequences. This chapter examined the development of AI, Blockchain, and cryptography in the contemporary computing environment, emphasizing their enormous potential and the revolutionary effects they have had on a range of businesses. Moving forward, we must carefully adopt new technologies, utilizing their potential for the benefit of society while proactively tackling the issues they raise.

References

1. Lyytinen, K., and Rose, G. M. 2003. The disruptive nature of information technology innovations: the case of internet computing in systems development organizations. *MIS quarterly*, 557–596.
2. Dooley, J. F. 2018. *History of cryptography and cryptanalysis: codes, ciphers, and their algorithms*. Springer.
3. Djenna, A., Harous, S., and Saidouni, D. E. 2021. Internet of things meet internet of threats: New concern cyber security issues of critical cyber infrastructure. *Applied sciences*, *11*(10), 4580.
4. Kimani, D., Adams, K., Attah-Boakye, R., Ullah, S., Frecknall-Hughes, J., and Kim, J. 2020. Blockchain, business and the fourth industrial revolution: Whence, whither, wherefore and how?. *Technological forecasting and social change*, *161*, 120254.
5. Xu, Y., Liu, X., Cao, X., Huang, C., Liu, E., Qian, S.,... and Zhang, J. 2021. Artificial intelligence: A powerful paradigm for scientific research. *The innovation*, *2*(4).
6. Duan, Y., Edwards, J. S., and Dwivedi, Y. K. 2019. Artificial intelligence for decision making in the era of Big Data–evolution, challenges and research agenda. *International journal of information management*, *48*, 63–71.
7. Hemamalini, V., Mishra, A. K., Tyagi, A. K., and Kakulapati, V. 2024. Artificial Intelligence–Blockchain-Enabled–Internet of Things-Based Cloud Applications for Next-Generation Society. *Automated secure computing for next-generation systems*, 65–82.
8. Beck, U. 2020. From industrial society to the risk society: Questions of survival, social structure and ecological enlightenment. In *risk management* (pp. 17–44). Routledge.
9. Allioui, H., and Mourdi, Y. 2023. Unleashing the potential of AI: Investigating cutting-edge technologies that are transforming businesses. *International journal of computer engineering and data science (IJCEDS)*, *3*(2), 1–12.
10. Hedabou, M. 2018. Cryptography for addressing cloud computing security, privacy, and trust issues. *Computer and cyber security*, 281–304.
11. Teng, L., Li, H., Yin, S., and Sun, Y. 2020. A Modified Advanced Encryption Standard for Data Security. *Int. J. Netw. Secur.*, *22*(1), 112–117.
12. He, S., Xing, X., Wang, G., and Sun, Z. 2023. A Data Integrity Verification Scheme for Centralized Database Using Smart Contract and Game Theory. *IEEE Access*.
13. Anderson, R. 2020. *Security engineering: a guide to building dependable distributed systems*. John Wiley & Sons.
14. Bauer, C. 2021. *Secret history: the story of cryptology*. CRC Press.
15. Mardon, A., Barara, G., Chana, I., Di Martino, A., Falade, I., Harun, R.,... and Varghese, N. 2021. Cryptography.
16. Kaur, J., Lamba, S., and Saini, P. 2021, March. Advanced encryption standard: attacks and current research trends. In *2021 International conference on advance computing and innovative technologies in engineering (ICACITE)* (pp. 112–116). IEEE.
17. Islam, M. A., Islam, M. A., Islam, N., and Shabnam, B. 2018. A modified and secured RSA public key cryptosystem based on "n" prime numbers. *Journal of computer and communications*, *6*(03), 78.
18. Makhtoum, H. E., and Bentaleb, Y. 2021. An improved IOT Authentication Process based on Distributed OTP and Blake2. *International journal of wireless and microwave technologies (IJWMT)*, *11*(5), 1–8.
19. Sarkar, A., Chatterjee, S. R., and Chakraborty, M. 2021. Role of cryptography in network security. *The" Essence" of network security: An end-to-end panorama*, 103–143.
20. Krombholz, K., Busse, K., Pfeffer, K., Smith, M., and Von Zezschwitz, E. 2019, May. " If HTTPS Were Secure, I Wouldn't Need 2FA"-End User and Administrator Mental Models of HTTPS. In *2019 IEEE Symposium on security and privacy (SP)* (pp. 246–263). IEEE.
21. Fang, W., Chen, W., Zhang, W., Pei, J., Gao, W., and Wang, G. 2020. Digital signature scheme for information non-repudiation in blockchain: a state of the art review. *EURASIP journal on wireless communications and networking*, *2020*(1), 1–15.
22. Thomas, K., Pullman, J., Yeo, K., Raghunathan, A., Kelley, P. G., Invernizzi, L.,... and Bursztein, E. 2019. Protecting accounts from credential stuffing with password breach alerting. In *28th USENIX Security symposium (USENIX Security 19)* (pp. 1556–1571).
23. Oppliger, R. 2023. *SSL and TLS: Theory and practice*. Artech House.

24. Kirillova, E. A., Pavlyuk, A. V., Mikhaylova, I. A., Zulfugarzade, T. E., and Zenin, S. S. 2018. Bitcoin, lifecoin, namecoin: The legal nature of virtual currency. *Journal of advanced research in law and economics*, *9*(1 (31)), 119–126.
25. Mckee, D. P. 2022. *Novel system compartmentalization and reverse engineering methods* (Doctoral dissertation, Purdue University Graduate School).
26. Centobelli, P., Cerchione, R., Del Vecchio, P., Oropallo, E., and Secundo, G. 2022. Blockchain technology for bridging trust, traceability and transparency in circular supply chain. *Information & management*, *59*(7), 103508.
27. Malik, S. 2021. *Transparent, trustworthy and privacy-preserving supply chains* (Doctoral dissertation, UNSW Sydney).
28. Moin, S., Karim, A., Safdar, Z., Safdar, K., Ahmed, E., and Imran, M. 2019. Securing IoTs in distributed blockchain: Analysis, requirements and open issues. *Future generation computer systems*, *100*, 325–343.
29. Nguyen, C. T., Hoang, D. T., Nguyen, D. N., Niyato, D., Nguyen, H. T., and Dutkiewicz, E. 2019. Proof-of-stake consensus mechanisms for future blockchain networks: fundamentals, applications and opportunities. *IEEE access*, *7*, 85727–85745.
30. Coglianese, C., and Lampmann, E. 2020. Contracting for algorithmic accountability. *Admin. L. Rev. Accord*, *6*, 175.
31. Bhutta, M. N. M., Khwaja, A. A., Nadeem, A., Ahmad, H. F., Khan, M. K., Hanif, M. A.,... and Cao, Y. 2021. A survey on blockchain technology: Evolution, architecture and security. *Ieee access*, 9, 61048–61073.
32. Niranjanamurthy, M., Nithya, B. N., and Jagannatha, S. J. C. C. 2019. Analysis of Blockchain technology: pros, cons and SWOT. *Cluster computing*, *22*, 14743–14757.
33. Ruiz, J. 2020. *Public-permissioned blockchains as common-pool resources* (Doctoral dissertation, Alastria Blockchain Ecosystem).
34. Yue, K. B., Kallempudi, P., Sha, K., Wei, W., and Liu, X. 2021, January. Governance attributes of consortium blockchain applications. In *Twenty-seventh americas conference on information systems, montreal, 2021.*
35. Alkhateeb, A., Catal, C., Kar, G., and Mishra, A. 2022. Hybrid blockchain platforms for the internet of things (IoT): A systematic literature review. *Sensors*, *22*(4), 1304.
36. Xiao, Y., Zhang, N., Lou, W., and Hou, Y. T. 2020. A survey of distributed consensus protocols for blockchain networks. *IEEE communications surveys & tutorials*, *22*(2), 1432–1465.
37. Biais, B., Bisiere, C., Bouvard, M., and Casamatta, C. 2019. The blockchain folk theorem. *The review of financial studies*, *32*(5), 1662–1715.
38. Duong, T., Chepurnoy, A., Fan, L., and Zhou, H. S. 2018, May. Twinscoin: A cryptocurrency via proof-of-work and proof-of-stake. In *Proceedings of the 2nd ACM workshop on blockchains, cryptocurrencies, and contracts* (pp. 1–13).
39. Bachani, V., and Bhattacharjya, A. 2022. Preferential Delegated Proof of Stake (PDPoS)—Modified DPoS with Two Layers towards Scalability and Higher TPS. *Symmetry*, *15*(1), 4.
40. Selvaraj, A. P., and Annamalai, S. 2022, December. Towards Reliable Medical Transactions for Blockchain based Healthcare Systems using Hybrid Consensus Method. In *Proceedings of the 4th international conference on information management & machine intelligence* (pp. 1–8).
41. Lao, L., Dai, X., Xiao, B., and Guo, S. 2020, May. G-pbft: a location-based and scalable consensus protocol for iot-blockchain applications. In *2020 IEEE international parallel and distributed processing symposium (IPDPS)* (pp. 664–673). IEEE.
42. B. Rawat, D., Chaudhary, V., and Doku, R. 2020. Blockchain technology: Emerging applications and use cases for secure and trustworthy smart systems. *Journal of cybersecurity and privacy*, *1*(1), 4–18.
43. Namasudra, S., Deka, G. C., Johri, P., Hosseinpour, M., and Gandomi, A. H. 2021. The revolution of blockchain: State-of-the-art and research challenges. *Archives of computational methods in engineering*, *28*, 1497–1515.
44. Mukhopadhyay, M. 2018. *Ethereum smart contract development: Build blockchain-based decentralized applications using solidity*. Packt Publishing Ltd.
45. Wang, S., Ouyang, L., Yuan, Y., Ni, X., Han, X., and Wang, F. Y. 2019. Blockchain-enabled smart contracts: architecture, applications, and future trends. *IEEE transactions on systems, man, and cybernetics: systems*, *49*(11), 2266–2277.

46. Vashistha, N., Hossain, M. M., Shahriar, M. R., Farahmandi, F., Rahman, F., and Tehranipoor, M. M. 2021. eChain: a blockchain-enabled ecosystem for electronic device authenticity verification. *IEEE transactions on consumer electronics*, *68*(1), 23–37.
47. Attaran, M. 2022. Blockchain technology in healthcare: Challenges and opportunities. *International journal of healthcare management*, *15*(1), 70–83.
48. Taş, R., and Tanrıöver, Ö. Ö. 2020. A systematic review of challenges and opportunities of blockchain for E-voting. *Symmetry*, *12*(8), 1328.
49. Mhatre, M., Kashid, H., Jain, T., and Chavan, P. 2023. BCPIS: Blockchain-based counterfeit product identification system. *Journal of applied security research*, *18*(4), 740–765.
50. Rogerson, M., and Parry, G. C. 2020. Blockchain: case studies in food supply chain visibility. *Supply chain management: an international journal*, *25*(5), 601–614.
51. Sunny, J., Undralla, N., and Pillai, V. M. 2020. Supply chain transparency through blockchain-based traceability: An overview with demonstration. *Computers & industrial engineering*, *150*, 106895.
52. Raschka, S., Patterson, J., and Nolet, C. 2020. Machine learning in python: Main developments and technology trends in data science, machine learning, and artificial intelligence. *Information*, *11*(4), 193.
53. Putri, T. D. 2021. Intelligent transportation systems (ITS): A systematic review using a Natural Language Processing (NLP) approach. *Heliyon*, *7*(12).
54. Yigitcanlar, T., Desouza, K. C., Butler, L., and Roozkhosh, F. 2020. Contributions and risks of artificial intelligence (AI) in building smarter cities: Insights from a systematic review of the literature. *Energies*, *13*(6), 1473.
55. Sarker, I. H. 2021. Deep learning: a comprehensive overview on techniques, taxonomy, applications and research directions. *SN Computer science*, *2*(6), 420.
56. Vargas, V. M., Rosati, R., Hervás-Martínez, C., Mancini, A., Romeo, L., and Gutiérrez, P. A. 2023. A hybrid feature learning approach based on convolutional kernels for ATM fault prediction using event-log data. *Engineering applications of artificial intelligence*, *123*, 106463.
57. Islam, M., Chen, G., and Jin, S. 2019. An overview of neural network. *American journal of neural networks and applications*, *5*(1), 7–11.
58. Khurana, D., Koli, A., Khatter, K., and Singh, S. 2023. Natural language processing: State of the art, current trends and challenges. *Multimedia tools and applications*, *82*(3), 3713–3744.
59. Szeliski, R. 2022. *Computer Vision: Algorithms and applications*. Springer Nature.
60. Che, E., Jung, J., and Olsen, M. J. 2019. Object recognition, segmentation, and classification of mobile laser scanning point clouds: A state of the art review. *Sensors*, *19*(4), 810.
61. Vlahos, J. 2019. *Talk to me: How voice computing will transform the way we live, work, and think*. Eamon Dolan Books.
62. Sarker, I. H., Hoque, M. M., Uddin, M. K., and Alsanoosy, T. 2021. Mobile data science and intelligent apps: concepts, ai-based modeling and research directions. *Mobile networks and applications*, *26*, 285–303.
63. Khayyam, H., Javadi, B., Jalili, M., and Jazar, R. N. 2020. Artificial intelligence and internet of things for autonomous vehicles. *Nonlinear approaches in engineering applications: Automotive applications of engineering problems*, 39–68.
64. Topol, E. J. 2019. High-performance medicine: the convergence of human and artificial intelligence. *Nature medicine*, *25*(1), 44–56.
65. Ahmed, A. A., and Echi, M. 2021. Hawk-eye: An ai-powered threat detector for intelligent surveillance cameras. *IEEE access*, *9*, 63283–63293.
66. Azadnia, R., Jahanbakhshi, A., Rashidi, S., and Bazyar, P. 2022. Developing an automated monitoring system for fast and accurate prediction of soil texture using an image-based deep learning network and machine vision system. *Measurement*, *190*, 110669.
67. Rawat, A., Suryavanshi, R. S., Nagar, V. and Yadav, D. (2023, November). Development of Blockchain-Enabled E-Healthcare Appointment System. In International Conference on Trends in Computational and Cognitive Engineering (pp. 359–369). Singapore: Springer Nature Singapore.
68. Rawat, A. and Rahman, D. H. U. 2023. Smart contract based approach for fixed deposit in blockchain networks with remix IDE. International Journal of Creative Research Thoughts (IJCRT), 11(5).
69. Rawat, A. and Ur Rahman, H. 2023. Implementation of algorithms for fixed deposit using smart contract. Journal of Emerging Technologies and Innovative Research (JETIR) JETIR, 10(6).

Chapter 2

Blockchain 1.0 to Blockchain 4.0—Evolution, Transformation and Comparison of Different Generations

Nikhil Pateria[1] and *Nikita Tiwari*[2,*]

1. Introduction

In the vast tapestry of technological innovation, few developments have sparked as much fascination, speculation, and transformative potential as blockchain technology. Birthed in the wake of the worldwide financial crisis of 2008, blockchain emerged to be a force for disruption, challenging traditional notions of trust, security, and decentralized systems [1]. Its journey from a simple ledger system underpinning the enigmatic Bitcoin cryptocurrency to a multi-faceted, far-reaching phenomenon encapsulates the essence of our digital age. Data security and trust have become crucial in today's environment, blockchain has stood out as a transformative force, promising transparency, immutability, and the elimination of intermediaries [2]. It has disrupted industries and challenged conventional governance, finance, and ownership notions. But, the story of blockchain is not one of a singular, unchanging technology; rather, it is a narrative of continuous innovation and adaptation [3].

Blockchain technology makes Cryptocurrencies (digital currencies secured with the aid of cryptography) like Bitcoin work similar to how the internet makes e-mail viable [6]. The blockchain is unalterable the record once created can't be modified because of which the blockchain has a reputation as immutable [4]. The distributed

[1] School of Computing Science Engineering, and Artificial Intelligence, VIT Bhopal University, Bhopal–Indore Highway, Kothrikalan, MP, Sehore, India, 466114.
Email: nikhil.pateria@vitbhopal.ac.in

[2] Department of Computer Science & Engineering, Pranveer Singh Institute of Technology, Kanpur, Uttar Pradesh, India.
Email: nikita.tiwari25@gmail.

ledger is the property of blockchain where a digital record of transactions or records which are saved in multiple devices on a computer network [5]. Blockchain has various use cases apart from Cryptocurrencies [6]. The d efinition of the blockchain according to the journal "Truth about blockchain" by Lansiti, Lakhani published in 2017 is as follows, "A reliable decentralized distributed ledger which verifiably as well as permanently stores records and transactions among the participants," with "efficient" indicating quick and scalable, "distributed" or decentralized implying the system is not controlled by a single entity or central body, and "open" meaning accessible to all. Permanent refers to the information's persistence, and verifiable refers to the ability of all blockchain participants to validate the accuracy of the information [7]. Blocks comprise the records that are stored in the blockchain, i.e., every transaction or record in the ledger is stored in a block [8]. On the Bitcoin blockchain, for example, blocks typically contain over 500 transactions of the Bitcoin. The records contained in a block depends on and are related to the information in a preceding block and over time form a chain of transactions, hence it has been given the name blockchain [9].

The narrative begins with Blockchain 1.0, where the pioneering concept of a decentralized digital currency laid the foundation for subsequent innovations. From this seminal point, as we explore the second-generation blockchain, witnessing the enactment of the decentralized applications and smart contracts, and then explore blockchain 3.0's scalability and interoperability solutions. Finally, we arrive at the forefront of technological innovation, Blockchain 4.0, where sustainability, security, and cross-industry integration reign supreme. This journey through the generations of blockchain technology is not just a historical account but a testament to human ingenuity and our relentless pursuit of decentralized, transparent, and efficient systems. As we unravel the complexities and nuances of each generation, we gain insights not only into the evolution of blockchain but also into the potential futures it promises. This chapter begins with a thorough investigation of the evolutionary trajectory of blockchain, illuminating pivotal milestones from its humble beginnings in Blockchain 1.0 to the sophisticated, interconnected networks of Blockchain 4.0. At its core, blockchain technology embodies the fusion of cryptography, computer science, and economics, forging a new paradigm of consensus mechanism and immutable record storage. Now that we are ready to explore the landscape across each blockchain generation, we will delve into the intricacies of their underlying mechanisms, unravel the diverse applications that have reshaped industries, and contemplate the profound implications on our global socio-economic fabric.

2. Basic Structure and its Terminologies

2.1 Fundamental Architecture of a Blockchain

A blockchain is a network of chains composed of information that is kept in blocks, as the name indicates. These blocks are containers that hold records of transactions on the blockchain. For Bitcoin, transactions are primarily transfers of Bitcoin. Some blockchains also contain various other information, such as computer program code [10].

A transaction cannot be removed from a block once it has been added. All information remains in the blockchain while the blockchain exists. Blocks are arranged linearly on top of each other. They create a chain containing all the past information or history of transactions within the blockchain. Depending on the specific type of blockchain, the structure of the blockchain may change [10–11].

This section is focused on describing the structure of a Bitcoin block, as this is the foundation on which many other blockchains are modeled. Based on the upper bound put forth in 2010 by Satoshi Nakamoto, the typical block size might be approximately one megabyte, with more than 500 transactions in it. However, now the block size might increase up to eight megabyte or higher. Preferably the block size is kept as large as it can be because the larger the block size, the greater the transactions it could process simultaneously [11].

Two components of the structure of the block are:

A. Header of the block
B. Block Body (Transaction List)

A. Block Header

The structure of the blockchain is defined as a set of blocks linked to one another sequentially to prevent the participants from performing any modifications once the transaction is stored. But it is just the header of each block that is connected together and not the entire blockchain [11–12]. There are many different fields in the blockchain header, which vary with the details of the implementation of the specific type of blockchain. The successive block's headers are linked in a series, therefore, t here must be an upgrade to the entire blockchain, if any changes are made anywhere in the blockchain [12]. The b lockchain header consists of six components. Block headers play a fundamental role in Blockchain as they link all blocks together:

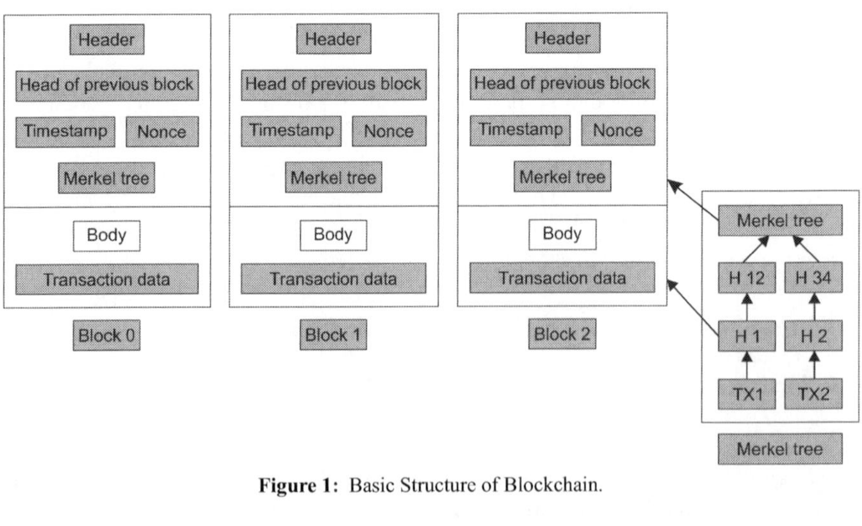

Figure 1: Basic Structure of Blockchain.

- **The software's version number:**
  In most cases software version is irrelevant. However, miners with specific version numbers can inform their decision of which protocol to support. The versions of the blockchain that are available to date are Blockchain 1.0, Blockchain 2.0, Blockchain 3.0, and Blockchain 4.0 [13].
- **Previous block hash:**
  Every block in the blockchain is linked to every other block because every new block contains the hash of every previous block. The chronology of each individual block will be lost if the previous block's hash value is not provided to the subsequent block [14].
- **Hash of Merkle tree root:**
  Since, it is difficult to maintain the hash of each individual transaction, the hash values of all the transactions were combined to create a single hash that is referred to as the Merkle tree's root hash [11–13]. These trees are the mathematical data structure consisting of various data blocks' hash values.
- **The Timestamp:**
  The timestamp contains information about the time when the transaction has been carried out. The time is given in seconds since 1.1.1970 [14].
- **Target Difficulty:**
  The target specifies how small the new hash must be in size to be validated. The size of the hash is measured in bits. The smaller the target, the higher the complexity of determining that hash. A hash with more leading zeros is smaller than a hash without zeros [15].
- **Nonce:**
  Nonce is a variable that increases in increments when an action is confirmed. Therefore, the miner determines a valid hash that is less than the difficulty target [16].

B. Block Body (List of Transactions)

The block body in a blockchain refers to the block section that contains the transactions. The Block body, containing transactions or smart contract code, forms the core data structure of a block [17]. This data is secured by cryptographic hashing and linked to the block header, ensuring the integrity and immutability of the blockchain's content. It's one of the two main components of a block, block header is another. The components of the Block Body are discussed below:

- **Transactions:**
  A list of transactions is the main information contained in a block body. Each transaction represents a record of data or an action carried out within the blockchain network. Transactions can vary widely based on the blockchain's purpose [18]. For instance, in a cryptocurrency blockchain like Bitcoin, transactions involve the transfer of currency between addresses. In other blockchains, transactions can represent various types of data interactions or smart contract executions [19].

- **Smart Contract Code:**
  In blockchains that support smart contracts (such as Ethereum), the block body might contain the actual code or bytecode of smart contracts [20]. Self-executing contracts with defined regulations stored within the blockchain are referred to as smart contracts. When smart contracts are executed or deployed, their code is included in the block body to facilitate their operation [21].
- **Types of Blockchain:**
  The blockchain has mainly four principle forms as shown in Fig. 4, namely public blockchain, private or non-public blockchain, consortium blockchain and hybrid blockchain [22]. As we cover later in this section, each type has unique advantages, drawbacks, and use cases. The targeted end users of different forms of blockchain may vary but the technology of the blockchain used behind each project always remains the same [23].

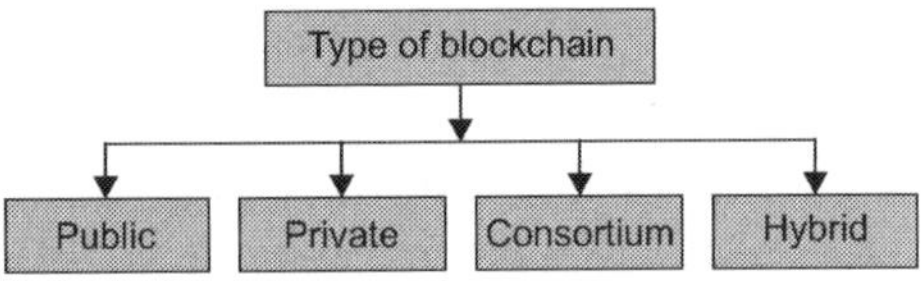

Figure 2: Types of Blockchain.

Public blockchains are unrestricted, and users are not required to take permission and anyone with internet access can register on the blockchain application to be an approved node. Once registered, the user can get access to the present as well as past transactions [24]. A registered user is also allowed to perform mining activities, which involve validation and addition of the transaction to the ledger by using complex calculations. A record (sometimes referred to as a transaction) once validated and added, then it cannot be altered. The source code is open source; therefore, any user within the network can review past transactions, find errors and might suggest corrections. The advantage of a public blockchain is they are not dependent on any authority.

Even in the case, that the organization that started the chain is no longer in existence, a public blockchain continues its working till computers remain connected. Another benefit of this type of blockchain is network transparency [25]. The security of the public blockchain is until users diligently follow their protocols and methods of security. The public blockchain does not pose scalability, as when a greater number of nodes is being added to the network, it becomes slow in processing. These blockchains are beneficial for organizations/companies that are built on trust and transparency, such as private organizations or social welfare organizations. As this network is public in nature it restricts private corporations from using them [26].

Private blockchain networks have a restricted environment and operate as closed networks. A single authority controls it as it is private in nature. This blockchain leverages decentralization and peer-to-peer connectivity, just like the public blockchain [27]. These blockchains have a smaller scope, as it does not allow anyone to participate and contribute. This blockchain has applications where small networks are required for example companies or organizations. Permissioned blockchains or enterprise blockchains are the other names of this blockchain. This kind of blockchain is more vulnerable to illegal access and cyberattacks [28]. To overcome the drawbacks of public and private blockchains, consortium and hybrid blockchains have been introduced. The consortium blockchain differs from the private blockchain, and it is authorized by several organizations instead of a single entity. Consortium blockchains can achieve higher decentralization as compared to private blockchains [29]. These blockchains are permissioned as they offer access to a limited group of people after the validation is carried out from the validator node, and due to high decentralization, they are more secure [30]. Forming a consortium requires cooperation among the participating organizations as it has both logistical challenges and potential antitrust risks, hence this makes creating a consortium a lengthy process [31]. Consortium blockchains offer less transparency and hence they are more suitable for banking and payments [32–33].

Hybrid blockchains come into existence when organizations need to implement the characteristics of private as well as public blockchains. These blockchains allow users to set up a network that combines the properties of private and public blockchains, i.e., a permissioned private system with an unauthorized public system, in which part of the specific record is made private, access of which is controlled by the system and some data is available publicly [34]. When a new user joins the network, their identity remains disclosed, however, once this user makes any transaction their identity is then shared with all other users in the network. Records and transactions in this blockchain are not made public but they are verifiable whenever needed [35]. This blockchain is highly scalable and it offers great protection and privacy while allowing the users to communicate with third parties. However, upgrading this blockchain is more challenging than other available blockchains [36].

3. Evolution of Blockchain

The advancement of this technology represents a fascinating journey through innovation, revolutionizing the way we envision trust, transparency, and decentralized systems. From its inception with Bitcoin to the emergence of diverse applications in Blockchain 4.0, this evolution mirrors a transformative saga reshaping industries and redefining digital interactions [37]. In its early stages, blockchain was the core technology underlying decentralized digital currencies such as B itcoin. The first iteration, Blockchain 1.0, laid the groundwork for safe, transparent transactions between peers, introducing the idea of an immutability, distributed ledger. With the advent of Blockchain 2.0, led notably by Ethereum, the technology expanded its horizons. Self-executing contracts, smart contracts, are embedded in blockchain technology and unlock a new realm of possibilities beyond currency [38]. This generation paved the way for decentralized applications (DApps) and innovative tokenization models.

Blockchain 3.0 brought forth solutions to scalability and interoperability challenges. Innovations aimed to enhance the technology's ability to handle a broader range of transactions while fostering compatibility between diverse blockchain networks. non-fungible tokens (NFTs) and Decentralized finance

Table 1: A comparison of the V arious B lockchain A rchitectures.

Parameters	Public Blockchain	Private Blockchain	Consortium Blockchain	Hybrid Blockchain
Access Control	Anyone can access; permissionless	Limited access; permissioned	Controlled access; permissioned	Variable access control; hybrid model
Governance	Decentralized	Centralized	Shared governance	Flexible governance structure
Transparency	High	Variable	Variable	Variable
Efficiency	Slower due to consensus mechanisms	Faster due to fewer nodes	Moderate	Variable
Use Cases	Cryptocurrencies, decentralized apps	Internal applications, enterprise	Industry-specific collaborations	Customizable applications
Scalability	Limited scalability	Potentially higher scalability	Moderate scalability	Variable scalability
Security	High security due to decentralization	Controlled by participants	Shared security responsibilities	Varies based on the hybrid model
Consensus Mechanism	Proof of Work (PoW), Proof of Stake (PoS)	Variable	Consortium agreement-based	Combination of consensus mechanisms
Examples	Bitcoin, Ethereum	Hyperledger Fabric, Corda	R3 Corda, Hyperledger	Quorum, Dragonchain

(DeFi) exemplify the diverse applications of this era [39]. As we transcend into Blockchain 4.0, the focus shifts to sustainability, scalability, and holistic ecosystem growth. This generation emphasizes enhanced sustainability measures through energy-efficient consensus mechanisms. Moreover, it envisions an interconnected landscape fostering interoperability among various blockchains and industries. Throughout its evolution, blockchain technology has permeated industries far beyond finance. From Governance to supply chain management, entertainment and healthcare, its transformative potential continues to be realized in diverse sectors [40]. The technology's applicability in changing traditional paradigms is supported by its capacity to promote efficiency, security, and transparency. The evolution of blockchain reflects a continuum of advancements—each phase building upon the previous, transcending limitations, and opening new frontiers. Understanding this evolution unveils not just where the technology has been, but also envisions the boundless possibilities that lie ahead, promising a future shaped by decentralized, secure, and interconnected systems [41].

3.1 Blockchain 1.0

The first generation, as they are frequently referred to, Blockchain 1.0, laid the groundwork for the pioneering concepts and functionalities that underpin modern blockchain systems. Blockchain 1.0's core technology is distributed ledger technology, or DLT, serving as the underlying framework for systems like Bitcoin [42]. A decentralized database, or DLT, records transactions by multiple participants or locations within a network. Blockchain 1.0, specifically, utilizes DLT uses a chronological, linear sequence of blocks to record and secure transactions. Using blockchain technology, Bitcoins are the very first and most widely recognized example. First mentioned in a whitepaper in 2008 by Satoshi Nakamoto, Bitcoin fundamentally altered the landscape of finance by introducing a decentralized digital currency [43]. This is a network-based peer-to-peer decentralized digital currency, or cryptocurrency, that enables users to transfer as well as receive the payment and it does not require the involvement of middlemen like banks or other financial organizations. Blockchain implemented for Bitcoin utilizes DLT to develop a decentralized, transparent distributed ledger of financial transactions. Transactions are added to blocks, cryptographically secured, and verified by network nodes through the consensus mechanism [44]. This section focuses on the fundamental ideas and characteristics of Blockchain 1.0, as well as its key components and the revolutionary changes it brought about in the realm of digital transactions.

3.1.1 Architecture

Blockchain 1.0 using Bitcoin established the groundwork for decentralized, trustless financial transactions without involving any intermediaries. Its architecture provides transparency, security, and resilience, laying the foundation for subsequent blockchain developments and inspiring innovations beyond cryptocurrencies [45]. The working of the blockchain 1.0 using B itcoin is illustrated in Fig. 3. The working of the bitcoin blockchain is explained below:

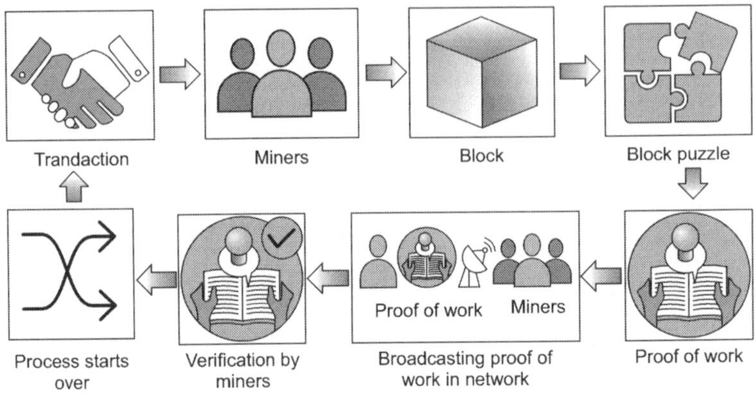

Figure 3: Working of B lockchain using Bitcoin.

A. Transaction Initiation:

- Initiating Transactions: Users initiate transactions by generating a digital message specifying the receiver's address, amount, and signature. Finally, broadcasting of the transaction is done over the whole network.
- Transaction Verification: Nodes in the network verify the transaction's validity by ensuring the sender's ownership of the bitcoins and the availability of funds. Subsequently, the transaction gets added to the mempool, which is a pool of pending or unconfirmed transactions.

B. Mining and Block Creation:

- Block Formation: A group of pending transactions is chosen by miners within the mempool to be included in a newly created block. They combine transactions with a header of the block which contains a reference of the preceding blocks' hash value, a timestamp, and a unique nonce.
- Proof of Work (PoW): Miners in a blockchain compete with each other to find solutions to computationally complex cryptographic challenges, known as the Proof of Work. Miners use different nonces to hash the block header continuously until they find a specified hash value (which is approximately lower than a target threshold). The new block is broadcast to the network by the very first miner who finds the valid hash.

C. Consensus and Block Addition:

- Block Verification: After receiving the recently mined block, other nodes in the network independently confirm that the associated transactions and the Proof of Work are valid.
- Consensus Mechanism: By acknowledging the longest valid chain—the one with the highest cumulative computational work—as a valid blockchain, nodes reach consensus. The process proceeds with, the approved block getting linked to the current blockchain.

D. Incentives:

- Block Rewards: Miners are incentivized to participate in the network through block rewards. When a miner is successful in adding a block to a blockchain, miners are credited with a set number of recently mined bitcoins. As of now, the reward is halved approximately every four years through a process called "halving."
- Transaction Fees: To encourage miners to prioritize their transactions, users might include transaction fees. Once the block is successfully added to the blockchain, the miner receives these fees.

E. Security and Cryptography:

- Cryptography: Bitcoin in its implementation includes cryptographic methods to maintain the blockchain's integrity, and to enhance the security of transactions including hashing and digital signatures.
- Decentralization: The network's decentralized architecture increases security and resilience against manipulation and attacks by preventing a single party from controlling the system.

F. Wallets and Addresses:

- Wallets: Digital wallets, which hold the private keys needed for accessing and approving transactions, are used by users to maintain their B itcoin holdings.
- Addresses: Bitcoin addresses are identifiers used for sending and receiving bitcoins that can be derived from public keys. These addresses are pseudonymous, providing a degree of privacy.

3.1.2 Areas of Application

Blockchain 1.0, was primarily associated with Bitcoin, and initially focused on the application of decentralized digital currencies. Because it provides transparency, security, and decentralization, its applications extend beyond financial transactions, showcasing its ability to completely transform various industries. The following are the key areas where Blockchain 1.0 has been applied:

- **Cryptocurrencies and Digital Assets**:

  Digital Currencies for example Bitcoin paved the way for various cryptocurrencies functioning as alternative forms of money, enabling secure, borderless transactions. Also, the assets tokenization, representing real estate or commodities on a blockchain, enables fractional ownership and facilitates easier trading [46].

- **Financial Services and Banking**:

  Blockchain removes the requirement for financial intermediaries, providing faster as well as cheaper overseas transactions. It provides an efficient and cost-effective way for individuals to send money internationally.

- **Logistics and Supply Chain**:

  Blockchain makes the supply chain transparent and traceable by providing immutable records at each stage, lowering fraud and counterfeit goods.

Blockchain provides provenance verification by verifying the authenticity of products and tracing their origins for industries like food, pharmaceuticals, and luxury goods [47].

- **Identity Verification and Authentication:**

  Utilizing blockchain for securing, decentralized identity verification, reduced identity theft and providing greater control over personal data. Blockchain provides credentialing storing educational qualifications, professional certifications, and licenses on a blockchain for easy verification.

- **Voting and Governance:**

  Blockchain has provided secure voting systems, i.e., utilizing blockchain technology to develop transparent as well as secure voting procedures that maintain electoral integrity and reduce fraud. Another application is to introduce decentralized governance, i.e., implementing decentralized decision-making processes in organizations, enhancing transparency and inclusivity [48].

3.1.3 Merits of Blockchain 1.0

- **Decentralization and Trustless Transactions:** This eliminates the requirements of intermediaries, allowing peer-to-peer transactions directly. A transparent and immutable ledger maintained by a decentralized network of nodes enhances trust in blockchain technology [49].
- **Security and Immutability:** Cryptographic techniques provide the security as well as the integrity of records over blockchain. The i mmutability of records prevents tampering and fraud, ensuring a trustworthy transaction history.
- **Transparency and Traceability:** Offers transparent and publicly accessible transaction records, promoting accountability. Facilitates traceability in supply chains, improving accountability and reducing fraud.
- **Efficiency and Cost Savings:** S treamlines processes, cutting down on the time and costs required using conventional intermediary companies in transactions. Potentially lower transaction fees in comparison to conventional banking systems, particularly for international transactions [50].
- **Financial Inclusion:** It helps those who are underbanked or unbanked by providing bank- related services in areas with very restricted banking facilities.
  Demerits of Blockchain 1.0.
- **Scalability Challenges:** Blockchain 1.0 faces scalability issues, resulting in slower transaction speeds as the network grows. Limited throughput, processing a lower number of transactions per second compared to traditional financial systems [51].
- **Energy Consumption:** Consensus mechanisms such as Proof of Work (PoW) used in this version require huge computational capacity, which results in higher energy consumption.
- **Regulatory Uncertainty:** Decentralization poses challenges in terms of regulatory compliance and adaptation to existing legal frameworks. Governments

and regulatory bodies face challenges in controlling and regulating decentralized systems.

- **Lack of Smart Contract Functionality:** Smart contracts in Blockchain 1.0 are limited in functionality, restricting the scope of automated and programmable contracts [52].
- **Environmental Concerns:** The energy-intensive mining process in Blockchain 1.0, especially with Proof of Work, raises concerns about its environmental impact and carbon footprint.
- **Limited Use Cases Beyond Currency:** Initial blockchain applications (Bitcoin) were primarily limited to financial transactions, restricting widespread adoption in other industries.

3.2 Blockchain 2.0

Blockchain 2.0 is popular for the evolution of this technology beyond its initial application in Bitcoin, which is often referred to as Blockchain 1.0. The f irst version has primarily focused on enabling digital currencies, offering a secure, decentralized means of enabling peer-to-peer transactions without the need for intermediaries [53]. Blockchain 2.0 introduced several advancements and enhances blockchain technology's applications more than just digital currencies. This version has widened the scope to encompass various industries and applications like smart contracts. Smart contracts became a cornerstone of Blockchain 2.0 [54]. These contracts are self-executing since their terms have been encoded within the source code. In numerous kinds of industries, including supply chain management, real estate, banking, and more, smart contracts offer enhanced efficiency, transparency, and dependability by automatically enforcing and carrying out the conditions agreed upon without requiring intermediaries. These contracts enable the automation of agreements and transactions based on predefined rules, without intermediaries. Smart contracts revolutionize traditional contract execution by automating processes, reducing the need for intermediaries, minimizing the risk of manipulation or fraud, and increasing reliability, efficiency and also transparency in many fields such as real state, supply chain management, finances, and more [55]. These advancements signify the maturation and diversification of blockchain technology beyond its original use case of digital currency, offering a broad spectrum of applications across various industries and sectors. Different projects and platforms contribute to the evolution of Blockchain 2.0, each with its unique features and approaches towards addressing the limitations and expanding the capabilities of blockchain technology [56].

3.2.1 Architecture

Smart contracts in blockchain are implemented using Ethereum. The decentralized blockchain platform Ethereum was introduced in 2015 after being conceived by Vitalik Buterin in late 2013. Smart contracts, or contracts that execute themselves with the conditions of the agreement explicitly put into code, are something that Ethereum is well-known for being able to perform [57]. The Ethereum network uses a specific programming language called Solidity to write smart contracts. They are deployed on the Ethereum blockchain and operate according to the r ules and conditions set within their code. Executing smart contracts on Ethereum requires a payment called "gas". The computational work done by the Ethereum network to process and execute transactions is measured in the unit called gas. U sers use ether (ETH) to pay for gas, Ethereum's native cryptocurrency, to compensate miners for their work [58]. DApps, or decentralized applications, are made possible by Ethereum's smart contracts. These are Ethereum-based apps that make use of smart contracts to enable a range of features, including decentralized exchanges (DEXs), non-fungible tokens (NFTs), and decentralized finance (DeFi) [59]. The w orking process of the Ethereum blockchain is illustrated in Fig. 4.

A. **Code Creation:** The smart contract's terms and conditions are drafted by a developer in a programming language appropriate for the blockchain network on which it will execute its processes. One popular language for Ethereum smart contracts is called Solidity.

B. **Deployment on the Blockchain:** As a transaction, the smart contract code is assembled and put into the blockchain network. Because it has a distinct address on the blockchain, everyone on the network can access this contract.

C. **Triggering Conditions:** Smart contracts contain predefined conditions or triggers. These conditions are coded into the contract and can be based on various parameters such as time, date, specific events, or actions.

D. **Transaction Execution:** When the predetermined conditions are coded into the smart contract, this smart contract self-executes. For example, in the case of an insurance smart contract, if a flight delay is reported, triggering the condition in the contract, it would automatically release the compensation funds to the affected party.

E. **Transaction Cost:** Every action performed by the smart contracts in the blockchain incurs a cost or transaction fee, known as gas in Ethereum. Gas fees are paid in cryptocurrency and cover the computational cost required to execute the contract.

F. **Verification and Validation:** By using consensus techniques such as proof-of-work or proof-of-stake, nodes (computers) on the blockchain network authenticate and verify the operations carried out by smart contracts, guaranteeing the precision and integrity of the contract execution.

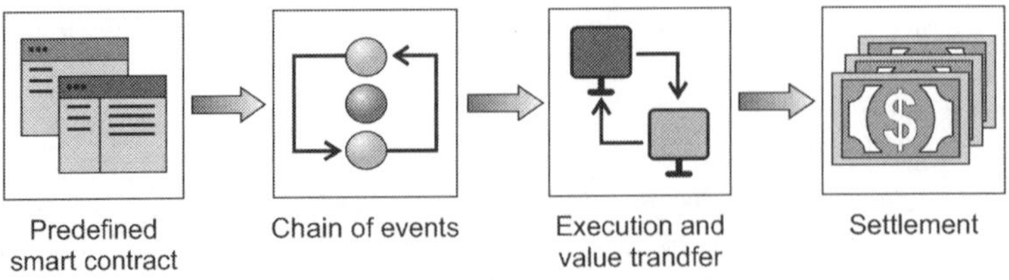

Figure 4: Working of Blockchain using Smart Contracts.

3.2.2 Areas of Application

Blockchain 2.0 introduced several advancements and enhanced the applications of blockchain technology more than just digital currencies. Some key features and concepts associated with Blockchain 2.0 include:

- **Smart Contracts:** These self-executing contracts contain explicit code encoding the terms of the agreement. In numerous sectors, such as supply chain management, real estate, banking, and more. Smart contracts promote efficiency, transparency, and dependability by automatically enforcing and carrying out the terms of the agreement without the need for mediators [60].
- **Diverse Use Cases:** Blockchain 2.0 widened the scope of applications beyond finance. It introduced the idea of using blockchain for various purposes, including voting systems, decentralized finance (DeFi), decentralized applications (dApps), identity verification, supply chain management, non-fungible tokens (NFTs), and many more applications.
- **Governance Models:** Blockchain 2.0 also addresses governance issues within decentralized networks. It explores different models for decision-making and consensus among network participants, seeking to create a more democratic and effective governance structure [61].
- **Privacy and Security:** Enhanced privacy features and improved security measures are integral to Blockchain 2.0. Solutions like privacy coins, zero-knowledge proofs as well and other cryptographic techniques focus on increasing privacy and protection by ensuring the security of transactions and data.

3.2.3 Merits of Blockchain 2.0

Blockchain 2.0, with its advancements beyond the original concept of blockchain technology, offers several merits and demerits:

- **Diverse Use Cases:** The applications of blockchain technology go beyond virtual currencies with Blockchain 2.0. It enables applications in non-fungible tokens (NFTs), voting systems, supply chain management, identity verification, and decentralized finance (DeFi) fostering innovation across various industries [62].
- **Smart Contracts:** The introduction of smart contracts automates and executes agreements without intermediaries, improving efficiency, reducing costs, and ensuring the transparent execution of predefined conditions.
- **Interoperability:** Some Blockchain 2.0 initiatives focus on interoperability between different blockchains. This connectivity allows assets and information to move seamlessly across various blockchain networks, enhancing usability and facilitating collaboration between different projects [63].
- **Enhanced Privacy and Security:** Newer blockchain iterations often implement improved privacy features such as zero-knowledge proofs and enhanced security measures. Such advancements aim to secure transactions and sensitive data while maintaining the decentralized nature of the network.

- **Decentralized Applications (dApps):** Blockchain 2.0 facilitates the development and deployment of decentralized applications, providing greater transparency, censorship-resistant, and a user-centred experience compared to traditional centralized applications.

3.2.4 Demerits of Blockchain 2.0

- **Scalability Issues:** Some Blockchain 2.0 platforms still face scalability challenges, struggling to manage an excessive number of transactions efficiently. Scaling solutions are in development but are yet to be universally implemented.
- **Complexity and Adoption:** The technology behind Blockchain 2.0 can be complex for non-technical users or traditional businesses to understand and adopt. Integrating blockchain into existing systems often requires significant adjustments and technological expertise [64].
- **Regulatory Challenges:** As blockchain technology expands its applications, regulatory frameworks struggle to keep pace. The adoption and development of blockchain-based solutions may be limited by unclear or stringent regulations in different jurisdictions.
- **Energy Consumption:** Consensus Mechanisms such as Proof-of-Work (PoW) which is integrated by some blockchains, like Bitcoin and Ethereum (currently), consume substantial energy. This has sparked questions about how sustainable these kinds of systems are and how they affect the ecosystem [65].
- **Security Concerns:** Although the security of the blockchain system is an advantage, vulnerabilities in smart contracts or protocols can still be exploited, leading to significant financial losses or system disruptions.
- Blockchain 2.0 represents a significant advancement in blockchain technology, offering expanded possibilities and addressing some limitations of earlier iterations. However, it also presents several issues that must be resolved before a broad acceptance and further evolution [66].

3.3 Blockchain 3.0

The third generation of blockchain technology is Blockchain 3.0, building upon the foundations of its predecessors (Blockchain 1.0 and 2.0). Its focus is on overcoming the limitations and shortcomings of the earlier versions while introducing advancements in scalability, interoperability, sustainability, and governance [67]. Blockchain 3.0 allows Decentralized Applications (DApps) to execute a large number of transactions per second (TPS). Decentralized Applications (DApps) are a fundamental component of this technology, leveraging its decentralized structure to develop a new breed of apps that runs on a peer-to-peer network rather than centralized servers [68]. Traditional blockchains like Scaling problems with Bitcoin and Ethereum led to increased fees and slower transaction times amid spikes in network activity. Third-generation blockchains use various techniques such as sharding, layer-2 solutions, and sidechains to improve scalability, enabling DApps to handle more users and transactions without congestion. DApps developed on Blockchain 3.0 platforms can communicate and make transactions across different blockchains, fostering interoperability. This means DAps aren't limited to a single blockchain network; they can leverage assets, data, and functionalities from multiple blockchains, expanding their scope and utility [69].

3.3.1 Architecture

The architecture of Blockchain 3.0 using decentralized applications is different from traditional applications. To ensure reliability, high security and privacy DApps require a unique system design. In this architecture the client application directly communicates to the blockchain instead of communicating with the database, like in a centralized system, i.e., in this system application state is not being stored in a centralized database [70]. Decentralized Applications (DApps) operate within a blockchain environment, leveraging its decentralized nature to function. Working of DApps within a blockchain is shown in figure [5] and discussed in detail below:

- **User Interactions:** Users interact with DApps through the frontend interface. They can perform various actions, such as sending/receiving tokens, accessing decentralized finance (DeFi) services, playing games, accessing content, or participating in governance, depending on the DApp's purpose.
- **Frontend Interface:** DApps have a user-facing frontend interface, such as a web app, mobile app, or desktop client. This interface interacts with the backend, enabling participants to access as well as interact with the DApp's services and information saved on the blockchain.
- **Smart Contracts:** As defined earlier in this chapter, Smart contracts are self-executing agreements with their terms recorded in digital code. Since they automate and enforce rules, these contracts are essential to DApps because they make it easier for users to communicate and transact within the application. Smart contracts are transparent and unchangeable; they reside within the blockchain network [71].
- **Ethereum Virtual Machine (EVM):** One essential part of the Ethereum blockchain network is the Ethereum Virtual Machine (EVM). This works as a runtime environment for Ethereum-based smart contracts, or decentralized apps (DApps). The Ethereum blockchain is used to implement smart contracts, and the EVM runs their code. To guarantee that smart contracts are executed consistently throughout the network, each Ethereum node runs a separate instance of the EVM [72].
- **Ethereum Blockchain:** Developers can create and implement decentralized apps (DApps) and smart contracts on the Ethereum blockchain, an open-source, decentralized platform. The native cryptocurrency of the Ethereum network is called ether. It has been utilized to cover transaction fees, reward participants (miners or validators) who execute calculations and validate transactions, and communicate with smart contracts [73].
- **Consensus Mechanism:** Nodes in the blockchain network validate the transaction per the particular blockchain's consensus process (Proof of Work, Proof of Stake, etc.). After verification, the transaction is added to a block and appended to the blockchain, forming a permanent and immutable record [74].

- **Transactions and Validation:** When a user initiates an action, such as a transaction or interaction with the DApp, it triggers a transaction in a blockchain. The transaction executed recently is validated and executed by the smart contracts according to predefined rules and conditions [75].
- **Immutable Record:** All transactions and data related to the DApp's activities are stored in a blockchain, creating a tamper-proof, and transparent as well as auditable history. Users can verify the integrity of the DApp's operations by accessing the blockchain's data [76].

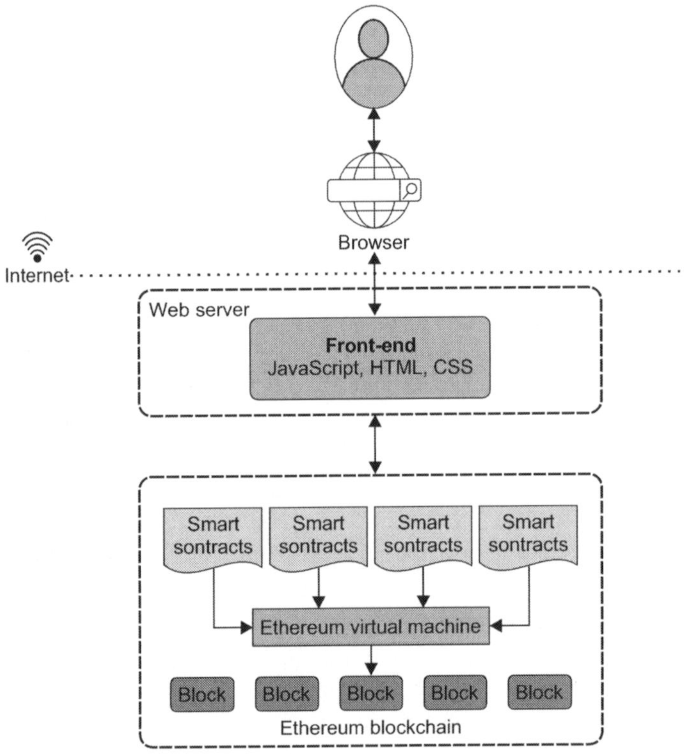

Figure 5: Blockchain 3.0 using DApps.

3.3.2 Areas of Application

Blockchain 3.0, with its focus on scalability, interoperability, sustainability, and advanced governance, extends the applicability of blockchain technology beyond financial applications [77]. Some notable areas where Blockchain 3.0 may make a big difference include:

- **Supply Chain Management:** This version offers improved traceability, transparency, and efficiency in supply chains. It enables the tracking of products at every stage, reducing counterfeiting, ensuring product authenticity, and enhancing overall supply chain logistics.
- **Healthcare:** Blockchain 3.0 can revolutionize healthcare by securely managing patient data, ensuring interoperability between different healthcare providers,

streamlining processes like insurance claims, and enabling the secure sharing of sensitive medical records while maintaining patient privacy.

- **Decentralized Finance (DeFi):** Building on the foundation of Blockchain 2.0, Blockchain 3.0 platforms enhance DeFi by offering more scalability, lower transaction costs, and improved interoperability among different financial protocols. This makes a greater variety of banking offerings possible, including asset management, decentralized exchanges, lending, and borrowing.
- **Identity Management:** Blockchain 3.0 introduces identity solutions that are more verified and secure. By decreasing identity theft and offering a decentralized, un-hackable identity verification mechanism, it gives people more control over their digital identities.
- **Energy and Sustainability:** Sustainability is a key focus of third -generation blockchain, aiming for reducing the environmental impact of this technology. Innovations include energy-efficient consensus mechanisms and platforms that promote green initiatives, such as tracking renewable energy production and carbon credit trading.
- **Cross-Chain Interoperability:** Blockchain 3.0 platforms facilitate seamless communication and interaction between different blockchains. This interoperability allows the exchange of assets and data across various blockchain networks, fostering a more connected and efficient decentralized ecosystem.
- **Gaming and Digital Assets:** Improved scalability and lower transaction costs in Blockchain 3.0 make it suitable for gaming applications, enabling the creation and trading of in-game assets as non-fungible tokens (NFTs). This creates new possibilities for ownership and monetization within the gaming industry.
- **Governance and Voting:** Blockchain 3.0 introduces more sophisticated governance models, enabling decentralized decision-making processes. It can be used to make voting systems tamper-proof and more transparent, enhancing organizational governance and electoral fairness and integrity.

The above-mentioned applications illustrate the capacity of Blockchain 3.0 to disrupt and innovate across various sectors beyond finance, providing solutions for enhanced security, efficiency, and trust in a wide variety of domains and applications.

3.3.3 Merits of Blockchain 3.0

- **Scalability:** Improved scalability is a significant advantage of Blockchain 3.0. Enhanced throughput and transaction speeds enable more efficient processing of transactions across the network, catering to a larger user base without congestion.
- **Interoperability:** Interaction and transactional capabilities between various blockchains represent a significant breakthrough. Interoperability creates a versatile and connected environment, allowing for seamless asset and data exchange between disparate platforms [78].
- **Enhanced Smart Contracts:** Secure, efficient and more intelligent smart contracts are a hallmark of Blockchain 3.0. These contracts offer expanded functionalities and capabilities, enabling complex automation of agreements and processes across various industries [79].

- **Improved Governance:** Novel governance models empower the community to participate in decision-making procedures, enhancing decentralization and enabling more democratic control over network upgrades and protocol changes [80].
- **Sustainability:** Efforts toward energy efficiency and environmental friendliness are commendable in Blockchain 3.0. Innovations aim to reduce the energy consumption required for consensus mechanisms, addressing issues of blockchain technology's effects on the environment [81].

3.3.4 Demerits of Blockchain 3.0

- **Complexity:** The introduction of new features and technologies in Blockchain 3.0 can lead to increased complexity. Developing and understanding these advanced systems might pose challenges for users and developers, potentially limiting widespread adoption.
- **Adoption and Integration:** Integrating and adopting these new blockchain advancements might face resistance or slow adoption due to established systems, interoperability issues, or the need for industry-wide standards.
- **Security Challenges:** As with any technological advancement, new security vulnerabilities might emerge with the implementation of complex systems. Ensuring the robustness and security of these sophisticated networks remains a critical concern.
- **Regulatory Uncertainty:** New features and functionalities might pose regulatory challenges in various jurisdictions. Regulatory frameworks often lag behind technological advancements, creating uncertainty and potential hurdles for Blockchain 3.0 adoption.
- **Transition Challenges:** Transitioning from earlier blockchain versions to Blockchain 3.0 might require significant upgrades, migrations, or protocol changes. These transitions can be time-consuming, and resource-intensive, and may face resistance from stakeholders invested in existing systems.

3.4 Blockchain 4.0

Blockchain 4.0 is a conceptual term used to describe the next evolution or iteration of blockchain technology beyond its predecessors, such as Blockchain 1.0 (Bitcoin), Blockchain 2.0 (Ethereum, smart contracts), and Blockchain 3.0 (scalability, interoperability). It encompasses several advancements and potential features that are envisioned for increasing the capabilities, scalability, interoperability, as well as utility of these networks. However, it's not a universally defined term, it generally refers to the potential developments that could further revolutionize the blockchain space [82]. Improving scalability is a significant focus in Blockchain 4.0. This involves enhancing the throughput of transactions to handle a larger volume without compromising network speed or efficiency. Solutions like sharding, layer-2 scaling solutions, and novel consensus mechanisms aim to address scalability issues. Interoperability is another prominent feature of Blockchain 4.0, enabling different blockchains to communicate and interact seamlessly is a crucial aspect [83]. This version of blockchain aims to break down the barriers between various blockchain networks, allowing them to share data and assets across platforms securely. Blockchain 4.0 also explores the different ways to integrate artificial intelligence with blockchain for improving data analysis, decision-making, and automation within decentralized systems. These features represent a vision of the future of this technology, aiming to overcome limitations and issues faced by earlier iterations. On the other hand, it's essential to remember that Blockchain 4.0 currently represents only an idea, and its realization depends on ongoing research, development, and collaborative efforts within the blockchain community [84].

3..4.1 Architecture

Blockchain 4.0 has innovation as its main goal. Blockchain 4.0 will prioritize usability, speed, and user experience for a wider audience [85]. Applications for Blockchain 4.0 can be split into two vertical categories:

- **Web 3.0 And Blockchain 4.0**

  The term "Web 3.0" describes the next stage of internet development, which aims to produce a web experience that is more transparent, decentralized, and user-centric. Unlike Web 1.0 (the early internet with static web pages) and Web 2.0 (User-generated content and social media on the social and interactive web), Web 3.0 envisions a more interconnected and decentralized internet [86]. In the context of Web 3.0, blockchain serves as a foundational technology that underpins non-fungible tokens (NFTs), smart contracts, decentralized applications, and decentralized finance (DeFi) are just a few examples. These blockchain-based applications aim to create more transparent, secure, and user-controlled digital experiences, aligning with the principles of Web 3.0 [87]. A key component of blockchain technology is in the realization of Web 3.0 due to its core characteristics that are discussed below:

i. **Decentralization:** Blockchains are decentralized ledgers where across a network of computers, or nodes, data is distributed. This aligns to decentralize the internet in Web 3.0, reducing reliance on centralized authorities [88].

ii. **Security:** The cryptographic principles of blockchains ensure data integrity and security, addressing concerns related to privacy and data tampering.

iii. **Trust:** The transparent and immutable nature of blockchain ledgers fosters trust among participants without the requirement for mediators or intermediaries.

iv. **Tokenization:** Blockchain makes it possible to create and manage tokens (digital assets) representing various real-world or digital entities, supporting new economic models and DApps [89].

Together, Web 3.0 and blockchain technology strive to reshape the internet and digital interactions by promoting decentralization, data ownership, and trust in online ecosystems.

- **Metaverse And Blockchain 4.0**

  The collective virtual shared space known as the metaverse, is often envisioned as a convergence of physical reality, Virtual reality (VR), augmented reality (AR), and other digital environments. It is a persistent, immersive, and interconnected digital universe that allows users inter action, a place to socialize, work, play, and transact with each other [90]. The amalgamation of Blockchain 4.0's advancements with the metaverse holds the potential to create a more vibrant, interconnected, and user-driven virtual universe. Whereas, it is essential to remember that both concepts are still in the development stage, and their convergence is subject to ongoing technological developments and collaborations within the tech community. Blockchain is integrated with the metaverse to enhance user experiences, foster economic activities, establish ownership and governance models, and create more secure and interconnected virtual environments [91]. As both blockchain and the metaverse continue to evolve, their combined applications are likely to expand, offering new possibilities and innovations for users and developers alike.

3.4.2 Areas of Application

- **Digital Asset Ownership:** Blockchain's role in the metaverse involves facilitating ownership and trade of digital assets within these virtual environments. Blockchain's ability to tokenize assets and establish ownership through non-fungible tokens (NFTs) allows users to possess unique digital items like virtual real estate, avatars, collectibles, and more [92]. This ownership is transparent, secure, and transferable across different metaverse platforms.
- **Decentralized Economy:** The creation of decentralized economies within the metaverse is enabled by blockchain. It allows for the development of in-world currencies, marketplaces, and economies in which users can create, purchase, sell, as well as trade goods and services. These transactions on blockchain are governed by smart contracts, ensuring transparency and security [93].
- **Interoperability and Standards:** Blockchain 4.0 aims to enhance interoperability between different blockchain networks. This can be crucial in the metaverse, where various platforms, games, and virtual worlds exist. Interoperability standards can enable seamless movement of data and assets among these distinct environments, developing a more unified metaverse experience [94].
- **Identity and Reputation Systems:** Identity solutions based on blockchain can provide users with self-sovereign identities within the metaverse. This allows for secure and portable digital identities across different virtual platforms. Reputation systems built on blockchain can also establish trust and reliability among users within the metaverse [95].
- **Decentralized Governance:** Blockchain's decentralized governance models, like DAOs (Decentralized Autonomous Organizations), are essential in governing certain elements of the metaverse. Community-driven decision-making processes can shape the rules, policies, and developments within virtual worlds [96].
- **Immersive Experiences and AI Integration:** The integration of AI technologies within the metaverse, coupled with blockchain, can create more immersive and responsive experiences. AI algorithms can personalize interactions and

environments based on user preferences and behaviours, enhancing the overall user experience [97].

3.4.3 Merits of Blockchain 4.0

- **Scalability Improvements:** Addresses the scalability issues of earlier blockchains, aiming to handle higher transaction throughput without compromising speed or efficiency.
- **Enhanced Interoperability:** Aims to facilitate seamless communication and interaction between different blockchains, fostering data sharing and asset transfer across platforms.
- **Improved Privacy and Security:** Introduces advanced privacy features, like zero-knowledge proofs and enhanced cryptography, to maintain confidentiality without sacrificing transparency. Additionally, heightened security measures bolster protection against hacks and fraud.
- **Advanced Smart Contracts:** Smart contracts become more versatile, secure, efficient, and capable of executing complex operations with improved auditing and safety.
- **Decentralized Governance Models:** Innovations in governance facilitate decentralized decision-making and community-driven participation in network governance, promoting fairness and inclusivity.
- **AI Integration and Emerging Technologies:** Integrating AI with various other emerging technologies could improve data analysis, automation, and decision-making within decentralized systems.
- **Tokenization of Assets:** Further expansion of tokenization allows a representation of a broader range of real-world assets and their trading over blockchain networks, unlocking new use cases across industries.

3.4.4 Demerits of Blockchain 4.0

- **Complexity:** Implementing advanced features can make blockchain systems more complex, potentially leading to increased development and operational challenges.
- **Adoption and Education:** As with any new technology, widespread adoption might face hurdles due to the need for education and familiarization with the advancements of Blockchain 4.0 among developers and users.
- **Regulatory Uncertainty:** Regulatory frameworks often struggle to keep up with rapid technological advancements. The distinctive features of Blockchain 4.0 can create difficulties for regulatory adaption and compliance.
- **Technological Risks:** The introduction of new features and technologies might bring unforeseen risks or vulnerabilities that need to be addressed through rigorous testing and ongoing development.

- **Energy Efficiency:** While advancements in consensus mechanisms aim for energy efficiency, some implementations might still have environmental impacts, necessitating a balance between functionality and sustainability.

4. Evolution Flow from Blockchain 1.0 to Blockchain 4.0

Each version builds upon the previous one, aiming to overcome limitations and enhance blockchain technology's capabilities for broader applications beyond cryptocurrencies. The following is an overview of the evolution of blockchain from its initial stages to the more advanced versions:

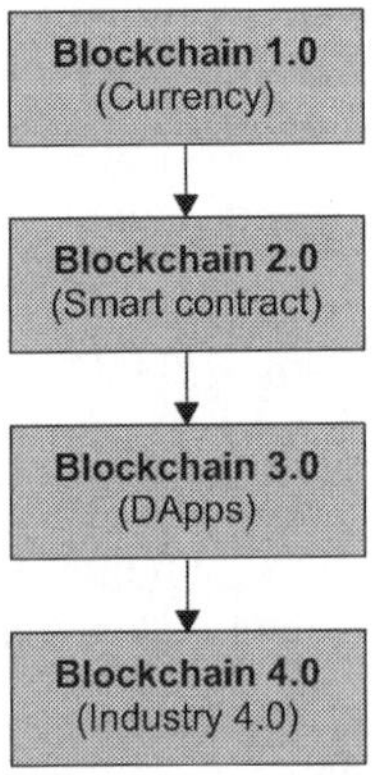

Figure 6: Evolution of Blockchain.

4.1 Blockchain 1.0—Bitcoin

- This phase begins with the creation of Bitcoin by Satoshi Nakamoto in 2009.
- Its main focus was on digital currency and decentralized transactions.
- The emergence of a peer-to-peer, decentralized electronic cash system was the primary innovation.

4.2 Blockchain 2.0—Smart Contracts and Ethereum

- Vitalik Buterin introduced Ethereum in 2015 and expanded the capabilities of blockchain beyond currency.
- It enabled the creation of self-executing contracts called "smart contracts," which have the conditions of the agreement explicitly encoded into the code.
- This phase introduced the concept of decentralized applications (dApps), allowing developers to build various applications on the blockchain.

4.3 Blockchain 3.0—Scalability and Interoperability

- This phase focused on addressing scalability and interoperability issues.

- Projects like Cardano, Polkadot, and Cosmos aimed to improve blockchain scalability, allowing for more transactions per second (TPS) without compromising decentralization.
- Interoperability protocols emerged to enable different blockchains to communicate and share data securely.

4.4 Blockchain 4.0—Advanced Features and Integration

- Blockchain 4.0 is an anticipated phase focusing on advanced features and further integration.
- It focuses on solving remaining problems such as scalability, sustainability, and governance.
- Features like sharding (breaking the blockchain into smaller parts for efficiency), advanced consensus mechanisms, AI integration, quantum resistance, and enhanced privacy are expected.
- Additionally, this phase might see increased integration with artificial intelligence (AI), the Internet of Things (IoT), and other cutting-edge technologies.

This table provides a broad overview and the evolution of blockchain technology across generations based on various parameters. Each generation brought advancements to overcome limitations and improve the capabilities of the preceding one.

5. Major Challenges to Deploy Blockchain

Deploying blockchain technology comes with several challenges, including:

i. **Scalability:** Networks of the blockchain, mainly early ones such as Bitcoin and Ethereum, struggle with scalability. As more transactions occur, the network can become slower and more expensive. Scaling solutions are essential to accommodate more throughput per second without compromising decentralization [98].

ii. **Interoperability:** Different blockchains often operate in isolation, leading to interoperability issues. Creating a seamless flow of data and value between different blockchain networks is an issue that needs to be addressed for widespread adoption and better communication between systems [99].

iii. **Security:** Although blockchain's decentralized structure and use of cryptographic methods make it notable for its security, it is not resistant to specific inefficiencies. 51% of attacks (in proof-of-work systems), smart contract bugs, also hacking incidents on exchanges are among the security concerns that need continuous attention and improvement [100].

iv. **Regulatory Compliance:** Blockchain technology operates across borders, which can create regulatory challenges. Varying regulations and legal frameworks around the world regarding cryptocurrencies, tokenization, and decentralized finance (DeFi) pose challenges for deployment and adoption [101].

Table 2: Comparison of different generations of blockchain.

Parameters	**First Generation**	**Second Generation**	**Third Generation**	**Fourth Generation**
Consensus Mechanism	Proof of Work (PoW)	Proof of Work (PoW)/Proof of Stake (PoS)	Proof of Stake (PoS)/Delegated Proof of Stake (DPoS)	Asynchronous Byzantine Fault Tolerance (aBFT)/Pure Proof of Stake (PPoS)
Smart Contracts	Limited scripting language	Turing-complete smart contracts	Advanced smart contract capabilities	Efficient and fast smart contract execution
Scalability	Limited scalability	Improved scalability through sharding and sidechains	Scalability solutions implemented, some with interoperability	Higher throughput and faster transaction speeds
Interoperability	Limited interoperability	Basic interoperability	Emphasis on cross-chain communication	Enhanced interoperability and compatibility
Governance	Decentralized	Some elements of decentralization	Focus on decentralized governance models	Enhanced decentralized governance models
Security	Generally secure	Vulnerable to certain exploits	Enhanced security measures and focus	Emphasis on security and resistance to attacks
Use Cases	Primarily cryptocurrency	Smart contracts and DApps	Diverse applications across industries	Wide-ranging applications, including finance, IoT, etc.
Energy Efficiency	Less energy-efficient	Moving towards energy efficiency	Efforts to improve energy efficiency	Focus on energy-efficient consensus mechanisms

v. **User Experience (UX):** Creating user-friendly interfaces and experiences for blockchain applications is crucial for mass adoption. The complexity of wallets, keys, and transactions can be intimidating for non-technical users. Improving the UX is essential to make blockchain more accessible [102].

vi. **Energy Consumption and Sustainability:** Some consensus mechanisms in blockchain, like proof-of-work (PoW), consume substantial amounts of energy. The sustainability of blockchain technology and its effects on the environment has been placed into serious consideration due to this. Developing energy-efficient consensus mechanisms is crucial [103].

vii. **Privacy:** While blockchains are transparent, maintaining privacy for sensitive information is essential. Balancing transparency with data privacy and confidentiality is a challenge, especially in public blockchains [104].

Research must be done continuously to address these issues, development, collaboration, and innovation in the blockchain domain. Many projects and initiatives are working on solutions to overcome these hurdles and make blockchain technology more scalable, secure, and user-friendly for widespread adoption.

6. Conclusion

The development of blockchain technology, traversing from its rudimentary form in Blockchain 1.0 to the sophistication of Blockchain 4.0, embodies a remarkable journey of innovation and transformative potential. This progression mirrors the maturation of an idea into a disruptive force, reshaping industries and societal norms. Blockchain 1.0 laid the groundwork with B itcoin, showcasing the power of decentralized currency and immutable ledgers. Its simplicity paved the way for Blockchain 2.0, marked by Ethereum's smart contracts, enabling programmable transactions and decentralized applications. This phase expanded possibilities, fostering a diverse ecosystem of decentralized finance, supply chain management, and more. Blockchain 3.0 saw scalability and interoperability become focal points, addressing limitations in speed, energy consumption, and cross-chain operability. Solutions emerged, seeking to reconcile the trade-offs between decentralization, security, and efficiency.

Now, blockchain 4.0 heralds a new era, integrating cutting-edge technologies like Artificial Intelligence, the Internet of Things, and Q uantum C omputing. It promises enhanced security, scalability, and governance models, setting the stage for unprecedented applications in healthcare, governance, and beyond. The emphasis shifts toward sustainability, inclusivity, and user-centric design, aiming to bridge the gap between blockchain technology and mainstream acceptance. The evolution of blockchain is not merely a technical narrative but a testament to human ingenuity and collaborative progress. Each iteration builds upon the last, addressing limitations and unlocking new potentials. As we stand on the cusp of blockchain 4.0, the journey continues, beckoning us to explore uncharted territories, stretching beyond the boundaries of what's achievable, and ultimately, shaping the future of decentralized innovation.

References

1. Zheng, Z., Xie, S., Dai, H-N., Chen, X., and Wang, H. 2018. Blockchain challenges and opportunities: A survey. *International journal of web and grid services 14*(4), 352–375.
2. Mukherjee, P., and Pradhan, C. 2021. Blockchain 1.0 to blockchain 4.0—The evolutionary transformation of blockchain technology. Blockchain technology: applications and challenges. *Cham: Springer International Publishing*, 29–49.
3. 'A timeline and history of blockchain technology' by Sheldon, R. 2021. https://whatis.techtarget.com/feature/A-timeline-and-history-of-blockchain-technology. Accessed 18 Jan 2022
4. Patel, M.M., Tanwar, S., Gupta, R., and Kumar, N. 2020. A deep learning-based cryptocurrency price prediction scheme for financial institutions. *J. Inf. Secur. Appl. 55*: 102583. https://doi.org/10.1016/j.jisa.2020.102583. ISSN 2214–2126.
5. 'Blockchain tutorial: history of blockchain' by java T point. 2021. https://www.javatpoint.com/history-of-blockchain. Accessed 18 Jan 2022
6. Gupta, R., Tanwar, S., Kumar,. N., and Tyagi, S. 2019. Blockchain-based security attack resilience schemes for autonomous vehicles in industry 4.0: a systematic review. *Comput. Electr. Eng. 86*. https://doi.org/10.1016/j.compeleceng.2020.106717
7. Kakkar, R., Gupta, R., Tanwar S., and Rodrigues, J.J.P.C. 2022. Coalition game and blockchain-based optimal data pricing scheme for ride sharing beyond 5G. IEEE Syst J. https://doi.org/10.1109/JSYST.2021.3126620
8. 'History of blockchain' by Binance Academy. 2021. https://academy.binance.com/en/articles/history-of-blockchain. Accessed 18 Jan 2022
9. 'Ethereum: decentralized applications and autonomous organizations' by Zeitgeist Lab. 2021. http://www.zeitgeistlab.ca/doc/Ethereum_Decentralized_Applications_Autonomous_Organizations.html. Accessed 18 Jan 2022
10. 'Ethereum' by Wikipedia. 2022. https://en.wikipedia.org/wiki/Ethereum. Accessed 18 Jan 2022
11. Akram, S.V., Malik, P., Singh, R., Gehlot, A., and Tanwar, S. 2020. Adoption of blockchain technology in various realms: opportunities and challenges. Secur Priv 3:e109. https://doi.org/10.1002/spy2.109
12. 'Blockchain evolution: from 1.0 to 4.0' by Unibright.io 2017. https://unibrightio.medium.com/blockchain-evolution-from-1-0-to-4-0-3fbdbccfc666. Accessed 20 Jan 2022
13. Ølnes, S., Ubacht, J., and Janssen, M. 2017. Blockchain in government: benefits and implications of distributed ledger technology for information sharing. *Gov. Inf. Q., 34*(3), 355–364. https://doi.org/10.1016/j.giq.2017.09.007. ISSN 0740-624X
14. 'A brief history of blockchain' by Future Learn. 2022. https://www.futurelearn.com/info/courses/introduction-to-blockchain-dlt/0/steps/250288. Accessed 20 Jan 2022
15. Kumari, A., Gupta, R., Tanwar, S., Tyagi, S., and Kumar, N. 2020. When blockchain meets smart grid: secure energy trading in demand response management. *IEEE Netw, 34*(5), 299–305. https://doi.org/10.1109/MNET.001.1900660
16. Narayanan, A., Bonneau, J., Felten, E., Miller, A., and Goldfeder, S. 2016. Bitcoin and cryptocurrency technologies: a comprehensive introduction. Princeton University Press, Princeton
17. MATH Google Scholar
18. Xu, M., Chen, X., and Kou, G. 2019. A systematic review of blockchain. *Financ. Innov., 5*: 27. https://doi.org/10.1186/s40854-019-0147-z
19. 'Blockchain transaction life-cycle' by GeeksforGeeks. 2021. https://www.geeksforgeeks.org/blockchain-transaction-life-cycle/. Accessed 20 Jan 2022
20. Ankalkoti, P., and Santhosh, S.G. 2017. A relative study on bitcoin mining. *Imp. J. Interdiscip. Res.* (IJIR) 3
21. Google Scholar
22. 'Blockchain 2.0 - from bitcoin transactions to smart contract applications' by Kehrli, J. 2016. https://www.niceideas.ch/roller2/badtrash/entry/blockchain-2-0-from-bitcoin. Accessed 20 Jan 2022
23. Bhutta, M. N. M. et al. 2021. A survey on blockchain technology: evolution, architecture and security. *IEEE Access,* 9: 61048–61073. https://doi.org/10.1109/ACCESS.2021.3072849
24. 'Blockchain 2.0, smart contracts and challenges' by von Haller Gronback, M. 2016. https://www.twobirds.com/en/news/articles/2016/uk/blockchain-2-0--smart-contracts-and-challenges. Accessed 21 Jan 2022

25. '101 smart contracts and decentralized apps in ethereum' by Cibraro, P. 2021. https://auth0.com/blog/101-smart-contracts-and-decentralized-apps-in-ethereum/. Accessed 21 Jan 2022
26. Chen, T., Li, X., Luo, X., and Zhang, X. 2017. Under-optimized smart contracts devour your money. In: 2017 IEEE 24th international conference on software analysis, evolution and reengineering (SANER), pp. 442–446. https://doi.org/10.1109/SANER.2017.7884650
27. 'Blockchain 3.0 and the future of the decentralized internet' by Sara Technologies Inc. 2018. https://saratechnologiesinc.medium.com/blockchain-3-0-the-future-of-the-decentralized-internet-63ba199e2a5. Accessed 21 Jan 2022
28. 'Solving the blockchain scalability issue: sharding vs sidechains' by Willemse, L. 2018. https://medium.com/swlh/solving-the-blockchain-scalability-issue-sharding-vs-sidechains-1b0c6d1f6c0d. Accessed 21 Jan 2022
29. 'Explaining directed acylic graph (DAG), the real blockchain 3.0' by Lee, S. 2018. https://www.forbes.com/sites/shermanlee/2018/01/22/explaining-directed-acylic-graph-dag-the-real-blockchain-3-0/?sh=3016383b180b. Accessed 21 Jan 2022
30. 'A complete guide to directed acyclic graph (DAG)' by Chamira, R. 2022. https://www.sofocle.com/blog/a-complete-guide-to-directed-acyclic-graph-dag/. Accessed 21 Jan 2022
31. 'Can blockchain 3.0 finally make the technology more mainstream' by Joshi, N. 2019. https://www.allerin.com/blog/can-blockchain-3-0-finally-make-the-technology-more-mainstream. Accessed 21 Jan 2022
32. 'InterValue the world's first practical blockchain 4.0' by Rahaman, A. 2018. https://medium.com/@aminurrahaman/intervalue-the-worlds-first-practical-blockchain-4-0-b9324878c262. Accessed 21 Jan 2022
33. 'InterValue's TestNet 2.0 release: implementation of HashNet consensus, committed to building the practical infrastructure of the Blockchain 4.0 era' by InterValue 201. https://www.globenewswire.com/news-release/2018/06/28/1531189/0/en/InterValue-s-TestNet-2-0-release-implementation-of-HashNet-consensus-committed-to-building-the-practical-infrastructure-of-the-Blockchain-4-0-era-.html. Accessed 21 Jan 2022
34. 'Blockchain 4.0' by Takyar, A. 2022. https://www.leewayhertz.com/blockchain-4-0/. Accessed 21 Jan 2022
35. Monrat, A. A., Schelén, O., and Andersson, K. 2019. A survey of blockchain from the perspectives of applications, challenges, and opportunities. *IEEE Access 7*: 117134–117151.
36. Belotti, M., Božić, N., Pujolle, G., and Secci, S. 2019. A vademecum on blockchain technologies: When, which, and how. *IEEE communications surveys & tutorials 21*(4): 3796–3838.
37. Nofer, M., Gomber, P., Hinz, O., and Schiereck, D. 2017. Blockchain. Business & Information Systems Engineering *59*(3), 183–187.
38. Wüst, K., and Gervais, A. 2018. Do you need a blockchain? In 2018 Crypto Valley Conference on Blockchain Technology (CVCBT), pp. 45–54. IEEE.
39. Yaga, D., Mell, P., Roby, N., and Scarfone, K. 2019. Blockchain technology overview. arXiv preprint arXiv:1906.11078 .
40. Li, X., Jiang, P., Chen, T., Luo, X., and Wen, Q. 2020. A survey on the security of blockchain systems. *Future generation computer systems 107*, 841–853.
41. Dinh, T.T.A., Liu, R., Zhang, M., Chen, G., Ooi, B.C., and Wang, J. 2018.Untangling blockchain: A data processing view of blockchain systems. *IEEE transactions on knowledge and data engineering*, 30(7), 1366–1385.
42. Pilkington, M. 2016. Blockchain technology: principles and applications. In Research handbook on digital transformations. Edward Elgar Publishing.
43. Tasatanattakool, P., and Techapanupreeda, C. 2018. Blockchain: Challenges and applications. In 2018 International Conference on Information Networking (ICOIN), pp. 473–475. IEEE.
44. Bodkhe, U., Tanwar, S., Parekh, K., Khanpara, P., Tyagi, S., Kumar, N., and Alazab, M. 2020. Blockchain for industry 4.0: A comprehensive review. *IEEE Access, 8*, 79764–79800.
45. Risius, M., and Spohrer, K. 2017. A blockchain research framework. *Business & Information Systems Engineering 59*(6), 385–409.
46. Golosova, J., and Romanovs, A. 2018. The advantages and disadvantages of the blockchain technology. In 2018 IEEE 6th workshop on advances in information, electronic and electrical engineering (AIEEE), pp. 1–6. IEEE.

47. Gatteschi, V., Lamberti, F., Demartini, C., Pranteda, C., and Santamaria, V. 2018. To blockchain or not to blockchain: That is the question. *It Professional, 20*(2), 62–74.
48. Xu, M., Chen, X., and Kou, G. 2019. A systematic review of blockchain. *Financial Innovation,* 5(1), 1–14.
49. Aste, T., Tasca, P., and Di Matteo, T. 2017. Blockchain technologies: The foreseeable impact on society and industry.
50. Maesa, D.D.F, and Mori, P. 2020. Blockchain 3.0 applications survey. *Journal of Parallel and Distributed Computing, 138*, 99–114.
51. Syed, T. A., Alzahrani, A., Jan, S., Siddiqui, M.S., Nadeem, A., and Alghamdi, T. 2019. A comparative analysis of blockchain architecture and its applications: Problems and recommendations. *IEEE access, 7*, 176838–176869.
52. Gorkhali, A., Li, L., and Shrestha, A. 2020. Blockchain: A literature review. *Journal of Management Analytics, 7*(3), 321–343.
53. Halaburda, H. 2018. Blockchain revolution without the blockchain? *Communications of the ACM, 61*(7), 27–29.
54. Al-Jaroodi, J., and Mohamed, N. 2019. "Blockchain in industries: A survey. *IEEE Access,* 7, 36500–36515.
55. Zheng, Z., Xie, S., Dai, H., Chen, X., and Wang, H. 2017. An overview of blockchain technology: Architecture, consensus, and future trends. In 2017 IEEE international congress on big data (BigData congress), pp. 557–564. Ieee.
56. Brilliantova, V., and Thurner, T.W. 2019. Blockchain and the future of energy. *Technology in Society, 57*, 38–45.
57. Sheth, H., and Dattani, J. 2019. Overview of blockchain technology. Asian Journal For Convergence In Technology (AJCT) ISSN-2350-1146.
58. Samaniego, M., Jamsrandorj, U., and Deters, R. 2016. Blockchain as a Service for IoT. In 2016 IEEE international conference on internet of things (iThings) and IEEE green computing and communications (GreenCom) and IEEE cyber, physical and social computing (CPSCom) and IEEE smart data (SmartData), pp. 433–436. IEEE.
59. https://www.euromoney.com/learning/blockchain-explained/the-rise-of-private-blockchains
60. https://openledger.info/insights/consortium-blockchains/
61. https://blockchain.intellectsoft.net/blog/how-the-consortium-blockchain-works/
62. Agbo, C. C., Mahmoud, Q.H., and Eklund, J.M. 2019. Blockchain technology in healthcare: a systematic review. *In Healthcare, 7*(2), 56. MDPI, 2019.
63. Wang, Q., and Su, M. 2020. Integrating blockchain technology into the energy sector—from theory of blockchain to research and application of energy blockchain. *Computer Science Review, 37*, 100275.
64. Berdik, D., Otoum, S., Schmidt, N., Porter, D., and Jararweh, Y. 2021. A survey on blockchain for information systems management and security. *Information Processing & Management, 58*(1), (2021): 102397.
65. Kshetri, N., and Voas, J. 2018. Blockchain in developing countries. *It Professional, 20*(2), 11–14.
66. Lin, I -C., and Liao, T-C. 2017. A survey of blockchain security issues and challenges. *Int. J. Netw. Secur., 19*(5), 653–659.
67. Politou, E., Casino, F., Alepis, E., and Patsakis, C. 2019. Blockchain mutability: Challenges and proposed solutions. *IEEE Transactions on Emerging Topics in Computing, 9*(4), 1972–1986.
68. De Filippi, P., Mannan, M., and Reijers, W. 2022. The alegality of blockchain technology. *Policy and Society, 41*(3), 358–372.
69. Feng, Q., He, D., Zeadally, S., Khan, M.K., and Kumar, N. 2019. A survey on privacy protection in blockchain system. *Journal of Network and Computer Applications, 126*, 45–58.
70. Mingxiao, D., Ma, X., Zhang, Z, Wang, X., and Chen, Q. 2017. A review on consensus algorithm of blockchain. In 2017 IEEE international conference on systems, man, and cybernetics (SMC), pp. 2567–2572. IEEE.
71. Yeoh, P. 2017. Regulatory issues in blockchain technology. *Journal of Financial Regulation and Compliance.*
72. Dasgupta, D., Shrein, J.M., and Datta Gupta, K.. 2019. A survey of blockchain from security perspective. *Journal of Banking and Financial Technology, 3*(1), 1–17.

73. Shahnaz, A., Qamar, U., and Khalid, A. 2019. Using blockchain for electronic health records. IEEE Access 7, 147782–147795.
74. Mohamed, N., and Al-Jaroodi, J. 2019. Applying blockchain in industry 4.0 applications. In 2019 IEEE 9th annual computing and communication workshop and conference (CCWC), pp. 0852–0858. IEEE.
75. Fernández-Caramés, Tiago M., and Fraga-Lamas, P. 2018. A Review on the Use of Blockchain for the Internet of Things. *Ieee Access,* 6, 32979–33001.
76. Crosby, M., Pattanayak, P., Verma, S., and Kalyanaraman, V. 2016. Blockchain technology: beyondbitcoin. *Appl. Innov.* 2(6–10), 71.
77. Dorri, A., Kanhere, S.S., and Jurdak, R. 2017. Towards an optimized blockchain for IoT. In 2017 IEEE/ACM Second International Conference on Internet-of-Things Design and Implementation (IoTDI), pp. 173–178. IEEE.
78. Kan, L., Wei, Y., Muhammad, A.H., Siyuan, W., Gao, L.C., and Kai, H. 2018. A multiple blockchains architecture on inter-blockchain communication. In 2018 IEEE international conference on software quality, reliability and security companion (QRS-C), pp. 139–145. IEEE.
79. Wang, X., Zha, X., Ni, W., Liu, R.P., Guo, Y. J., Niu, X., and Zheng, K. 2019. Survey on blockchain for Internet of Things. *Computer Communications, 136*: 10–29.
80. Kuo, T -T., Kim, H-E., and Ohno-Machado, L. 2017. Blockchain distributed ledger technologies for biomedical and health care applications. *Journal of the American medical informatics association, 24*(6), 1211–1220.
81. McGhin, T., Choo, K-K.R., Liu, C.Z., and He, D. 2019. Blockchain in healthcare applications: Research challenges and opportunities. *Journal of network and computer applications, 135,* 62–75.
82. Yang, Q., Zhao, Y., Huang, H., Xiong, Z., Kang, J., and Zheng, Z. 2022. Fusing blockchain and AI with metaverse: A survey. *IEEE Open journal of the computer society, 3,* 122–136.
83. Alam, A. 2022. "Platform Utilising Blockchain Technology for eLearning and Online Education for Open Sharing of Academic Proficiency and Progress Records. In Smart Data Intelligence, pp. 307–320. Springer, Singapore.
84. Yue, K., Zhang, Y., Chen, Y., Li, Y., Zhao, L., Rong, C., and Chen, L. 2021. A survey of decentralizing applications via blockchain: The 5g and beyond perspective. *IEEE Communications Surveys & Tutorials, 23*(4), 2191–2217.
85. Pilkington, M. 2016. Blockchain technology: principles and applications. In: Research Handbookon Digital Transformations. Edward Elgar Publishing.
86. Zheng, Z., Xie, S., Dai, H., Chen, X., and Wang, H. 2017. An overview of blockchain technology:architecture, consensus, and future trends. In: 2017 IEEE International Congress on Big Data(BigData Congress), pp. 557–564. IEEE.
87. Yaga, D., Mell, P., Roby, N., and Scarfone, K. 2019. Blockchain technology overview. arXiv preprintarXiv:1906.11078.
88. Nofer, M., Gomber, P., Hinz, O., and Schiereck, D. 2017. *Blockchain. Bus. Inf. Syst. Eng. 59*(3), 183–187.
89. Preneel, B.1994. Cryptographic hash functions. Eur. Trans. *Telecommun.* 5(4), 431–448.
90. Bakhtiari, S., Safavi-Naini, R., and Pieprzyk, J. 1995. Cryptographic hash functions: a survey, vol. 4. Technical Report 95-09, Department of Computer Science, University of Wollongong.
91. Carlozo, L. 2017. What is blockchain? *J. Account. 224*(1), 29.
92. Nakamoto, S. 2008. Bitcoin: peer-to-peer electronic cash system.
93. Underwood, S. 2016. Blockchain beyond bitcoin.
94. Urquhart, A. 2016. The inefficiency of Bitcoin. *Econ. Lett. 148*, 80–82.
95. Baliga, A. 2017. Understanding blockchain consensus models. *Persistent 2017*(4), 1–14.
96. Mohanta, B.K., Panda, S.S., and Jena, D. 2018. An overview of smart contract and use cases inblockchain technology. In: 2018 9th International Conference on Computing, Communicationand Networking Technologies (ICCCNT), pp. 1–4. IEEE.
97. Cachin, C. 2006. Architecture of the hyperledger blockchain fabric. In: Workshop on DistributedCryptocurrencies and Consensus Ledgers, *310*(4).
98. Abrar, W. 1990. Untraceable electronic cash with Digicash,
99. Friis, J.B. 2003. Digicash Implementation. University of Aarhus.

100. Haber, S., and Stornetta, W.S. 1990. How to time-stamp a digital document. In: Conference on the Theoryand Application of Cryptography, pp. 437–455. Springer, Berlin, Heidelberg.
101. Bayer, D., Haber, S., and Stornetta, W.S. 1993. Improving the efficiency and reliability of digital time-stamping. In: Sequences Ii, pp. 329–334. Springer, New York, NY.
102. Singh, S., Sanwar Hosen, A.S.M., and Yoon, B. 2021. "Blockchain security attacks, challenges, and solutions for the future distributed iot network." IEEE Access 9 : 13938–13959.
103. Zyskind, G., Nathan, O. et al. 2015 'Decentralizing privacy: Using blockchain to protect personal data', Security and Privacy Workshops (SPW), 2015 IEEE, IEEE, pp.180–184.
104. Zheng, Z., et al. 2018. "Blockchain challenges and opportunities: A survey." *International journal of web and grid services* 14.4 : 352–375.

CHAPTER 3

Traditional AI vs Modern AI

Rishika Aggarwal,[1,*] *Shachi Sachan,*[1] *Rajat Verma,*[1] and *Namrata Dhanda*[2]

1. Introduction

The early phases of Artificial Intelligence (AI) [19] development may be seen in Traditional AI, which is characterized by rule-based systems [8] and expert knowledge encoding. This period, which mostly covered the latter half of the 20th century, saw AI researchers concentrating on developing computer systems that could replicate human intelligence [46] by adhering to specific rules and directives. These systems established the field and provided the ground for later developments, despite their limitations in comparison to contemporary AI methods. The focus of conventional AI was on manually creating rules and logical representations to model certain problem areas. Expert systems, a well-known example of conventional AI, sought to replicate the knowledge of human experts in a certain topic. To make choices and offer answers to challenging issues, these systems integrated an inference engine with a knowledge store of facts and rules. Using a set of criteria based on symptoms and medical expertise, for instance, an expert [7] system in medical diagnosis may recommend probable disorders. Traditional AI systems showed some degree of success in specific, well-defined fields despite their limitations. They demonstrated how formal logic and symbolic thinking may be used to simulate human competence. These systems, however, frequently have trouble scaling to address larger settings and resolving ambiguity and uncertainty. They were inflexible when confronted with novel or unexpected situations because of their rigorous dependence on manually coded rules.

Modern AI [16] emerged when more data-driven and adaptable methods replaced previous AI techniques as technology and research [28] advanced. AI research has increasingly emphasized machine learning, especially deep learning [37], which enables computers to learn from data and carry out tasks that were

[1] Department of Computer Science and Engineering, Pranveer Singh Institute of Technology, Kanpur, Uttar Pradesh, India.

[2] Department of Computer Science and Engineering, Amity University, Uttar Pradesh, Lucknow Campus, India.

* Corresponding author

difficult or impossible for rule-based systems in the past. In conclusion, classical AI, which is defined by rule-based systems and the encoding of expert knowledge, is an early stage in the evolution of AI. Although these methods had inherent drawbacks, they paved the way for the development of AI into more sophisticated and flexible forms. Researchers began utilizing the potential of data-driven learning to construct intelligent systems with increased autonomy and capabilities, Signaling a fundamental change from classic AI to contemporary AI.

2. Traditional Artificial Intelligence

This also known as classical artificial intelligence [29] or symbolic artificial intelligence, is the early stage of artificial intelligence research and development that concentrates on developing computer [32] programs and systems that imitate human intellect by using explicit rules, logical representations, and symbolic reasoning. In conventional AI, algorithms and rule-based systems use human knowledge and skill to solve certain issues and carry out tasks.

2.1 Attributes of Traditional Artificial Intelligence

Traditional Artificial Intelligence (AI) is a term that describes the early methods and procedures employed in AI research before deep [35] learning and neural networks became popular. These strategies are distinguished by their dependence on symbolic thinking, rule-based frameworks, and subject-matter expertise. Some of the main characteristics of conventional AI are as follows:

- **Explicit Programming:** Explicit programming of rules and logic statements is the foundation of traditional AI systems. Human specialists use a set of preset rules and conditions to encode their knowledge and problem-solving techniques. These regulations specify how the system ought to react to certain inputs and circumstances.
- **Symbolic Representation:** Traditional AI frequently uses symbolic language to describe knowledge, with ideas, connections, and rules being represented by symbols and logical expressions. This makes it possible to represent complicated information and thought processes systematically.
- **Domain Knowledge:** Traditional AI relies on the domain knowledge of human experts. The rules and knowledge base that the AI system will utilize to make judgments are defined by this expert.
- **Rule-Based Inference:** Inference engines in conventional AI conclude presuppositions and rules that are provided. These systems are excellent at making logical inferences and adhering to prescribed inference rules.
- **Narrow Task Focus:** Traditional AI systems cannot generalize or transfer their expertise to new or related tasks because they are intended for specialized tasks only. They are frequently created for challenges [27] with clear objectives and limited scopes.

- **Scalability Issues:** As problems become more complex, managing and maintaining a large number of rules and exceptions becomes more difficult for classical AI systems. These systems can become cumbersome and impractical to scale.
- **Lack of Common Sense Reasoning:** Conventional AI struggles to recognize context outside of the clear rules embedded in the system and to use common sense reasoning. It might struggle with handling ambiguous

2.2 Types of Traditional Artificial Intelligence

Traditional AI included a variety of AI system types that were each created to tackle certain problem areas utilizing rule-based methodologies and explicit human expertise. The f ollowing are some instances of conventional AI of several types:

- **Expert Systems:** Computer programs known as "expert systems" are designed to mimic the decision-making abilities of a human expert in a certain subject. Their components consist of an inference engine that uses a knowledge base of facts and rules to form conclusions or offer recommendations [23]. Examples include:
    - **MYCIN:** A pioneering expert system for identifying bacterial illnesses and making antibiotic recommendations.
    - **DENDRAL:** A program that used mass spectrometry data to infer the structure of organic compounds.
- **Rule-Based Systems:** To accomplish logical reasoning and decision-making, rule-based systems employ a set of if-then rules. These regulations codify human knowledge and skill and control how the system behaves. Examples comprise:
    - **XCON:** A rule-based computer system configuration tool used by Digital Equipment Corporation.
    - **R1/XCON:** A more comprehensive version of XCON is used to set up VAX systems.
- **Systems for planning and scheduling:** These systems concentrate on creating plans or schedules to accomplish particular objectives. They entail analyzing constraints, dependencies, and actions through reasoning. Examples comprise:
    - A planning system designed to resolve symbolic planning issues is called STRIPS (Stanford Research Institute Problem Solver).
    - The Navy's Optimization for Allocation of Hardware, or NOAH, which is used to optimize resource distribution throughout Navy onboard systems.
- **Systems for Processing Natural Language (NLP):** Conventional NLP systems sought to process and comprehend human language. To perform tasks like parsing and translation, these computers frequently used syntactic and semantic rules. Examples comprise:
    - **SHRDLU:** An application that allows users to provide commands in plain language to move blocks around in a virtual block world.

- **ELIZA:** An early chatbot [42] that used pattern-matching methods to have brief chats with users.
- **Pattern Recognition Systems:** Systems that detect and categorize patterns [49] in data, such as pictures, sounds, or signals.
- **Programs that Play Games:** Traditional AI also generated programs that played games according to predetermined rules and strategies. Examples comprise:
  - World chess champion Garry Kasparov who was beaten in 1997 by the computer program Deep Blue.
  - Software that plays the board game checkers (draughts) against real players.

2.3 Approaches of Traditional Artificial Intelligence

Traditional AI called "Good Old-fashioned AI" (GOFAI) or "Symbolic AI," traditional artificial intelligence is a primitive form of technology [48] that depends on the explicit programming of rules and symbols. In contrast, learning from data is a key component of contemporary machine-learning approaches. The following are some crucial methods used in traditional AI:

- **Systems with Rules:** Rule-based systems are constructed using a predetermined set of explicit rules or logical assertions. These guidelines specify how the system ought to operate in certain circumstances. An example of a rule-based system is an expert system, which simulates a human expert's decision-making process in a particular field.
- **Reasoning and Representation of Knowledge:** Conventional AI systems concentrate on using logical reasoning to reach conclusions and structured knowledge representation. Examples of knowledge representation techniques that are used to arrange data so that the system can reason about relationships and entities are semantic networks and frames.
- **Algorithms for Search:** Search algorithms are employed to methodically investigate potential solutions to an issue. These algorithms search for a solution by navigating through a problem space. In traditional AI, for instance, common algorithms for problem-solving include depth-first search, breadth-first search, and A* search.
- **Organizing and Scheduling:** Creating systems that can plan and schedule tasks based on predetermined guidelines and objectives is a component of traditional AI approaches. As an illustration, consider the early planning system STRIPS (Stanford Research Institute Problem Solver), which represented problems as states and actions.

  The foundation for creating systems that could carry out particular tasks based on clear rules and knowledge representations was laid by traditional AI approaches. However they had trouble managing uncertainty, and they lacked the flexibility and capacity for learning that characterize more recent machine learning and artificial intelligence systems.

2.4 Limitations of Traditional Artificial Intelligence

Traditional AI, sometimes called symbolic or rule-based AI, had several drawbacks that prompted the creation of more advanced AI techniques like machine learning and deep learning. The following are some of the main drawbacks of conventional AI:

- **Scalability and Complexity:** Historically, rules and knowledge representations had to be created by hand for traditional [14] AI systems. It got more challenging to develop and maintain complete rule sets for all conceivable circumstances as issue complexity arose. Due to this, classical AI was unable to manage the complexity of the actual world.
- **Knowledge Acquisition:** To create and update the knowledge base for conventional AI systems, subject-matter experts had to manually enter facts and rules. This procedure was time-consuming, costly, and frequently predicated on subjective evaluations. It was not ideal for fields where knowledge changed quickly or where there was a shortage of expertise.
- **Handling Uncertainty:** Traditional AI found it difficult to reason and predict with uncertain data. Numerous real-world situations contain ambiguity, missing data, or contradictory evidence, which classic AI systems find difficult to handle using rigid rule-based techniques. The capacity to learn from new experiences or data was lacking in traditional AI systems. They weren't built to adapt to changing circumstances or get better at what they did via learning; instead, they were made to follow predefined rules.
- **Lack of Generalization:** Despite being created for a specific context, traditional AI systems sometimes struggle to apply their understanding to brand-new or different circumstances. They were unable to identify traits or patterns that might not have been expressly included in their rule sets.
- **Inefficient for Big Data:** Big datasets are beyond the capabilities of traditional AI systems. Large data sets can be laborious to process and extract useful insights from, and this can take time and resources.
- **Absence of Autonomous Learning:** Conventional AI systems are incapable of autonomous learning. Their rules and knowledge bases are not updated by human specialists, which makes them unsuitable for ongoing learning and development.
- **Limited Robustness:** Conventional AI systems are susceptible to unforeseen changes in input. Inaccurate or unexpected results can result from slight modifications in the input.
- **Restricted Application in Unstructured Data:** Conventional AI performs well in settings with structured data but has trouble processing unstructured data, such as photos, sensor data, and natural language.

Even though older AI systems could give clear rule-based reasons for their judgments, these explanations were frequently convoluted and challenging for non-experts to comprehend. Traditional AI might develop complex thinking, making it difficult to understand and troubleshoot the system's behavior.

2.5 Goals of Traditional Artificial Intelligence

Traditional AI, sometimes referred to as restricted AI or symbolic AI, aims to:

- **Intelligently Solve Certain Tasks:** Conventional AI is made to carry out particular activities that are generally seen as intelligent, such as playing chess, identifying objects in pictures, or translating languages. It accomplishes this by making judgments and acting based on pre-established rules, algorithms, and knowledge bases.
- **Perform Repetitive Work:** Data input, sorting, and classification are just a few of the repetitive, time-consuming operations that traditional AI can perform. People may be able to concentrate on more strategic [31] and creative work as a result.
- **Improve Decision-making:** Conventional AI is capable of analyzing vast volumes of data and spotting patterns that are hard or impossible for people to notice. This can assist companies in making more informed judgments about everything from marketing plans to product development.
- **Increase Productivity and Efficiency:** Process and workflow optimization using traditional AI can result in notable increases in productivity and efficiency.
- **Offer Advice and Insights:** Conventional AI is capable of data analysis and insight generation that may assist companies in making wiser decisions. AI may be used, for instance, to spot fraud, suggest new items to consumers, and recognize patterns in their behavior.

3. Current Trends and Future Directions for Traditional Artificial Intelligence

Although it uses symbolic or rule-based methods, traditional artificial intelligence (AI) has achieved significant advances in the field. It's crucial to remember that more contemporary AI methods like machine learning and deep learning have mostly supplanted traditional AI. However, there are still several fields and possible avenues for future development where classical AI finds use:

- **Hybrid Approaches:** Combining cutting-edge AI approaches with conventional methodologies is a developing trend. To make use of the advantages of both techniques, researchers are looking at ways to integrate machine learning with rule-based systems and symbolic reasoning.
- **Explainable AI (XAI):** Research toward improving the explainability of conventional AI systems is being driven by the requirement for transparency [40] and interpretability in AI systems. This is especially crucial in industries where trust and accountability are paramount, such as healthcare and finance.
- **AI for Common Sense Reasoning:** To help AI systems comprehend and reason about context and daily circumstances more effectively, advances in common-

sense reasoning are being made. For applications [5] like chatbots and virtual assistants, this is essential.

- **Knowledge Graphs:** Complex information is organized and represented using traditional artificial intelligence methods such as knowledge graphs and ontologies. They are used in many different fields, such as data integration, recommendation systems, and search engines.
- **Expert Systems Evolution:** To enhance decision-making and knowledge acquisition, expert systems, a classical AI classic, are being updated with contemporary AI elements like machine learning.
- **Natural Language Processing (NLP):** Conventional AI methods, such as rule-based systems, are occasionally combined with cutting-edge NLP strategies. When paired with machine learning models, rule-based systems may produce and comprehend natural language writing with greater comprehension.
- **Rule-based Chatbots:** For certain jobs and frequently asked questions, certain chatbots, and virtual assistants continue to rely on rule-based systems. When a consistent set of inquiries and answers is anticipated, these systems may work well.
- **Cognizant Computing:** Systems using cognitive computing combine standard artificial intelligence with human-like mental processes. Such tasks as recommendation systems, data analysis, and decision-making include the use of systems.

While deep learning, machine learning, and neural networks remain essential areas of concentration for AI research and development, classical AI still has a role in these domains. Several fields, including data analysis, computer vision [34], natural language processing, and reinforcement learning, have witnessed remarkable outcomes using these techniques. As a result, cutting-edge and rapidly developing AI applications mostly concentrate on these modern AI methodologies.

Traditional AI might still be useful in certain specialized or niche applications as AI research develops, but the field's main trend is definitely toward more advanced machine learning and neural network models that can handle a larger variety of challenging issues and adjust to shifting environments.

4. Development Stage of Modern Artificial Intelligence

Symbolic AI, limited AI, or traditional AI, has gone through numerous major stages of development.

- **Phase I (1950s–1970s):** The development of foundational ideas and methods, including search algorithms, symbolic reasoning, and knowledge representation, defined the early phases of classical AI. During this time, the Turing test was developed, the first expert system for medical diagnosis was created, and the first chess-playing computer program was created.
- **Middle Stage (1970s–1990s):** Expert systems, knowledge-based systems, and probabilistic reasoning emerged during the middle period. Applications of AI

have spread to fields including natural language processing (NLP), robotics, and planning. However, the possibilities of classical AI were limited by constraints on computer power and data availability.

- **Contemporary Stage (1990s–Today):** Traditional artificial intelligence has come back into vogue thanks to the development of machine learning, bigger datasets, and more potent computers. The accuracy and scalability of AI models have increased because of methods like random forests, decision trees, and support vector machines. Nowadays, traditional AI is widely employed across many sectors and applications.
- **Present Situation:** Conventional AI is still a well-developed and established area with many successful applications in several fields. It keeps developing, adding new features and methods, and growing in scope. However, attention has turned to more sophisticated AI techniques like deep learning and reinforcement learning [50] as a result of the quest for artificial general intelligence (AGI).

5. Modern Artificial Intelligence

The integration of data-driven learning algorithms and the capacity of systems to adapt, generalize, and carry out challenging tasks independently describe modern artificial intelligence (AI), which marks a transformational moment in the field. Modern artificial intelligence (AI) uses cutting-edge techniques like machine learning and deep learning to enable computers to learn from data and make decisions based on patterns and statistical probabilities, in contrast to its predecessor, traditional AI, which heavily relied on rule-based systems and explicit human knowledge. The idea of machine learning, which enables AI systems to get better over time, lies at the core of contemporary AI. Large data sets are analyzed by machine learning algorithms to find significant patterns, trends, and correlations. AI systems are more flexible in a variety of settings because of this data-driven approach, which enables AI systems to automatically modify their behavior and decision-making processes.

5.1 Attributes of Modern Artificial Intelligence

- **Data-Driven:** To discover patterns, correlations, and insights from data, modern AI relies on vast databases. This contrasts with the rule-based AI of classical AI, where human specialists explicitly encode information.
- **Machine Learning Algorithms:** Modern AI employs a range of machine learning algorithms, including supervised, unsupervised [45/20], and reinforcement learning [22], to automatically learn from data and make predictions.
- **Learning and Adaptation:** Modern AI systems may learn from new data and modify their behavior over time, in contrast to traditional AI systems. By improving their models [12] in response to input, they can perform better.
- **Feature Extraction:** Automatic feature extraction [33] from raw data is a common step in contemporary AI approaches. Because the algorithms learn to recognize pertinent aspects on their own, there is no longer a requirement for human specialists to describe specific features.

- **Generalization:** Modern AI is built to transfer the knowledge it learns from training data to new, unheard-of examples. This enables it to generate precise predictions or judgments based on previously unexplored facts.
- **Neural Networks and Deep Learning:** Deep learning, a subclass of machine learning, employs multilayer neural networks to model complex patterns and representations. This has led to advancements in computer vision, natural language processing, and speech recognition [39], among other fields.
- **High Computational Power:** To train complicated models and process and analyze massive datasets effectively, modern AI frequently needs a lot of computing power.

5.2 Types of Modern Artificial Intelligence

A wide range of cutting-edge methods and techniques are included in modern AI, which uses data-driven learning to complete challenging tasks. The following list of contemporary AI types includes examples:

- **Computer learning:** In the discipline of computer science or machine learning, the phrase computer learning is not often used. It might be a mistake or an uncommon way to say machine learning or computer-based learning. Let's investigate both options:
  - **Supervised learning:** To generate predictions or categorize data, models are trained using labeled examples. An illustration of this is image classification, where a model is taught to recognize objects in photos using labeled training data.
  - **Unsupervised Learning:** Without labeled data, unsupervised learning entails grouping and dimensionality reduction. An illustration of this would be customer segmentation for marketing, which groups comparable consumers based on behavior.
- **Deep learning:**
  - **Convolutional Neural Networks (CNNs):** CNNs [41] are excellent at jobs that need spatial data and visuals. Examples include object detection, picture production, and image categorization.
  - **Recurrent Neural Networks (RNNs):** RNNs analyze data sequences, making them appropriate for jobs requiring sequences like time series analysis and language processing. Using text data to analyze sentiment as an example.
  - **Transformer Models:** By identifying contextual linkages in big text datasets, transformers have revolutionized natural language processing jobs. Examples include text summarization, chatbots, and language translation.
  - Generator and Discriminator Models: A generator and a discriminator battle to create accurate data in G enerative A dversarial N etworks (GANs) [25]. Examples include creating convincing visuals, convincing films, and style emulation.

- **NLP: Natural Language Processing:**
  - Text: Text [24] may be translated between multiple languages using AI-powered translation tools like Google Translate.
  - **Sentiment analysis:** NLP models can ascertain the sentiment included in text, such as the positive or negative nature of a review.
- **Computer Vision:**
  - **Object D etection:** AI systems can find and recognize things in photos and movies. For instance, self-driving automobiles can recognize other vehicles and people.
  - **Face Recognition:** Identity verification and access control both employ face recognition technology.
  - Image Segmentation: Image segmentation entails grouping every pixel in a picture into a certain category. An MRI scan image analysis example would be able to identify various tissues.
- **Intelligent Systems:**
  - Autonomous Cars: Autonomous cars driven by AI can navigate and make judgments while driving thanks to sensors and deep learning algorithms.
  - Drones: Drones that fly in the air can independently map a region, examine a piece of land, or deliver items.

5.3 Approaches of Modern Artificial Intelligence

These are some important techniques used in contemporary AI. Modern AI encompasses a variety of approaches that leverage advanced computational techniques and data-driven methodologies. Here are some key approaches within modern AI.

- **Machine Learning (ML):** Machine learning (ML) [47] algorithm development enables computers to identify patterns in data and make inferences from it. This broad category includes several subfields, such as supervised learning, unsupervised learning, and reinforcement learning. As an illustration, supervised learning uses labeled training data to teach algorithms how to predict or categorize newly discovered information.
- **Deep Learning (DL):** Modeling complicated patterns with multi-layered neural networks, or deep neural networks, is the basis of deep learning, a subfield of machine learning. Applications such as photo and speech recognition using deep learning are particularly successful. For example, sequence data is processed by recurrent neural networks whereas image identification [36] is processed by convolutional neural networks.
- **Networks of Neurals:** The structure and functions of the human brain serve as the model for neural networks, which are computer systems. They are composed of layers of networked nodes, or neurons, and are capable of learning hierarchical representations. For instance, feedforward neural networks—in

which data flows only in one direction [10]—are frequently employed in a wide range of settings.

- **Learning by Reinforcement (RL):** In reinforcement learning (RL), agents acquire decision-making skills through interactions with their surroundings. In the form of incentives or sanctions, the agent is given feedback. As an illustration, consider AlphaGo, a computer program that uses reinforcement learning to develop superhuman skills in the board game Go.

5.4 Goals of Modern Artificial Intelligence

The goal of artificial intelligence (AI) as it exists now is to develop intelligent machines that are capable of doing activities that would typically need human intelligence. These goals reflect the ambition of the AI community to develop systems that can comprehend, learn from, and adjust to various environments. Among the primary goals of modern AI are:

- **Machine Learning and Deep Learning:** Establishing deep learning and machine learning models and algorithms so that computers may learn from data and progressively become more proficient at what they do. Through the use of multi-layered neural networks, a subset of machine learning known as "deep learning" enhances data abstraction and representation.
- **Natural Language Processing (NLP):** By employing natural language processing, or NLP, it is feasible to enable machines to understand, interpret, and generate human language. Applications like sentiment analysis, language translation, and chatbots require natural language processing (NLP).
- **Computer vision:** Give computers the ability to interpret and grasp visual data from the outside world, such as images and videos. This is necessary for applications like photo recognition, object detection, and autonomous [38] cars.
- **Robotics:** Use AI to give robotic devices the ability to see their surroundings, act, and carry out manual jobs. Applications in manufacturing, healthcare, and autonomous systems are included in this.
- **Autonomous Systems:** Develop AI systems that can behave and make decisions on their own without the need for human interaction. This includes drones that operate on their own, self-driving cars, and other intelligent agents.
- **Reinforcement Learning:** Develop systems that can pick up new skills through interaction with their surroundings and feedback in the form of incentives or punishments. For agents to be trained to make sequential judgments in dynamic contexts, this is essential.
- **Fairness and Ethics in AI:** Discuss ethical issues surrounding AI, such as algorithmic prejudice, data privacy, and the responsible creation and application of AI technologies. An essential objective is to guarantee fairness and prevent discrimination in AI systems.
- **Explainable AI (XAI):** Improve the interpretability and transparency of AI models so that users can comprehend the reasoning behind the models' conclusions. This is known as explainable AI or XAI. Establishing confidence

in AI systems is crucial, particularly in delicate areas. Together, these objectives advance AI research and development, enabling it to be a potent tool for solving difficult problems and enhancing many facets of our everyday lives. It's crucial to remember that ethical issues must be taken into account when pursuing these objectives because AI technologies continue to have a big impact on society.

6. Limitations of Modern Artificial Intelligence

Although extremely capable and adaptable, modern AI nevertheless has several limits and difficulties that academics and practitioners are actively attempting to resolve. Some of the main drawbacks of contemporary AI include:

- **Lack of Common Sense:** AI systems frequently lack the ability to reason using common sense, which leaves them susceptible to errors or giving incomprehensible responses in circumstances that humans would readily comprehend.
- **Dependency on Data:** The majority of AI models, particularly deep learning models, need a lot of high-quality data to function well. When data is lacking or skewed, they can have trouble, and they might even confirm biases already present in the data.
- **Narrow Domain Expertise:** Many AI systems lack universal intelligence and are instead specialized for particular tasks. Effective knowledge transfer from one domain to another is impossible for them.
- **Interpretability:** Neural networks and other deep learning models are frequently referred to as "black boxes" [43]. It's difficult to comprehend why individuals choose certain actions, which poses a serious obstacle in crucial domains like law or healthcare.
- **Ethical Concerns:** AI has the potential to unintentionally notice and reinforce biases found in training data, producing unfair or discriminating results. A significant difficulty is ensuring justice and moral behavior in AI systems.
- **Risks to security [26]:** AI systems are susceptible to adversarial assaults, in which little modifications to the input data can result in inaccurate outputs or even malevolent behavior.

The field of artificial intelligence is continually conducting research and development to address these constraints and difficulties. Researchers and practitioners are attempting to reduce the downsides of AI while enhancing its robustness [45/20], interpretability, ethics, and accessibility.

7. Current trends and Future Direction for Modern AI

The following are some topics of interest and possible future paths:

- **Developments in Deep Learning:** As more sophisticated brain architectures [1] and attention mechanisms are created, deep learning keeps advancing, allowing artificial intelligence systems to manage tasks that are becoming increasingly complex.

- **Generative Models:** Image generation, text-to-image synthesis, and data augmentation are examples of tasks that employ generative models, such as GANs (Generative Adversarial Networks) and VAEs (Variational Autoencoders).
- **Self-Supervised Learning:** These learning strategies are gaining popularity. These methods reduce the need for huge labeled datasets by training AI models with unlabeled data.
- **Edge AI:** Directly deploying AI models on edge devices, such as smartphones and Internet of Things gadgets, is becoming increasingly common. Edge AI reduces latency, conserves bandwidth, and enhances privacy by processing data locally.
- **AI That Is Explainable and Interpretable:** As AI systems get more sophisticated, there is an increasing focus on ensuring that they are both interpretable and explainable. Scholars are now developing methods to offer perspectives on the decision-making process of models.
- **AI in Healthcare:** For activities like drug development, predictive analytics, and medical [30] imaging analysis, modern AI—including deep learning—is making major strides. Future healthcare solutions should be increasingly more precise and individualized.
- **AI in Autonomous Vehicles:** Deep learning in particular, a contemporary kind of AI, is essential to the advancement of autonomous vehicle technology. The development of autonomous [13] vehicles and associated technologies will remain a priority.
- **Advances in Reinforcement Learning:** As reinforcement learning continues to grow, its uses are being applied to more complicated decision-making situations, robotics, gaming, and recommendation systems.
- **AI for Natural Language Processing:** Persistent advancements in natural language processing, emphasizing sentiment analysis, multilingual assistance, and context awareness.
- **AI and Sustainability:** AI is being used to optimize energy use, cut waste, and improve resource management, among other environmental and sustainability concerns.
- **Artificial Intelligence in Finance:** Algorithmic trading, fraud detection, risk assessment, and financial advising services are just a few of the ways that AI is being utilized in the finance sector.
- **5556+Quantum AI:** Continued research into quantum computing and its use in AI might result in important advances in the understanding and solution of challenging machine learning and optimization issues.

8. Development Stage of Modern

A few pivotal phases can be used to broadly classify the evolution of contemporary AI. The general stages through 2022 are as follows:

- **Symbolic AI:** Also known as Good Old-fashioned AI, or GOFAI, this phase began in the middle of the 20th century. Symbolic reasoning and logic, which used explicit rules to represent knowledge, were the foundations of early AI systems. These systems, however, were unable to learn from data and had difficulty managing uncertainty.
- **Machine Learning (ML):** As processing power grew, machine learning techniques became more important. At this stage, algorithms that can recognize patterns in data and forecast outcomes are being developed. The three main paradigms in machine learning are supervised learning, unsupervised learning, and reinforcement learning.
- **Deep Learning (DL):** Deep learning, a branch of machine learning, models complex patterns by using multiple-layered neural networks, or deep neural networks. The 2010s saw a major increase in popularity for this stage, which has been crucial to innovations [15] in speech and picture recognition.
- **Narrow AI (Weak AI):** The majority of the applications of AI that are in use today are classified as narrow AI or weak AI. A single task or a group of related tasks can be completed by a narrow AI. Recommendation algorithms [44], image recognition software, and virtual assistants are a few examples.
- **Pre-trained Models and Transfer Learning:** Recent developments include the application of pre-trained models and transfer learning. Massive datasets are used to train large neural networks for generic tasks, after which they are tuned for particular uses. This methodology has demonstrated efficacy across multiple fields, including natural language processing.
- **AI Ethics and Bias Mitigation:** As AI is used more frequently across a range of industries, ethical issues and biases in AI systems are becoming more widely recognized. Practitioners and researchers are hard at work creating approaches and frameworks to deal with these problems.

It's crucial to remember that AI development is a continuous process and that since our last update, new phases and innovations may have appeared. For the most recent developments, it is important to monitor publications and developments in the field.

9. Evolution of Traditional AI to Modern AI

Modern AI differs from classic AI in that it uses different methods, tools, and underpinning technology. The significant differences are broken down as follows:

- **Complexity M anagement:**
    - **Traditional AI:** Due to its inflexible rule-based methodologies, traditional AI found it difficult to handle complex and unpredictable circumstances.

- **Modern AI:** Due to deep learning models' ability to identify nuanced patterns and correlations in vast and complicated datasets, modern AI excels at handling complexity.

- **Flexibility and I ndependence:**
  - **Traditional AI:** The autonomy and scalability of traditional AI systems were constrained by the need for human specialists to establish rules and make modifications.
  - **Modern AI:** Modern AI systems are more adaptable and scalable because they can work with more autonomy, always learning from new data and circumstances, and doing so.
- **Domain Particularity:**
  - **Traditional AI:** AI systems from the past were frequently constrained to certain, well-defined areas for which rules could be explicitly stated.
  - **Modern AI:** Modern AI is more adaptable and may be used in a variety of fields, such as robotics and healthcare, as well as image identification and natural language processing.
- **Dependency on Data:**
  - **Traditional AI:** Due to the laborious encoding of information by human specialists, traditional AI did not rely on enormous volumes of data.
  - **Modern AI:** Using the potential of big data [11] to improve performance, modern AI relies on enormous datasets for training and developing models.
- **Performance:**
  - **Traditional AI:** These systems frequently excelled in certain, well-defined domains where it was possible to set clear rules.
  - **Modern AI:** In tasks like speech and picture recognition, language translation, and more, modern AI systems have made amazing performance gains, frequently exceeding conventional approaches.

From conventional AI to modern AI, the development of AI has involved constant innovation and adaptation. Even while current AI has accomplished amazing things, it still has issues with bias, interpretability, ethics, and other issues. Further advancements in AI technology and a better comprehension of how to use AI's potential while resolving its drawbacks and ethical issues are likely to characterize the field's future. AI will keep transforming different industries and have a significant impact on how society and technology develop in the future.

10. Advantages of Modern AI Over Traditional AI

Several essential elements and benefits that current AI offers have fueled the transition from conventional AI to its new avatar. Here are a few explanations for this change:

- **Data Scale and Availability:** Modern AI makes use of the vast amounts of digital data that are already accessible. There is a lot of data available that may be utilized for training and enhancing AI models as a result of the expansion of the internet, social media, sensors, and other data sources. Large and diverse datasets were a challenge for traditional AI systems.

- **Performance Enhancements:** Modern AI methods, particularly deep learning, have demonstrated impressive performance enhancements in a variety of applications, including image identification and natural language processing. In many fields, these methods outperform conventional rule-based methods.
- **Adaptability and Generalization:** Conventional AI systems were frequently constrained by the particular information and rules they had been designed with, rendering them unresponsive to novel or unanticipated circumstances. Modern AI models can generalize their knowledge and learn from data, which enables them to adapt to new situations and tasks.
- **Automation of Feature Engineering:** In traditional AI, designing and extracting useful features from data frequently needed substantial human work. Deep learning in particular in modern AI may automatically discover and extract characteristics from unprocessed data, minimizing the need for human participation.
- **Handling Complexity:** Complex patterns and relationships are a common feature of many real-world issues. Modern AI is more suited to managing these difficulties due to its capacity to capture and represent these complicated interactions.
- **Autonomous Learning:** Modern AI systems have the capacity to constantly learn and develop over time even in the absence of explicit programming. They can stay current and relevant by changing their behavior in response to the latest information.
- **Accessibility of Computing Power:** Modern AI models' training and inference procedures are now completed much more quickly because of technology advancements like Graphics Processing Units (GPUs) and specialized hardware accelerators. Contemporary

AI approaches have significantly improved and advanced many real-world instances, demonstrating the transition from C lassic AI to C ontemporary AI. Here are a few noteworthy instances:

- **Healthcare:** Medical Image Analysis In radiology [18], conventional image analysis methods rely on rule-based algorithms to identify certain characteristics. Convolutional neural networks (CNNs), a type of deep learning model, have transformed medical image analysis through the automatic detection [2] of patterns and abnormalities in X-rays, MRIs, and CT scans.
- **Language Translation:** Early language translation systems frequently translate text using rule-based methodologies and established dictionaries. To create more accurate and natural translations, modern AI uses neural machine translation, a type of sequence-to-sequence modeling.

11. Case Studies and Examples of Traditional AI

Conventional Artificial Intelligence (AI), known as classical or symbolic AI, is a method based on rule-based systems and symbolic reasoning. Here are a few case studies and conventional AI examples:

- **System Experts:** One of the first expert systems created for bacterial illness diagnosis was MYCIN (1976). It arrived at diagnoses and suggested treatments using a backward chaining inference engine and a knowledge base of rules.
- **Natural Language Processing ELIZA (1966):** A mong the initial chatbots was ELIZA. It mimicked interactions with a Rogerian psychotherapist using pattern-matching techniques.
- **Automation and Robotic Reasoning:** Shakey (1966–1972), The Stanford Research Institute created Shakey, a groundbreaking mobile robot. It was one of the first artificial intelligence (AI) systems for robotics, using a symbolic reasoning mechanism to navigate and carry out tasks in a block environment.
- **Playing Games:** 1988 saw the release of Deep Thought, a chess-playing computer that employed heuristic and rule-based strategies. Even though it wasn't as advanced as more recent chess engines like Deep Blue or AlphaZero, it represented a significant advancement in artificial intelligence gaming.
- **Organizing and Timetables:** STRIPS (1971), Using formal language to represent the beginning state, desired state, and actions, the Stanford Research Institute Problem Solver (STRIPS) was an early planning system that employed symbolic reasoning to solve issues.
- **Interpretation of Natural Language:** CYC (1984), The goal of the ambitious CYC project (1984) is to turn common sense reasoning and knowledge into a computer program. It uses a broad body of knowledge and inference principles to comprehend and make sense of natural language content.
- **Identification and Resolution:** Dendral (1965), was an expert system created for the processing of chemical mass spectrometry data. Based on information from mass spectrometry, it identified chemical compounds using a set of guidelines and heuristics.
- **Medical Evaluation:** An early expert system intended for medical diagnosis; Internist-1 was created at the University of Pittsburgh. To aid in disease diagnosis, it made use of a rule-based inference engine and a knowledge store of medical information.

12. Case Studies and Examples of Modern AI

Contemporary AI, known as modern AI, refers to a broad spectrum of applications and technology that have undergone substantial recent evolution. Here are some case studies and contemporary AI examples:

- **Neural Networks and Deep Learning:** ImageNet Classification (2012), Deep learning and convolutional neural networks (CNNs) gained prominence with the ImageNet Large Scale Visual Recognition Challenge. The accuracy of picture classification was significantly increased by models such as Res-Net, Inception, and later, Alex-Net.

- **Natural Language Processing (NLP):** GPT-3 (2020) OpenAI's GPT-3 is a potent language model that can produce text that resembles that of a human. It can be used for text summarization, content creation, and chatbots.
- **Vision of Computers:** Self-Driving Cars: Autonomous vehicles are made possible by businesses like Waymo and Tesla, which employ computer vision and machine learning. These devices navigate and make decisions in real-time using cameras and sensors.
- **Medical Care:** IBM Watson for Oncology, Lenovo an AI program called Watson for Oncology helps medical professionals identify and treat cancer. It analyzes patient records and medical literature using machine learning and natural language processing.
- **Reward-Based Learning:** AlphaGo (2016) Lee Sedol, the world champion go player, was defeated by DeepMind's AlphaGo. It mastered the challenging game of Go by combining deep neural networks with reinforcement learning. The capabilities of contemporary AI, such as deep learning, reinforcement learning, and large datasets, enable computers to do jobs that were previously thought to be the exclusive domain of humans. The use of artificial intelligence is spreading throughout many areas of our lives and enterprises, and its applications are only going to get bigger and more varied in the future.

13. Challenges and Considerations of Traditional AI

Traditional AI, which is built on rule-based systems, expert knowledge, and symbolic reasoning, has its own set of issues to deal with. The following are some of the main issues and factors related to conventional AI:

- **Knowledge Acquisition:**
    - **Challenge:** Manually encoding expert information into rule-based systems is a significant part of traditional AI. This procedure could need constant updates and might be time-consuming.
    - **Consideration:** Develop effective methodologies and tools for knowledge acquisition to aid in the integration of expertise into AI systems.
- **Scalability:**
    - **Challenge:** Scaling and handling extremely vast and complicated datasets or domains may be difficult for traditional AI systems.
    - **Consideration:** Consider investigating techniques for scaling up, such as parallel processing and knowledge representation optimization.
- **Lack of Adaptability:**
    - **Challenge:** Traditional AI systems are frequently made for certain tasks and domains and may not be flexible enough to handle new or changing circumstances.

  - **Consideration:** Develop ways for system adaptability and take into account hybrid AI strategies that mix conventional AI with additional data-driven tactics.
- **Interpretability:**
  - **Challenge:** As rule-based systems expand in size and complexity, they can become complicated and challenging to understand.
  - **Consideration:** Maintain concise and well-documented rule sets, and take into account the use of visualization tools to help interpret system behavior.
- **Limited Data Learning:**
  - **Challenge:** Compared to current machine learning techniques, traditional AI systems are less capable of learning from data.
  - **Consideration:** Look at ways to apply statistical and machine learning methods to improve learning and adaptability.
- **Handling Uncertainty:**
  - **Challenge:** Making probabilistic judgments and addressing uncertainty may be difficult for traditional AI.
  - **Consideration:** Look into ways to include probabilistic reasoning and uncertainty modeling in rule-based systems.
- **Expertise Dependency:**
  - **Challenge:** To create and maintain knowledge bases, traditional AI mainly relies on domain specialists.
  - **Consideration:** Develop tactics for knowledge exchange and teamwork between specialists and those who create AI systems.
- **Combinatorial Explosion:**
  - **Challenge:** When faced with a huge number of rules and potential combinations, some rule-based systems may experience the combinatorial explosion problem.
  - **Consideration:** To reduce combinatorial complexity, use effective algorithms and pruning methods.
- **Real-Time Requirements:**
  - **Challenge:** In dynamic contexts, real-time decision-making for some classic AI systems might be difficult.
  - **Consideration:** To satisfy real-time requirements, optimize rule execution and inference procedures.
- **Integration with Modern AI:**
  - **Challenge:** For better performance, traditional AI may need to be combined with more advanced machine learning and deep learning methods.
  - **Consideration:** Take into account hybrid AI strategies [21], which combine the advantages of classical AI with data-driven techniques.

Traditional AI sometimes requires a strong grasp of the particular domain and job at hand to navigate these difficulties and concerns. To maintain the AI system relevant and efficient in changing situations, it could also need continual upkeep and upgrades.

14. Challenges and Considerations of Traditional AI

Modern AI, which frequently combines machine learning and deep learning methods, has resulted in revolutionary [9] improvements across several fields. However, it also has a unique set of difficulties and things to think about. The following are some important issues and difficulties related to contemporary AI:

- **Integration with Modern AI:**
  - **Challenge:** For better performance, traditional AI may need to be combined with more advanced machine learning and deep learning methods.
  - **Consideration:** Take into account hybrid AI strategies, which combine the advantages of classical AI with data-driven techniques
- **Interpretability and Explainability:**
  - **Challenge:** The difficulty is that many contemporary AI models, particularly deep learning models, are sometimes seen as "black boxes," making it difficult to comprehend how they make decisions.
  - **Consideration:** To win user trust and make debugging easier, provide methods and tools for model interpretability and explainability.
- **Bias and Fairness:**
  - **Challenge:** Biases in training data can produce biased AI models, which can produce unfair and discriminating results.
  - **Consideration:** Take into account regularly assessing and reducing biases in AI models. Implement assessment metrics and algorithms that are fairness-aware.
- **Generalization and Robustness:**
  - **Challenge:** Modern AI models can be excellent at some tasks, but they may find it difficult to generalize to new data or to cope with aggressive or noisy environments.
  - **Consideration:** Think about using methods like data augmentation, regularization, and adversarial training to improve model generalization. Investigate adversarial assaults and robustness testing to evaluate model resilience.
- **Computation and Resource Requirements:**
  - **Challenge:** The difficulty is that huge AI models might require a lot of computing time and resources to train and deploy.
  - **Consideration:** Think about streamlining model designs and training procedures. Investigate methods like quantization and model compression.

- **Data Privacy and Security:**
  - **Challenge:** Data privacy and security [17] issues might arise when sensitive data is used in machine learning.
  - **Consideration:** Implement data anonymization techniques, access limits, and robust encryption. according to data privacy laws.
- **Ethical Considerations:**
  - **Challenge:** The use of modern AI technology might bring up moral issues with responsibility, transparency, permission, and privacy.
  - **Consideration:** Establish ethical standards and norms for the advancement of AI. Conduct in-depth moral evaluations of AI application cases
- **Human-Machine Collaboration:**
  - **Challenge:** It can be difficult to integrate AI into human workflows and decision-making processes.
  - **Consideration:** To achieve successful synergy, pay attention to user experience, training, and collaboration between people and AI systems.
- **Continuous Learning:**
  - **Challenge:** The difficulty is that AI models require systems for ongoing learning and adaption as fresh data becomes available.
  - **Consideration:** To promote continual development, implement online learning strategies, reinforcement learning, and transfer learning.
- **Regulatory Compliance:**
  - **Challenge:** Depending on the industry and the area, different laws and compliance standards may apply to AI applications.
  - **Consideration:** Ensure that AI systems abide by all applicable laws and standards, such as the GDPR in Europe or HIPAA in the healthcare industry.
- **Environmental Impact:**
  - **Challenge:** Training big AI models requires a substantial amount of energy and has an impact on the environment.
  - **Consideration:** Investigate energy-efficient AI training techniques and support environmental initiatives.

A multidisciplinary approach incorporating skills in AI, ethics, legislation, and domain-specific knowledge is necessary to navigate these issues and considerations in contemporary AI. Organizations and developers should place a priority on ethical AI practices and keep up with changing ethical and legal requirements in the AI industry.

15. Conclusion

Concluding, the transition from conventional AI to contemporary AI was a revolutionary process that completely changed the landscape of artificial intelligence. The foundation for grasping fundamental ideas was created by traditional AI, which was built on rule-based systems and expert knowledge, but it frequently suffered with scalability and adaptation to complex and dynamic contexts. The power of data and computing, on the other hand, has been used by current AI, with its emphasis on machine learning and deep learning, to construct models that can learn from examples and make judgements that were previously thought to be beyond the scope of automation [4].

Traditional AI demonstrated the potential of human problem-solving skills, but current AI has surpassed these abilities by uncovering complex patterns from enormous datasets. This change has sparked improvements in several fields, including robotics, computer vision, healthcare, finance, and more. The issues that have emerged as a result of this advancement include data bias, ethics, transparency, and accountability. It's critical to understand that both conventional and current techniques have advantages and disadvantages in the dynamic field of artificial intelligence. While current AI excels at handling complicated tasks where data-driven insights and adaptation to changing settings are vital, traditional AI still has value in applications where clear rules and domain [3] expertise are crucial.

Researchers, practitioners, and policymakers must find a balance between utilizing the wisdom of conventional approaches and the potential of cutting-edge technologies as AI develops. We can negotiate the ethical, societal, and technological obstacles of AI while releasing its full potential to reshape industries, improve human capacities, and influence the future in unimaginable ways by grasping the advantages and disadvantages of both paradigms.

References

1. Khan, A., Sohail, A., Zahoor, U., and Qureshi, A. S. 2020. A survey of the recent architectures of deep convolutional neural networks. *Artificial Intelligence Review, 53*(8), 5455–5516. https://doi.org/10.1007/s10462-020-09825-6
2. Jena, B., Nayak, G. K., and Saxena, S. 2021. Convolutional neural network and its pre-trained models for image classification and object detection: A survey. *Concurrency and Computation: Practice and Experience, 34*(6). https://doi.org/10.1002/cpe.6767
3. Taillefer, C., Bonnin, M., Gilli, M., and Civalleri, P. P. 2021. A mixed time-frequency domain approach for the qualitative analysis of a hysteretic oscillator. *Synchroinfo Journal, 7*(4), 26–29. https://doi.org/10.36724/2664-066x-2021-7-4-26-29
4. Speck, D., Dornhege, C., and Burgard, W. 2017. Shakey 2016—How Much Does it Take to Redo Shakey the Robot? IEEE Robotics and Automation Letters, 1-1. https://doi.org/10.1109/lra.2017.2665694
5. Abdullah, M., Madain, A., and Jararweh, Y. 2022. ChatGPT: Fundamentals, Applications, and Social Impacts. 2022 Ninth International Conference on Social Networks Analysis, Management and Security (SNAMS). https://doi.org/10.1109/snams58071.2022.10062688
6. Myers, D., Mohawesh, R., Chellaboina, V. I., Sathvik, A. L., Venkatesh, P., Ho, Y.-H., Henshaw, H., Alhawawreh, M., Berdik, D., and Jararweh, Y. 2023. Foundation and large language models: Fundamentals, challenges, opportunities, and social impacts. Cluster Computing. https://doi.org/10.1007/s10586-023-04203-7
7. Benet, H., Ojasvi, G., and Neelu, J.A. 2016. Semiotics expert system: An integrative approach towards the maintenance of community peace. *International Journal of Peace and Development Studies, 7*(6), 50–61. https://doi.org/10.5897/ijpds2015.0256
8. Bai, X., and Yu, S. 2023. Rule-Based Machine Translation. *Routledge Encyclopedia of Translation Technology,* 193–207. https://doi.org/10.4324/9781003168348-11

9. Kinshuk, Chen, N.-S., Cheng, I.-L., and Chew, S. W. 2016. Evolution Is not Enough: Revolutionizing Current Learning Environments to Smart Learning Environments. *International Journal of Artificial Intelligence in Education, 26*(2), 561–581. https://doi.org/10.1007/s40593-016-0108-x
10. Merriam, S. B. 2001. Andragogy and Self-Directed Learning: Pillars of Adult Learning Theory. New Directions for Adult and Continuing Education, 2001(89), 3–14. https://doi.org/10.1002/ace.3
11. Indrakumari, R., Vijayalakshmi, S., Sagar, S., and Poongodi, T. 2021. Affiliation of Blockchain, IoT, and Big Data. Convergence of Blockchain, AI, and IoT, 1–18. https://doi.org/10.1201/9781003081180-1
12. Lu, Y. 2019. Artificial intelligence: A survey on evolution, models, applications and future trends. *Journal of Management Analytics, 6*(1), 1–29. https://doi.org/10.1080/23270012.2019.1570365
13. Slimani, K., Khoulji, S., Mortreau, A., and Kerkeb, M. L. 2023. From tradition to innovation: The telecommunications metamorphosis with AI and advanced technologies. *Journal of Autonomous Intelligence, 7*(1). https://doi.org/10.32629/jai.v7i1.1099
14. Fomin, V. V. 2020. The Shift from Traditional Computing Systems to Artificial Intelligence and the Implications for Bias. *Smart Technologies and Fundamental Rights,* 316–333. https://doi.org/10.1163/9789004437876_015
15. Verganti, R., Vendraminelli, L., and Iansiti, M. 2020. Innovation and Design in the Age of Artificial Intelligence. *Journal of Product Innovation Management, 37*(3), 212–227. https://doi.org/10.1111/jpim.12523
16. Luger, G. F. 2021. Modern AI and How We Got Here. Knowing Our World: An Artificial Intelligence Perspective, 49–74. https://doi.org/10.1007/978-3-030-71873-2_3
17. Bostrom, N. 2018. Strategic Implications of Openness in AI Development. Artificial Intelligence Safety and Security, 145–164. https://doi.org/10.1201/9781351251389-11
18. Hosny, A., Parmar, C., Quackenbush, J., Schwartz, L. H., and Aerts, H. J. W. L. 2018. Artificial intelligence in radiology. *Nature Reviews Cancer, 18*(8), 500-510. https://doi.org/10.1038/s41568-018-0016-5
19. Chamola, V., Hassija, V., Gupta, V., and Guizani, M. 2020. A Comprehensive Review of the COVID-19 Pandemic and the Role of IoT, Drones, AI, Blockchain, and 5G in Managing its Impact. *IEEE Access, 8*, 90225-90265. https://doi.org/10.1109/access.2020.2992341
20. Xu, K., Liu, S., Chen, P.-Y., Sun, M., Ding, C., Kailkhura, B., and Lin, X. 2020. Towards an Efficient and General Framework of Robust Training for Graph Neural Networks. *ICASSP 2020–2020 IEEE International Conference on Acoustics, Speech and Signal Processing (ICASSP).* https://doi.org/10.1109/icassp40776.2020.9054465
21. Mwase, C., Jin, Y., Westerlund, T., Tenhunen, H., and Zou, Z. 2022. Communication-efficient distributed AI strategies for the IoT edge. *Future Generation Computer Systems, 131*, 292–308. https://doi.org/10.1016/j.future.2022.01.013
22. Jaiswal, A., Babu, A. R., Zadeh, M. Z., Banerjee, D., and Makedon, F. 2020. A Survey on Contrastive Self-Supervised Learning. *Technologies, 9*(1), 2. https://doi.org/10.3390/technologies9010002
23. Yao, T., Yi, X., Cheng, D. Z., Yu, F., Chen, T., Menon, A., Hong, L., Chi, E. H., Tjoa, S., Kang, J. (Jay), and Ettinger, E. 2021. Self-supervised Learning for Large-scale Item Recommendations. Proceedings of the 30th ACM International Conference on Information & Knowledge Management. https://doi.org/10.1145/3459637.3481952
24. Zhang, M., Li, C., and Zhou, Z. 2020. Text to image synthesis using multi-generator text conditioned generative adversarial networks. *Multimedia Tools and Applications, 80*(5), 7789–7803. https://doi.org/10.1007/s11042-020-09965-5
25. Wang, H., Lin, G., Hoi, S. C. H., and Miao, C. 2021. Cycle-Consistent Inverse GAN for Text-to-Image Synthesis. Proceedings of the 29th ACM International Conference on Multimedia. https://doi.org/10.1145/3474085.3475226
26. Hu, Y., Kuang, W., Qin, Z., Li, K., Zhang, J., Gao, Y., Li, W., and Li, K. 2021. Artificial Intelligence Security: Threats and Countermeasures. *ACM Computing Surveys, 55*(1), 1–36. https://doi.org/10.1145/3487890
27. Picoito, J. 2020. Adoption of public interventions for adolescent alcohol use in Portugal: Challenges and opportunities. IACAPAP ArXiv. https://doi.org/10.14744/iacapaparxiv.2020.20003

28. Dwivedi, Y. K., Hughes, L., Ismagilova, E., Aarts, G., Coombs, C., Crick, T., Duan, Y., Dwivedi, R., Edwards, J., Eirug, A., Galanos, V., Ilavarasan, P. V., Janssen, M., Jones, P., Kar, A. K., Kizgin, H., Kronemann, B., Lal, B., Lucini, B., Medaglia, R., Le Meunier-FitzHugh, K., Le Meunier-FitzHugh, L. C., Misra, S., Mogaji, E., Sharma, S. K., Singh, J. B., Raghavan, V., Raman, R., Rana, N. P., Samothrakis, S., Spencer, J., Tamilmani, K., Tubadji, A., Walton, P., and Williams, M. D. 2021. Artificial Intelligence (AI): Multidisciplinary perspectives on emerging challenges, opportunities(Myers, et al., 2023), and agenda for research, practice, and policy. *International Journal of Information Management, 57*, 101994. https://doi.org/10.1016/j.ijinfomgt.2019.08.002
29. Marijuan, P. C. 1991. Intelligence as adaptive behavior: An experiment in computational neuroethology. *Biosystems, 24*(4), 321–323. https://doi.org/10.1016/0303-2647(91)90051-l
30. Currie, G. M. 2023. Academic integrity and artificial intelligence: Is ChatGPT hype, hero, or heresy? Seminars in Nuclear Medicine, 53(5), 719–730. https://doi.org/10.1053/j.semnuclmed.2023.04.008
31. Krogh, G. von, Ben-Menahem, S. M., and Shrestha, Y. R. 2021. Artificial Intelligence in Strategizing: Prospects and Challenges. *Strategic Management*, 625–646. https://doi.org/10.1093/oso/9780190090883.003.0035
32. Su, J., Ng, D. T. K., and Chu, S. K. W. 2023. Artificial Intelligence (AI) Literacy in Early Childhood Education: The Challenges and Opportunities. *Computers and Education: Artificial Intelligence*, 4, 100124. https://doi.org/10.1016/j.caeai.2023.100124
33. Thorp, C., Sisti, S., and Bennette, W. 2022. Feature Extraction for Out of Distribution Detection via Self-Supervised Learning. *2022 21st IEEE International Conference on Machine Learning and Applications (ICMLA)*. https://doi.org/10.1109/icmla55696.2022.00153
34. Hwang, G.-J., Xie, H., Wah, B. W., and Gašević, D. 2020. Vision, challenges, roles and research issues of Artificial Intelligence in Education. *Computers and Education: Artificial Intelligence*, 1, 100001. https://doi.org/10.1016/j.caeai.2020.100001
35. Bengio, Y. 2009. Learning Deep Architectures for AI. https://doi.org/10.1561/9781601982957
36. Karmakar, S., and Karmakar, S. 2023. Improved Image Recognition for AI Based Various Scientific Applications. *2023 IEEE World AI IoT Congress (AIIoT)*. https://doi.org/10.1109/aiiot58121.2023.10174451
37. Van Le, D., and Tham, C.-K. 2018. A deep reinforcement learning based offloading scheme in ad-hoc mobile clouds. *IEEE INFOCOM 2018 - IEEE Conference on Computer Communications Workshops (INFOCOM WKSHPS)*. https://doi.org/10.1109/infcomw.2018.8406881
38. Gacanin, H. 2019. Autonomous Wireless Systems With Artificial Intelligence: A Knowledge Management Perspective. *IEEE Vehicular Technology Magazine*, 14(3), 51–59. https://doi.org/10.1109/mvt.2019.2920162
39. Chen, L., Mao, X., Xue, Y., and Cheng, L. L. 2012. Speech emotion recognition: Features and classification models. *Digital Signal Processing*, 22(6), 1154–1160. https://doi.org/10.1016/j.dsp.2012.05.007
40. Larsson, S., and Heintz, F. 2020. Transparency in artificial intelligence. *Internet Policy Review*, *9*(2). https://doi.org/10.14763/2020.2.1469
41. Zhou, W., Wang, H., and Wan, Z. 2022. Ore Image Classification Based on Improved CNN. *Computers and Electrical Engineering, 99*, 107819. https://doi.org/10.1016/j.compeleceng.2022.107819
42. Chen, T.-Y., Chiu, Y.-C., Bi, N., and Tsai, R. T.-H. 2021. Multi-Modal Chatbot in Intelligent Manufacturing. *IEEE Access, 9*, 82118–82129. https://doi.org/10.1109/access.2021.3083518
43. Zhao, P., Liu, S., Chen, P.-Y., Hoang, N., Xu, K., Kailkhura, B., and Lin, X. 2019. On the Design of Black-Box Adversarial Examples by Leveraging Gradient-Free Optimization and Operator Splitting Method. *2019 IEEE/CVF International Conference on Computer Vision (ICCV)*. https://doi.org/10.1109/iccv.2019.00021
44. Bobadilla, J., Ortega, F., Hernando, A., and Gutiérrez, A. 2013. Recommender systems survey. *Knowledge-Based Systems, 46*, 109–132. https://doi.org/10.1016/j.knosys.2013.03.012
45. Xu, S., Zhang, X., Wu, Y., Wei, F., and Zhou, M. 2020. Unsupervised Extractive Summarization by Pre-training Hierarchical Transformers. *Findings of the Association for Computational Linguistics: EMNLP 2020*. https://doi.org/10.18653/v1/2020.findings-emnlp.161
46. Hayes-Roth, B. 1996. An architecture for adaptive intelligent systems. *Computational Theories of Interaction and Agency*, 329–366. https://doi.org/10.7551/mitpress/2026.003.0010

47. Haass, J. C. 2022. Cyber Threat Intelligence and Machine Learning. *2022 Fourth International Conference on Transdisciplinary AI (TransAI)*. https://doi.org/10.1109/transai54797.2022.00033
48. Mutingi, M., and Matope, S. 2013. Dynamics of information technology adoption in a complex environment. *2013 IEEE International Conference on Industrial Technology (ICIT)*. https://doi.org/10.1109/icit.2013.6505888
49. Robles-Kelly, A., and Ran Wei. 2016. Semi-supervised image labeling using barycentric graph embeddings. *2016 23rd International Conference on Pattern Recognition (ICPR)*. https://doi.org/10.1109/icpr.2016.7899852
50. Chhipa, P. C., Chopra, M., Mengi, G., Gupta, V., Upadhyay, R., Chippa, M. S., De, K., Saini, R., Uchida, S., and Liwicki, M. 2023. Functional Knowledge Transfer with Self-supervised Representation Learning. *2023 IEEE International Conference on Image Processing (ICIP)*. https://doi.org/10.1109/icip49359.2023.10222142

CHAPTER 4

Enhancing the Security of Resource Constraints Devices Using Blockchain

Abhishek Rawat, Rajat Verma, Ankita* and *Nayancy*

1. Introduction

This chapter, introduced the vast and convoluted world of blockchain technology, delving deeply into its enormous potential to revolutionize and strengthen the security of devices with limited resources. As the demand for smart devices that are connected grows, there is an increasing need to protect these devices from the ever-present threat of hostile attacks that aim to exploit vulnerabilities. Traditional security solutions, while successful in some situations, are insufficient when confronted with the unique problems given by limited resources devices, which are limited in computational power and memory capacity. However, a ray of light appears as Blockchain, which offers a viable and innovative solution that holds the key to unlocking unparalleled levels of security for these devices. The fundamental function of blockchain technology is to offer a decentralized platform that is more flexible than conventional security measures [1]. Blockchain technology, by leveraging the power of worldwide networks and algorithms for encryption, provides a safe basis for resource-constrained devices, allowing them to battle the ever-evolving threats that lurk in the digital arena. Because blockchain technology is decentralized, it eliminates the risks associated with depending on centralized authorities [2]. Blockchain creates an indelible record of transactions and data through the immutability of its ledger, protecting it from manipulation or unauthorized access. This chapter will examine how blockchain might rise to the occasion and meet the unique issues encountered by resource-constrained devices. Blockchain technology, by definition, is fundamentally

Department of Computer Science and Engineering, Pranveer Singh Institute of Technology, Kanpur, Uttar Pradesh, India.
* Corresponding author

compatible with the resource constraints that frequently afflict these devices. Its lightweight and efficient architecture enables smooth integration into the limited contexts in which these devices operate, ensuring that they are not burdened by excessive computing requirements.

Secure data storage is w here Blockchain Technology improves the security of resource-constrained devices. Sensitive data can be maintained in a way that provides uncompromised security, even in the face of constant assaults, thanks to the use of blockchain [3]. The use of encryption techniques and decentralized file systems strengthens data protection by adding an extra layer of defense against potential breaches. The Blockchain benefits transaction verification, which is an important part of device security. Transactions may be confirmed in a trustless and transparent manner by exploiting the consensus mechanisms inherent in blockchain, instilling confidence and integrity into the broader security architecture [4,52]. Smart contracts, which are powered by blockchain, automate and streamline security measures, ensuring that devices stay resilient and resistant to unauthorized activity.

However, problems and issues exist on the way to general acceptance, as with any emerging technology. Scalability and performance emerge as critical barriers, necessitating creative solutions to meet blockchain's fundamental constraints in resource-constrained contexts. Blockchain protocol optimization becomes critical to ensure that the technology runs smoothly within the limits imposed by limited resources. Another barrier to overcome is integration with existing systems. The coexistence of Blockchain Technology with traditional security measures needs a precise balance to ensure that the adoption of Blockchain does not disrupt the established security architecture. Collaboration and interoperability frameworks are required to integrate these different parts into a unified and strong security ecosystem.

1.1 Overview of Resource–Constrained Devices

Devices classified as resource-constrained have low amounts of RAM, storage, and computational power. These devices frequently operate in areas with limited resources and encounter performance, energy efficiency, and scalability difficulties. Devices with limited resources include activity trackers, coffee makers, refrigerators, microcontrollers, toys, smart thermometers, controllers, even wireless sensor network nodes [5]. The functionality and capacities of these devices are limited due to resource constraints. When compared to standard computer devices, they often have less computational power and memory. This constraint needs software and hardware optimization to enable efficient functioning within these constraints. Devices with limited resources are popular in IoT applications, where they serve as the backbone of networked systems, receiving and delivering data for a variety of functions.

Figure 1 illustrates the usage of diverse IoT devices that are somehow related to the smart industries via different techniques of network layer, that also acts as interface between the application layer and sensor layer of the three layered IoT architecture.

These gadgets' limited resources present significant security challenges. Because of their restricted processing power and memory, traditional security techniques may not be appropriate for resource-constrained devices. On the other hand, blockchain

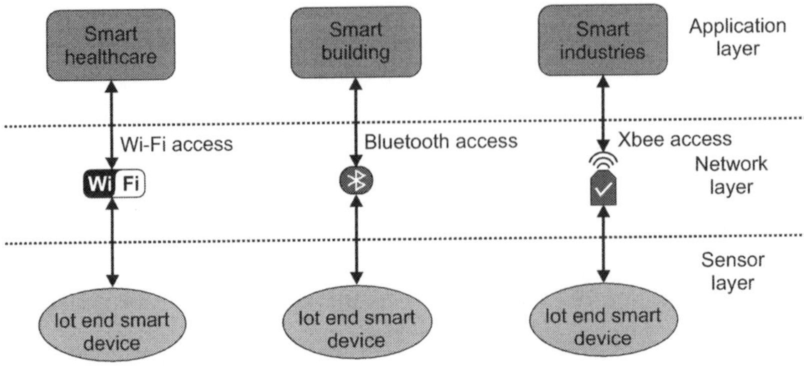

Figure 1: Connected Devices in IoT Spectrum.

offers a viable substitute to enhance the security of such gadgets. By utilizing blockchain's decentralized and irreversible nature, devices with limited resources can accomplish secure data storage, verification of transactions, including identity management. To summarize, resource-constrained devices have restricted processor power, memory, and storage capacity. They are often used in IoT applications and confront specific security challenges. By exploiting its decentralized and unchangeable nature, Blockchain can increase these devices' security.

1.2 Challenges in Terms of Security Due to their Inherent Limitations

Since resource-constrained devices have inherent constraints, they provide several security concerns. Table 1 highlights these difficulties [6].

Table 1 presents the security challenges associated with resource-limited devices, particularly those used in IoT applications. It outlines the constraints related to limited computing power and memory, energy optimization, communication limitations, physical vulnerability, lack of standardization, scalability and management difficulties, firmware and software vulnerabilities, and the absence of user interfaces. These challenges encompass issues such as energy usage optimization, communication effectiveness, physical security measures, and the complexities of managing diverse security methods across a large number of devices. Furthermore, it emphasizes how Blockchain Technology's decentralized and irreversible nature might improve security in devices with low resources.

1.3 The Limitations of Traditional Security Measures

While traditional security methods are successful in many situations, they frequently have substantial limits when applied to resource-constrained systems [7]. These restrictions include the following:

- **Resource-Intensive:** Conventional security protocols could need a lot of memory and processing power, which could be challenging for devices with limited resources. Using complicated encryption methods or executing resource-

Table 1: Security Issues with Devices with Limited Resources.

Limited Computing Power and Memory	They are often used in IoT applications and confront specific security challenges. By exploiting its decentralized and unchangeable nature, Blockchain can increase these devices' security.
Energy Constraints	These devices are usually battery-powered or have limited access to power sources. As a result, to extend battery life, energy usage must be optimized. Putting in place strong security measures while minimizing energy use is a huge task.
Communication Limitations	Wireless communication methods with limited bandwidth and range are frequently used by resource-constrained devices. This constraint makes it difficult to effectively transmit and receive security-related updates, patches, or authentication mechanisms.
Physical Vulnerability	These devices are frequently installed in physically accessible locations, leaving them vulnerable to manipulation or theft. Physical security measures must be considered to prevent unauthorized access or manipulation of the devices.
Lack of Standardization	The lack of standardized standards and protocols for resource-limited devices makes security implementation and interoperability more difficult. Different devices may have different security methods or may lack basic security measures entirely, leaving the total system vulnerable
Scalability and Management	Managing a large number of resource-limited devices can be difficult. Maintaining consistent security protections, upgrades, and monitoring across all devices becomes difficult, especially as the deployment grows.
Firmware and Software Vulnerabilities	Manufacturers may prioritize usefulness over security during the creation of firmware and software due to limited resources. This prioritization may result in weaknesses that attackers could exploit.
Lack of User Interface	Many low-resource devices are devoid of user interfaces or have restricted visual capabilities. This constraint makes interacting with security-related alerts or warnings difficult for users, possibly leading to security breaches.

intensive security protocols might put a burden on the device's capabilities, resulting in performance degradation or even device failure.

- **Scalability Issues:** Using a lot of devices with limited resources can make typical security solutions unsuitable for large-scale deployments. System inefficiencies can arise from the costs of managing security measures like encryption and authentication when the number of devices increases.
- **Lack of Flexibility:** Traditional security techniques are sometimes one-size-fits-all, which may not be appropriate for resource-constrained devices working in a variety of contexts. These devices have unique limits and requirements that must be taken into account when implementing security solutions. Traditional techniques' inflexibility can limit their effectiveness and adaptation to the special needs of resource-constrained devices.
- **Vulnerabilities in Legacy Systems:** Devices with limited resources may function in an ecosystem of outdated systems with security flaws. Traditional security solutions may not address these weaknesses properly, leaving the devices vulnerable to possible assaults. Interoperability issues between legacy systems and resource-limited devices can exacerbate these security threats.

- **Lack of Over-the-Air Updates:** To handle new threats and vulnerabilities, traditional security solutions frequently rely on timely security upgrades. However, due to limited connectivity or the inability to conduct sophisticated update operations, resource-constrained devices may experience difficulties obtaining over-the-air updates. This limitation may expose devices to known security flaws.
- **Limited User Interaction:** Many low-resource devices lack user interfaces or have limited display capabilities. For tasks like authentication or responding to security prompts, traditional security mechanisms frequently rely on human engagement. Without suitable user interfaces, it becomes difficult to successfully execute some security measures, thus jeopardizing the device's overall security.
- **Inadequate Physical Security:** Traditional security solutions may fail to address the physical vulnerabilities of resource-constrained devices completely. These devices are frequently installed in physically accessible locations, leaving them vulnerable to manipulation or theft. Traditional security measures may not adequately incorporate physical security factors such as secure enclosures or tamper-resistant designs.
- **Limited Security Expertise:** Devices with few resources are frequently designed by non-security professionals or small teams with limited security knowledge. Because of this constraint, best practices may be overlooked and inferior security measures used. Due to a lack of comprehensive security knowledge dedicated to these devices, they may be more vulnerable to assaults.

A customized strategy that considers the unique qualities and limitations of these devices is needed to address the restrictions of standard security measures with resource-constrained devices. Alternative security solutions, such as lightweight encryption algorithms, energy-efficient protocols, and decentralized alternatives like blockchain technology, can assist in overcoming these constraints and providing more effective security measures for resource-constrained devices.

1.4 Vulnerabilities Posed By Centralized Systems

Centralized systems, which centralize control and information in a single authority or body, introduce weaknesses that attackers can exploit [8]. These vulnerabilities are highlighted in Table 2.

This table outlines the vulnerabilities and challenges associated with centralized systems. It highlights the risks of single points of failure, such as susceptibility to attacks and potential consequences of compromise, including unauthorized access, data breaches, and service outages. Large volumes of sensitive data storage, scalability problems, user reliance on trust, low operational transparency, and the difficulties of regulatory compliance are some of the other concerns it solves. These factors collectively underscore the potential consequences of centralized system weaknesses, ranging from financial losses and brand damage to privacy concerns and legal implications. Decentralized and decentralized systems, such as the blockchain system, have become known as possible remedies for these weaknesses. Control and

Table 2: Vulnerabilities of Resources Constrained Devices.

Single Point of Failure	Centralized systems are vulnerable to attacks because they rely on a single source for command or authority. Compromising the main network can lead to unauthorized access, data breaches, or service outages, exposing the entire system.
Data Breaches	Centralized systems often store large amounts of sensitive data, making them vulnerable to breaches that could expose information to unauthorized individuals. The consequences of a breach include financial losses, brand damage, and privacy concerns.
Lack of Redundancy	Centralized systems are vulnerable to interruptions due to a lack of redundancy and failover methods. If the central system fails, whether due to hardware issues, software bugs, or cyber attacks, the entire system may become inaccessible, causing service interruptions and disrupting key operations.
Limited Scalability	As the number of users or devices increases, centralized systems may fail to scale successfully. The centralized architecture may be unable to keep up with the increasing demand, resulting in performance degradation, delays, or system breakdowns. During moments of high traffic or overload, this constraint may provide an opportunity for attackers to exploit system flaws.
Dependency on Trust	Users of centralized systems frequently must invest their trust in the central authority or entity. If the central authority is corrupted or acts maliciously, this trust can be exploited. Users may have limited control over their data and rely on the central system to protect it, leaving them open to abuse or unauthorized access.
Lack of Transparency	The operations and choice-making processes of centralized systems may be opaque. Users may have little insight into how their data is handled or shared, making it impossible to examine the central authority's security and privacy practices. The absence of transparency can undermine confidence and raise the likelihood of unauthorized data access or exploitation.
Regulatory Compliance Challenges	Compliance with data protection and privacy rules may be difficult for centralized systems, particularly when handling sensitive or private data. A large amount of data in just one place can make meeting regulatory standards more complex, raising the risk of non-compliance and related legal implications.

data are distributed among numerous nodes in these systems, decreasing reliance on one centralized figure and reducing some of the risks inherent in centralized systems.

2. Blockchain

Due to its potential to revolutionize multiple industries and remove inherent flaws in centralized systems, Blockchain Technology has garnered tremendous attention and interest. Organizations can benefit from increased security, openness, and speed in their operations by implementing blockchain technology. The ability of Blockchain to provide a safe and unchangeable system for recording and verifying transactions is one of its main advantages. Because blockchain technology is decentralized, there is no need for intermediaries, which reduces the risk of fraud and exploitation [9]. Each transaction is kept track of in a block, which is subsequently chained together using cryptographic hashes as shown in Fig. 2. Following the recording of a transaction, it becomes nearly immutable, making it highly resistant to unauthorized changes or tampering. Transparency is one of blockchain's main advantages. All network participants have access to identical information, and all transactions are transparent to all. Because anybody can check the validity and integrity of transactions, this transparency improves responsibility and confidence. It also lowers dependency on centralized authorities because blockchain's consensus mechanism ensures that transactions have been verified by a network of participants.

The blockchain architectural diagram is depicted in Fig. 2, which shows connected data blocks connected in a sequential chain. Every block is safeguarded by cryptographic hashing and contains encrypted data. Because the blocks are dispersed among a network of nodes and are decentralized, they guarantee tamper resistance, transparency, and immutability. Blockchain technology has several uses outside of cryptocurrencies. Supply chain managers may verify the validity and provenance of products by utilizing blockchain technology, which offers end-to-end visibility and traceability. By reducing the risk of identity theft, blockchain's ability to safely store and validate personal data can aid with identity management [10]. Data security, privacy, and integrity can all be improved by blockchain's decentralized architecture, which benefits data management. Still, a few concerns need to be resolved before Blockchain Technology becomes extensively used. Since present blockchain networks find it difficult to process many transactions quickly, scalability is still a problem [11]. Another area of study is regulatory compliance, as the decentralized nature of blockchain may contradict existing legislation. Integrating blockchain into current systems and infrastructure takes careful design and attention as well. Despite these obstacles, Blockchain Technology can alter sectors while also providing robust security and transparency. As companies continue to explore and grow with

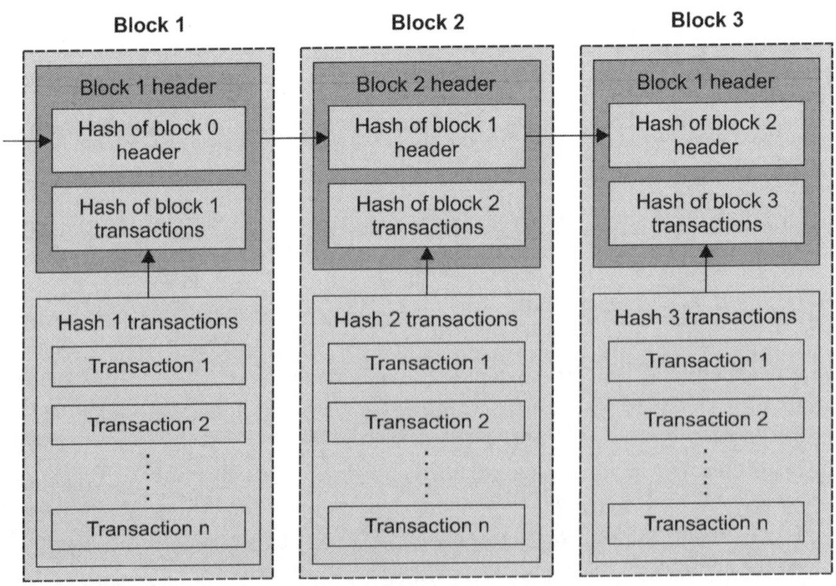

Figure 2: Blocks Architecture in Blockchain.

blockchain, it is expected to have a significant impact on many industries and provide a solution to the limitations of centralized systems.

2.1 Blockchain and its Core Principles

Recently, there has been a lot of interest in blockchain technology, a breakthrough technology. Blockchain records transactions or data securely and publicly. It is essentially a distributed, decentralized ledger. It is operated via a node network, which consists of computers that store a copy of the overall blockchain [12]. It is immune to censorship and manipulation since it does not require a centralized authority. Immutability is a fundamental feature of blockchain. It is exceedingly difficult to change or tamper with data after it has been saved on the blockchain. Every block has a distinct cryptographic hash it is generated based on the data within the block and the hash of the block before it [13]. This guarantees the integrity and reliability of the data because each modification in the data requires a new hash calculation for that block as well as all subsequent blocks.

Another fundamental component of blockchain is transparency. Any user on the network can view any transaction that is entered into the distributed ledger. Transparency makes it simpler to identify and stop fraudulent activity, which raises confidence and responsibility. Any modifications or modifications to the blockchain must be approved by network participants, who confirm and authenticate the transactions [14]. Blockchain also employs cryptographic methods to maintain data protection. Transactions are digitally signed using cryptographic keys, allowing participants' identities to be verified and ensuring that only authorized individuals can make modifications to the blockchain. Additionally, blocks are connected by cryptographic hashing, which makes it computationally difficult to alter the contents without being detected [15].

Blockchain accomplishes consensus amongst network users through a variety of consensus techniques, including proof of Work (PoW) and Proof of Stake (PoS) [16]. These methods ensure that all participants have a common understanding of the status of the Blockchain and prevent malevolent actors from influencing the system. For data verification and transaction recording, blockchain technology generally provides a decentralized, transparent, safe, and immutable foundation. With its ability to improve security, lower fraud, and increase participant confidence, it has the potential to drastically transform several businesses. Blockchain applications go beyond cryptocurrency, with possible use cases in SCM, healthcare, banking, and other industries.

2.2 Key Features that Make Blockchain Suitable for Enhancing Device Security

The functionality and advantages of blockchain technology are supported by a set of fundamental ideas [17]. Among these guidelines are:

- **Distributed Ledger:** Through a distributed ledger, blockchain technology gives each network user access to the complete database as well as every of its history.

Because of its decentralized structure, shared data control, transparency, and the elimination of centralized authority are all made possible.

- **P2P Network:** Blockchain, at its heart, is a peer-to-peer network that allows users to connect directly with each other without the need for intermediaries. This allows for immediate transactions and interaction among parties, which lowers costs and increases efficiency.
- **Mechanism of Consensus:** Blockchain uses a consensus process to confirm and determine what state the ledger is in at any given time. Before a transaction is uploaded to the blockchain, all network members can agree that it is valid. This process ensures the data's integrity and immutability.
- **Cryptographic Security:** To secure the information recorded on the blockchain, cryptographic techniques are used. Each block is tamper-resistant because it has a distinct cryptographic hash that represents the data inside. Furthermore, public-key cryptography is frequently used to authenticate and authorize transactions.
- **Transparency and Auditability:** Blockchain makes transactions on the network transparent by allowing anybody to see and verify them. Because the complete transaction history is public and auditable, this transparency increases confidence and accountability.
- **Intelligent Contracts:** Blockchain technology can be used to carry out autonomous contracts with predetermined rules and conditions, or smart contracts. Smart contracts streamline procedures and remove the requirement for intermediaries, ensuring that agreements are executed accurately and efficiently.

2.3 Blockchain Role in Device Security

Blockchain innovation has emerged as an effective technique for improving device security. It provides safe record-keeping, incorporates hardware security primitive beings, and improves overall safety online due to its decentralized and transparent nature. Blockchain Technology provides multiple advantages that lead to a safer digital environment, from safeguarding healthcare equipment to increasing IoT security. The most significant benefit of Blockchain in device security is its ability to enable safe and transparent record-keeping. Bitcoin, in contrast to conventional centralized systems, keeps a transparent and impenetrable decentralized record of transactions [18]. As a result, the information saved on the blockchain cannot be easily viewed or changed without the necessary authority. Blockchain preserves the integrity of data by utilizing cryptographic techniques, lowering the danger of unauthorized access or alteration.

Blockchain technology plays a crucial role in protecting connected devices within the Internet of Things (IoT) framework. Non-fungible tokens (NFTs), secure booting, smart contracts, dependable hardware, and other aspects help to safeguard the integrity and security of IoT data when combined with blockchain technology. Devices may securely communicate and authenticate transactions using blockchain, lowering the danger of unauthorized access or manipulation [19]. This paves the way for a more secure and dependable Internet of Things ecosystem. Another area where Blockchain Technology has considerable possibilities is healthcare device security. Creating a thin, safe medical blockchain for the Internet of Things medical devices will ensure the confidentiality and integrity of all medical data. Blockchain technology safeguards patient data by prohibiting illegal access or modification and by producing an auditable, tamper-proof record of healthcare transactions [20]. In the healthcare industry, where safeguarding the confidentiality and integrity of highly sensitive patient information is critical, this is especially important.

2.3.1 Decentralization: Reducing Reliance on Centralized Authorities

The goal of decentralization is to eliminate dependency on centralized authorities, which is a key premise of blockchain technology. Blockchain distributes power and decision-making across a network of nodes, improving security, transparency, and resilience. Due to the lack of a single point of failure and the transparency and immutability of the blockchain ledger, which provides an impermeable record of transactions, distributed systems are more resilient to attacks [21]. Decentralization improves device security by encouraging trust, removing the need for intermediaries, and protecting against unauthorized access or manipulation, making it an important component in the growth of secure and trustworthy digital ecosystems.

2.3.2 Immutable Ledger: Ensuring Data Integrity and Tamper Resistance

A key element of blockchain technology that guarantees the accuracy of information and resistance to tampering is an immutable ledger. Blockchain creates an immutable, permanent record of transactions using cryptography and consensus procedures. Without the network's approval, it is very difficult to alter or remove a transaction after it is registered on the blockchain [22]. Because of its immutability, which ensures that data is safe, dependable, and immune to unauthorized changes or manipulation, the blockchain ledger offers a transparent and trustworthy source of truth for a variety of applications, including supply chain management and financial transactions.

2.3.3 Consensus Mechanisms: Verifying transactions in a Trustless Environment

Consensus techniques are critical in Blockchain Technology for confirming transactions in a trustless environment. With the use of these techniques, even in the lack of a central authority, all participants in a decentralized network can agree on the validity and order of transactions. To verify transactions and add these to the blockchain, members of the network, known as miners, are required to solve difficult mathematical problems using the widely used proof-of-work (PoW) consensus technique [23]. Another popular technique is called Proof-of-stake (PoS), where users can approve transactions and add new blocks according to how many coins they have and are willing to "stake" as security [24]. By allowing network members to independently evaluate and agree upon the authenticity of transactions, these consensus techniques maintain the security thus integrity of the distributed ledger and create a trustless environment. While PoW is known for its high energy consumption, PoS provides scalability and speedier transaction acceptance. The choice of consensus mechanism is determined by the blockchain network's specific objectives and aims.

3. Enhancing Device Security with Blockchain

Secure data storage is a critical part of improving device security, and Blockchain provides unique solutions in this area. Data may be safely stored and accessed in a trustless environment by exploiting the blockchain's nature. The unchangeable ledger of blockchain assures data integrity and resistance to tampering. It is extremely difficult to alter or remove a transaction from the blockchain without consensus from the network [25]. Because of this feature, the blockchain is much more secure, which makes it a great option for storing sensitive data. Aside from the inherent security given by the blockchain ledger, various techniques and technologies can be used to improve secure data storage. Encryption techniques, for example, can protect data from unauthorized access. SMB Encryption, for example, can provide end-to-end privacy and integrity guarantees between file servers and clients. Data is rendered unreadable and useless by unauthorized persons even in the event of interceptions thanks to encryption [26].

Furthermore, making use of secure storage media and protocols, such as hardware-encrypted external drives, can add another degree of security to critical data. These gadgets are designed to safeguard information and stop unwanted access, offering data protection even if the storage device itself is compromised. It is crucial to realize that secure data storage extends beyond technology. Individuals and organizations must also follow best practices and rules for data retention and disposal. To comply with regulations like the Payment Cards Industry Security Standard (PCI DSS), organizations must securely destroy data that is no longer required for business and legal purposes [27]. Proper data disposal practices, including secure destruction of physical media, are critical for preventing data breaches of stored data.

3.1 Storing Private Information on Devices with Limited Resources Using Blockchain

Using blockchain to store sensitive data on resource-constrained devices is a safe and efficient alternative. Sensitive data may be safely kept and accessed by using the decentralized nature of blockchain technology, even on devices with limited processing power and storage capacity. The blockchain ledger's immutable and tamper-resistant nature assures the integrity and privacy of the stored data [28]. Furthermore, the adoption of encryption techniques improves the protection of sensitive information. Blockchain allows resource-constrained devices to securely store and process sensitive data, enabling secure transactions and interactions in a variety of domains such as IoT devices, healthcare systems, and financial applications while minimizing the need for considerable computational resources.

3.2 Encryption Techniques and Decentralized File Systems for Added Security Verification

When storing sensitive data, encryption techniques and decentralized file systems might give additional security. Decentralized file systems, such as IPFS, spread data across a worldwide network, ensuring efficient and secure storage. These systems use Blockchain Technology to establish a tamper-resistant and immutable ledger, improving data security [29]. Additionally, the security of sensitive data depends heavily on encryption. Only authorized parties will be able to access and decrypt data thanks to encryption mechanisms like the Advanced Encryption Standard, also known as AES, that may encrypt data before it is put on the blockchain or distributed file system [30]. An effective solution for safely storing private information, regardless of devices with constrained resources, is the combination of distributed file systems with encryption software.

3.3 Leveraging Blockchain's Consensus Mechanisms for Secure Transaction Verification

The consensus processes of blockchain are critical to providing secure transaction verification. By utilizing these ideas, blockchain networks can foster consensus and trust among users without the need for a central authority. In the well-known consensus method known as proof-of-work (PoW), miners compete to solve challenging mathematical puzzles to confirm transactions and add those to the blockchain [31]. Adversaries find it harder to change the data using this strategy since it uses a lot of computing resources. Proof-of-stake (PoS) is another popular tactic. In PoS, validators are chosen based on how many coins they have and are willing to "stake" as a guarantee. Validators are selected according to how many coins they stake, and they are picked to verify deals and create new blocks. In comparison to PoW, this approach not only improves security but also saves energy usage. Other consensus mechanisms, like as DPoS and PBFT, provide other approaches to secure transaction verification. o prevent fraud, double spending, and unauthorized alterations, these consensus procedures guarantee that transactions are verified and published to the blockchain safely and transparently [32]. By leveraging the power of consensus processes, blockchain technology provides a solid foundation for secure verification of transactions in a variety of applications, including financial and supply chain management.

3.4 Smart Contracts and their Role in Automating Security Protocols

In blockchain technology, smart contracts play a key role in automating security procedures. These self-executing contracts eliminate the need for middlemen and reduce the possibility of human error by being programmed to carry out predefined actions when specific conditions are met [33]. Security protocols can be automated using smart contracts, ensuring consistent and dependable implementation without the need for user intervention. The ability of smart contracts to increase transaction security is a key benefit.. Smart contracts are powered by the distributed, immutable ledger known as the blockchain. As a result, once a transaction is recorded on the blockchain, it cannot be changed or tampered with, protecting the process' transparency and integrity.

Smart contracts can also automate security measures like access control and authentication. A smart contract, for example, can require multi-factor authentication before allowing access to specific resources or systems. This automation decreases the likelihood of unauthorized access and improves overall security posture [34]. Furthermore, smart contracts can make it easier to design safe and auditable procedures. Organizations can ensure that important processes are regularly followed by automating the execution of security protocols, minimizing the likelihood of security breaches or vulnerabilities. The blockchain's transparency and immutability allow for simple auditing and verification of security mechanisms.

It is crucial to remember that although smart contracts offer many benefits, there are drawbacks as well. The intricacy of designing smart contracts necessitates careful study to eliminate vulnerabilities and ensure the contract's security. To detect and minimize potential hazards, thorough testing and auditing methods are required [35]. Finally, smart contracts are critical in automating security protocols within blockchain technology. They allow for the secure and dependable execution of predetermined activities, improve transaction security, automate access control and authentication, and make auditable workflows possible. Organizations can use smart contracts to streamline security processes, eliminate human error, and improve the overall security of their systems and transactions.

3.5 Blockchain–based Methods for Safe Identity Confirmation and Access Control

Blockchain Technology provides a variety of methods for safe identity confirmation and access management. These solutions improve security and privacy by exploiting the blockchain's decentralized and unchangeable nature [36]. Here are some crucial points:

- **Decentralized Identity:** Blockchain makes it easier to create decentralized identity systems where people are in charge of their data. Users can manage access to their data by storing their identification information in private identity stores known as identity hubs. This method of self-sovereign identity decreases the danger of data breaches and unauthorized access.
- **Elimination of Multiple Login Accounts:** Users using centralized identification systems are frequently required to create several accounts and passwords for various websites and applications. This raises the danger of security breaches and makes it difficult for users to manage several login credentials. Blockchain-based identification solutions can help to speed up this process by offering a single, secure digital identity that can be used across numerous platforms, eliminating the need for multiple accounts.
- **Efficient Know Your Customer (KYC) Processes:** KYC processes, which entail confirming individuals' identities, may be time-consuming and costly

for businesses. By securely storing and distributing verified identification information, blockchain-based authentication solutions can streamline and speed up KYC processes. This lowers the need for repeated verification and improves onboarding process efficiency.

- **Decentralized Identifiers (DIDs):** Decentralised identifiers (DIDs) are used in blockchain-based identity systems instead of traditional identifiers such as usernames and passwords. DIDs are cryptographically secure unique identifiers that can be used to authenticate and verify identities without relying on centralized authorities.
- **Secure Access Control:** Blockchain technology can be used to create secure access control systems. By setting access controls and permissions, smart contracts guarantee that only authorized individuals or entities can view or perform particular tasks. By reducing the likelihood of unwanted access and enforcing strict access control procedures, this enhances security.
- **Tamperproof Audit Trails:** Because blockchain technology is transparent and immutable, it may be used to create audit trails that are impossible to tamper with. Every transaction as well as access request can be recorded on the blockchain, resulting in an accurate record of every single activities. Organizations may now track and audit identity-related events, improving accountability and security.

Overall, blockchain-based solutions for safe identity verification and access management provide greater privacy, less reliance on centralized authorities, simplified operations, and higher security. These technologies have the power to revolutionize supply chain management, healthcare, finance, and identity management in many other areas.

3.6 Enhancing the Authorization and Authentication Procedures for Devices

To provide secure resource access, device authorization and authentication procedures must be improved. Authentication is the process of authenticating a user's or device's identity before giving access, whereas authorization specifies the specific permissions and level of access allowed to the authenticated entity. Here are several blockchain-powered methods for improving device authentication and authorization [37].

- **Decentralized Identity Management:** Blockchain makes it easier to create decentralized identification systems in which individuals control their data and may securely authenticate themselves without depending on centralized authority. Decentralized identifiers (DIDs) and verifiable credentials can be utilized in peer-to-peer authentication and trust establishment.
- **Immutable Identity Records:** An identity record cannot be altered or tampered with once it is placed on the blockchain thanks to its immutability. This offers a secure and impenetrable way to store and confirm the identities of devices.
- **Smart Contracts for Access Control:** Access control rules and permissions can be defined using smart contracts. These contracts ensure that only those with

permission can access resources or carry out specific tasks by automatically enforcing access controls based on predetermined conditions.

- **Secure Key Management:** The digital keys that are used for device authentication can be safely maintained and stored using blockchain technology. The blockchain can hold private keys, providing for safe and decentralized key management.
- **Consensus Mechanisms for Trust:** Blockchain consensus mechanisms, including proof-of-work and proof-of-stake, guarantee that only valid and approved transactions are stored there. This helps to prevent unauthorized devices from connecting to the network.
- **Auditability and Transparency:** Blockchain records device authentication and authorization events in a transparent and auditable manner. Every transaction and access request may be recorded on the blockchain, making the authentication process straightforward to audit and verify.

Device authentication and authorization processes can be reinforced by employing blockchain technology, enabling improved security, privacy, and confidence in resource access. In several industries, including supply chain, cyber security, and the Internet of Things, these solutions offer a decentralized and impenetrable method to guarantee safe device interactions.

4. Case Studies and Implementations: Platforms for IoT Security base on Blockchain

IoT security solutions built on blockchain are becoming more and more popular as a means of enhancing the privacy and security of IoT data and devices. The following are a few examples, while there are more case studies and implementations in this sector.

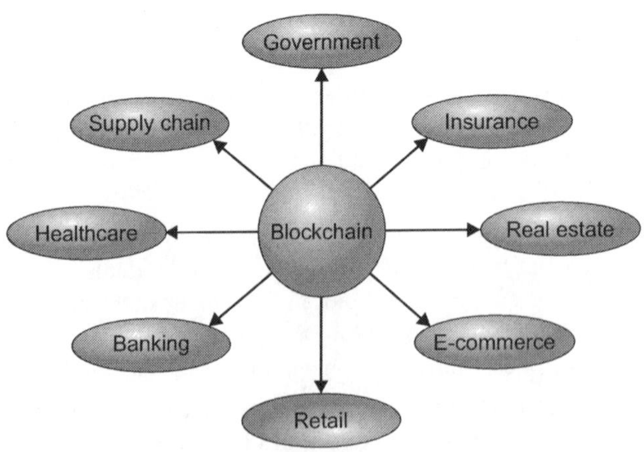

Figure 3: Different Blockchain Use Case.

[38]. Figure 3 illustrating different blockchain use cases, including financial transactions, supply chain management, healthcare records, and digital identity verification. Each use case demonstrates the versatility of Blockchain Technology in facilitating secure, transparent, and decentralized applications across various industries.

- **Blockchain Technology and Maritime Shipping:** A study that looked into the application of a platform built on blockchain for a demonstration project in maritime transport demonstrated how blockchain technology can enhance security and traceability within the shipping sector.
- **Blockchain-Based Frameworks for Food Traceability:** Several case studies have focused on integrating blockchain for food traceability, ensuring supply chain transparency and security. These deployments have proven blockchain's promise for improving food safety and minimizing fraud in the food business.
- **Blockchain-Enabled Smart Contracts for IoT Security:** IoT security can be enhanced by using smart contracts provided by blockchain, according to research. These smart contracts can automate security protocols, providing secure and dependable execution without the use of middlemen.

These examples demonstrate blockchain technology's promise to improve IoT security by providing decentralized and tamper-resistant solutions for verification, accountability, and access control. However, it is crucial to note that every implementation may have unique design concerns and obstacles that must be addressed for the deployment to be effective.

4.1 Exploring Existing Platforms that Utilize Blockchain for Securing Resource—Constrained Devices

Examining current blockchain-based resource-constrained device security platforms provides innovative approaches for enhancing the safety of devices with little memory and processing power. These systems protect data transactions, enforce access control, and foster trust by utilizing the decentralized and immutable features of blockchain technology. For instance, DAG structures like the "Tangle" are used by platforms like IOTA to facilitate safe communication and feeless microtransactions between IoT devices while maintaining data integrity and thwarting manipulation [39]. Similarly, platforms like IoTeX offer lightweight blockchain solutions developed expressly for resource-constrained IoT devices, with capabilities like secure firmware updates, decentralized identity management, and verifiable data integrity [40]. These platforms show how blockchain may be used to secure resource-constrained devices, enable trustworthy interactions, and guard against vulnerabilities in IoT contexts.

4.2 Protocols and Standards for Device Security: Examples from the Real World and Success Stories

These are a few real-world instances and triumphs about standards and methods for safeguarding devices [41].

- **Bluetooth:** Bluetooth is a commonly used wireless communication system that connects and exchanges data between devices over short distances. It was primarily designed for wireless headphones but has now grown to support a wide range of applications. Bluetooth includes security mechanisms to safeguard data transmission and prevent unauthorized access.
- **Mobile Device Management (MDM):** MDM solutions offer protocols and standards for managing and protecting mobile devices in the company. Organizations may manage device configurations, enforce security standards, and remotely track and regulate devices with MDM solutions like Microsoft Intune and Office 365.
- **Project for Open Web Application Safety (OWASP):** A community-driven group called OWASP is committed to enhancing the security of application software. They offer standards, best practices, and testing tools to assist developers in developing safe applications. Their security resources cover a wide range of topics, including device security.
- **Web Content Accessibility Guidelines (WCAG):** While WCAG is primarily concerned with accessibility, it also incorporates security recommendations. These standards ensure that all users have access to web material while still maintaining security safeguards to protect sensitive information.

These examples show how protocols and standards can be used in practice to ensure device security. They emphasize the significance of installing strong security measures, keeping to industry standards, and sticking to best practices to protect devices and data.

4.3 Evaluating their Effectiveness and Adoption Rates

Various elements must be considered when assessing the success and acceptance rates of industry standards and protocols that include blockchain for security enhancement. Adoption of open-source standards and reuse of existing protocols are critical components in assuring cost-effective and secure use of information and communications technology [42]. Blockchain-based solutions have demonstrated potential in enabling faster, more cost-effective, and customized issuance of digital securities, resulting in higher market adoption.

The development of effective and lightweight machine learning and artificially intelligent models for Internet of Things (IoT) devices, the exploration of novel security models and protocols to safeguard data exchange and communication among diverse IoT devices and systems, and the integration of flexible and expandable Blockchain Technology to facilitate safe and transparent sharing of information and management between IoT devices and stakeholders are some of the future instructions in device security [43]. Although blockchain technology has the potential to improve security, different industries have different adoption rates for these protocols and standards. For example, artificial intelligence usage in the building industry is still quite low, despite recognition of its potential to increase profitability, effectiveness, safety, and security. Similarly, the use of blockchain-based finance and AI techniques in finance has the potential to increase efficiency gains, however, the level of acceptance will vary.

The usefulness and effectiveness of blockchain technology in enhancing security and transparency are demonstrated by real-world achievements, such as the use of blockchain in the development of transparent supply chains [44]. Furthermore, encryption and data erasure techniques are critical components of data security, giving a competitive advantage for businesses and increasing consumer trust. Overall, the efficacy and pace of acceptance of industry standards and protocols that use blockchain for security enhancement differ among industries and application cases. Even while there are examples of achievement and the advantages are becoming more widely recognized, more study, innovation, and cooperation are needed to fully achieve the potential of blockchain technology to improve security.

5. Conclusion and Future Scope

Resource-constrained devices, like Internet of Things (IoT) gadgets and edge devices, face unique challenges when it comes to incorporating blockchain technology. These difficulties focus on computational power, storage capacity, and energy efficiency. There are, however, intriguing future avenues that can address these issues and enable wider adoption of blockchain in resource-constrained contexts. The resource-intensive nature of blockchain networks is one of the key issues. Traditional blockchain systems, such as Bitcoin or Ethereum, necessitate a substantial amount of processing power and storage space to maintain the distributed ledger and validate transactions. This is a challenge for devices that have restricted processing and memory capacity. The strict requirements of blockchain algorithmic consensus and transaction verification procedures might be too much for these devices to handle. The significant energy usage of the blockchain miner and validation processes is another problem. Blockchain networks frequently use Proof-of-Work (PoW) consensus algorithms, which require a lot of computing power and consume a lot of energy [45]. The operating lifespan or overall performance of devices with scarce resources that run on limited electrical power can be significantly impacted by this energy inefficiency.

Furthermore, the scalability of blockchain networks poses a barrier for devices with limited resources. The processing and storage requirements of a blockchain network expand in tandem with the number of transactions and participants. It is critical to ensure that blockchain solutions for resource-constrained devices are efficient and scalable to accommodate their limits and enable smooth integration into existing infrastructures [46]. To solve these challenges and prepare the way for the future adoption of blockchain in resource-constrained environments, several intriguing avenues are being explored. One path includes the creation of lightweight and efficient consensus algorithms tailored to resource-constrained devices. These algorithms attempt to lower processing requirements and energy consumption while maintaining the blockchain network's security and integrity. Another path is to investigate off-chain solutions and layer-two protocols. These solutions offload some of the blockchain network's core operations to lighten the load on devices with limited resources. Resource-constrained devices can offload computational activities and minimize storage requirements by leveraging technologies such as sidechains, state channels, or payment channels, while still benefiting from the security and transparency provided by blockchain technology. Furthermore, advances in hardware technology, such as low-power processors and energy-efficient architectures, can aid in mitigating the energy consumption issue. Future hardware advancements may allow resource-constrained devices to conduct blockchain-related tasks more efficiently, hence extending their operational lives and overall capabilities.

5.1 Addressing the Scalability Limitations of Blockchain in Resource—Constrained Environments

Addressing blockchain's scalability limits in resource-constrained contexts is a critical component of enabling its effective application. While resource-constrained devices, such as IoT devices, have computing power, storage capacity, and energy efficiency restrictions, there are potential solutions and future directions to address these difficulties [47]. The creation of lightweight consensus algorithms specifically built for resource-constrained devices is one method for tackling scaling issues. These algorithms seek to minimize the computational and energy requirements associated with standard blockchain networks. For example, in resource-constrained distributed systems, the Proof-of-Authentication (PoAh) consensus method has been presented as a solution for quick and scalable private blockchain networks.

Another path is to investigate off-chain solutions and layer-two protocols. These solutions offload some of the blockchain network's core operations to lighten the load on devices with limited resources. Resource-constrained devices can offload computational duties and minimize storage requirements while still benefiting from the security and transparency provided by Blockchain Technology by leveraging technologies such as sidechains, state channels, or payment channels. Furthermore, developments in hardware technology can play an important role in overcoming scaling constraints. Low-power CPUs and energy-efficient architectures can enable resource-constrained devices to conduct blockchain-related operations more efficiently, extending their operational lives and overall capabilities [48]. Collaborations and standardization efforts across sectors and research communities are also critical for addressing scaling issues. Stakeholders can collaborate to design efficient and scalable blockchain solutions that fit the needs of resource-constrained environments by defining shared protocols and optimizing resource allocation. While there are continuous initiatives to address scalability issues, blockchain acceptance, and deployment in resource-constrained contexts may still face hurdles [49]. Complexity, technical skill, and compatibility difficulties are all potential barriers to wider adoption. However, attempts to clarify blockchain's intricacies through user-friendly interfaces, quicker onboarding processes, and instructional resources can help lower entry barriers and promote broader use.

5.2 Optimizing Blockchain Protocols for Efficient Device Security

Optimizing blockchain protocols for efficient device security is a critical component of assuring effective Blockchain Technology implementation in resource-constrained contexts. Several approaches and techniques can be used to improve the efficiency of blockchain protocols while still maintaining strong security safeguards.

One way is to use lightweight consensus mechanisms. These protocols aim to lower blockchain networks' computational and energy requirements, making them more appropriate for resource-constrained devices. For instance, leveraging consensus methods that have been tuned, like Proof-of-Authority (PoA) or Proof-of-Stake (PoS), can significantly improve blockchain network efficiency without compromising security [50]. Another thing to think about is optimizing the cryptography algorithms used in blockchain protocols. To reduce the computational cost associated with cryptographic computations while retaining an elevated degree of security, for example, extremely efficient cryptographic techniques such as ECC (Elliptic Curve Cryptography) can be adopted. Additionally, the effectiveness of blockchain-based security methods can be increased with the application of machine learning techniques. Security protocols, network efficiency, and resource allocation can all be enhanced by machine learning techniques. Machine learning-based load-balancing techniques, for example, can aid in the efficient distribution of computational jobs among resource-constrained devices. Furthermore, developing optimized blockchain models that are specifically suited for resource-constrained environments is a potential area. These models consider device restrictions and tailor blockchain architecture and protocols accordingly. These models can increase the overall efficiency of blockchain systems by optimizing data storage, transaction processing, and consensus procedures.

5.3 Challenges and Considerations for Integrating Blockchain into Current Device Security Infrastructure

Integrating blockchain into current device security infrastructure presents several challenges and considerations [51]. Here are some key points to consider.

- **Resource Constraints:** IoT devices, for example, frequently have limited computing power, storage capacity, and energy efficiency. Integrating Blockchain Technology into these devices necessitates the optimization of protocols and algorithms to reduce resource usage while retaining security and usefulness.
- **Scalability:** Scalability problems appear on blockchain networks as the volume of transactions and users increases. When blockchain is incorporated into devices with limited resources, this issue gets worse. Ensuring effective consensus procedures and transaction processing that can handle the constraints of the device is essential.
- **Privacy and Security:** Although Blockchain Technology has built-in security features, integrating it with device security architecture requires careful consideration of security and privacy issues. The need for anonymity in some applications may conflict with blockchain's transparency. To protect sensitive data, further precautions like encryption along with access controls could be needed.
- **Interoperability:** Integrating blockchain into existing device security infrastructure necessitates guaranteeing interoperability with other systems

and protocols. Compatibility and seamless integration with older systems and protocols are critical for avoiding disruptions and ensuring smooth operations.

- **Standardization:** The lack of standardized protocols and frameworks for integrating blockchain into device security infrastructure can create difficulties. Common standards and protocols can improve interoperability, increase security, and promote widespread adoption.
- **Cost and Complexity:** Implementing Blockchain Technology can be expensive and complicated. During the integration process, organizations must examine the financial consequences, technical knowledge necessary, and potential interruptions. To address these problems, proper planning, training, and support are required.
- **Regulatory and Legal Considerations:** Blockchain Technology operates inside a regulatory framework that differs from nation to jurisdiction. When integrating blockchain into a device security architecture, compliance with data protection, privacy, and other relevant requirements is critical. It is critical to understand the legal ramifications and to ensure compliance.

5.4 Predictions for the Future of Securing Resource—Constrained Devices using Blockchain

Resource-constrained devices, such as the edge and Internet of Things (IoT) devices, face unique challenges when it comes to incorporating blockchain technology. Restrictions in energy efficiency, processing power, and storage capacity present significant obstacles. The future of protecting these gadgets with blockchain, on the other hand, looks promising. Efforts are being undertaken to optimize blockchain protocols, create lightweight consensus algorithms, and investigate efficient cryptographic techniques that consume fewer resources while preserving security.

The integration of pooled learning techniques with blockchain is expected to improve machine learning security and privacy on resource-constrained devices. Federated learning based on blockchain ensures distributed computing stability, allowing for secure and private data processing. Additionally, blockchain technology will be crucial in enhancing IoT device security. Through decentralized and irreversible transactions, IoT devices may safely share data, verify transactions, and foster trust amongst devices and stakeholders. Blockchain technology has potential applications in the area of edge computing security. Blockchain technology provides safe data storage, tamper-proof audit trails, and access control to protect the confidentiality and safety of sensitive data handled by edge devices. Efforts to establish interoperability standards and protocols for integrating blockchain into device security infrastructure will enable seamless integration with existing systems, promoting widespread adoption. Privacy protection is a significant focus in using blockchain to secure resource-constrained devices. To use the security and transparency advantages of blockchain technology, methods including homomorphic encryption, differential privacy, and zero-knowledge proofs are being researched as ways to protect sensitive data. Collaborative security mechanisms, such as blockchain-based collaborations and consortiums, are also emerging. To improve the security of resource-constrained devices, organizations and stakeholders are collaborating to exchange threat intelligence, develop best practices, and establish trust frameworks. Advances in hardware technology, such as low-power CPUs and energy-efficient architectures, are expected to help blockchain-based security protocols run efficiently on resource-constrained devices. The blockchain solutions' performance and energy efficiency will be further enhanced by these hardware-level improvements.

In conclusion, while there are issues in securing resource-constrained devices with blockchain technology, the future holds promise. Optimized protocols, lightweight consensus algorithms, federated learning integration, improved IoT security, interoperability standards, privacy-preserving solutions, collaborative security measures, and hardware-level optimizations are all expected to play important roles in securing resource-constrained devices with blockchain.

References

1. Politou, E., Casino, F., Alepis, E., and Patsakis, C. 2019. Blockchain mutability: Challenges and proposed solutions. *IEEE Transactions on Emerging Topics in Computing*, *9*(4), 1972–1986.
2. Zarrin, J., Wen Phang, H., Babu Saheer, L., and Zarrin, B. 2021. Blockchain for decentralization of internet: prospects, trends, and challenges. *Cluster Computing*, *24*(4), 2841–2866.
3. Verma, R.., Dhanda, N.. and Nagar, V. 2023. Analysing the Security Aspects of IoT using Blockchain and Cryptographic Algorithms. *International Journal on Recent and Innovation Trends in Computing and Communication, 11*(1s), pp. 13–22. doi: 10.17762/ijritcc.v11i1s.5990.
4. Rawat, A., and Rahman, D. H. U. 2023. Smart Contract Based Approach for Fixed Deposit in Blockchain Networks with Remix IDE. *International Journal of Creative Research Thoughts (IJCRT)*, *11*(5).
5. Verma, R., Dhanda, N. and Nagar, V. 2022. Enhancing & Optimizing Security of IoT Systems using Different Components of Industry 4.0. *International Journal of Engineering Trends and Technology, 70*(7), pp.147–157. doi:10.14445/22315381/ijett-v70i7p216.
6. Kornaros, G. 2022. Hardware-assisted machine learning in resource-constrained IoT environments for security: review and future prospective. *IEEE Access*, *10*, 58603–58622.
7. Yaacoub, J. P. A., Salman, O., Noura, H. N., Kaaniche, N., Chehab, A., and Malli, M. 2020. Cyber-physical systems security: Limitations, issues and future trends. *Microprocessors and Microsystems*, *77*, 103201.
8. Butun, I., Österberg, P., and Song, H. 2019. Security of the Internet of Things: Vulnerabilities, attacks, and countermeasures. *IEEE Communications Surveys & Tutorials*, *22*(1), 616–644.
9. Rawat, A., and Ur Rahman, H. 2023. Implementation of Algorithms for Fixed Deposit Using Smart Contract. *Journal of Emerging Technologies and Innovative Research (JETIR) JETIR*, *10*(6).
10. Verma, R., and Dhanda, N. 2023. Application of Supply Chain Management in Blockchain and IoT - A Generic Use Case. 2023 13th International Conference on Cloud Computing, Data Science & Engineering (Confluence). https://doi.org/10.1109/confluence56041.2023.10048815
11. Hafid, A., Hafid, A. S., and Samih, M. 2020. Scaling blockchains: A comprehensive survey. *IEEE Access*, *8*, 125244–125262.
12. Aggarwal, S., and Kumar, N. 2021. Basics of blockchain. In *Advances in Computers* (Vol. 121, pp. 129–146). Elsevier.
13. Politou, E., Casino, F., Alepis, E., and Patsakis, C. 2019. Blockchain mutability: Challenges and proposed solutions. *IEEE Transactions on Emerging Topics in Computing*, *9*(4), 1972–1986.
14. Yiannas, F. 2018. A new era of food transparency powered by blockchain. *Innovations: Technology, Governance, Globalization*, *12*(1–2), 46–56.
15. Le Nguyen, B., Lydia, E. L., Elhoseny, M., Pustokhina, I., Pustokhin, D. A., Selim, M. M.,... and Shankar, K. 2020. Privacy preserving blockchain technique to achieve secure and reliable sharing of IoT data. *Computers, Materials & Continua*, *65*(1), 87–107.

16. Application of Truffle Suite in a Blockchain Environment. Proceedings of Third International Conference on Computing, Communications, and Cyber-Security, 693–702. https://doi.org/10.1007/978-981-19-1142-2_54
17. Hughes, L., Dwivedi, Y. K., Misra, S. K., Rana, N. P., Raghavan, V., and Akella, V. 2019. Blockchain research, practice and policy: Applications, benefits, limitations, emerging research themes and research agenda. *International Journal of Information Management, 49*, 114–129.
18. Ayvaz, S., and Cetin, S. C. 2019. Witness of Things: Blockchain-based distributed decision record-keeping system for autonomous vehicles. *International Journal of Intelligent Unmanned Systems*, *7*(2), 72–87.
19. Arcenegui, J., Arjona, R., Román, R., and Baturone, I. 2021. Secure combination of IoT and blockchain by physically binding IoT devices to smart non-fungible tokens using PUFs. *Sensors*, *21*(9), 3119.
20. Verma, R., Dhanda, N., and Nagar, V. 2023. The modern internet of things with blockchain: IoT-B. Emerging Trends in IoT and Computing Technologies, 420–425. https://doi.org/10.1201/9781003350057-61
21. Wang, Y., and Chen, H. 2022. Blockchain: A potential technology to improve the performance of collaborative emergency management with multi-agent participation. *International Journal of Disaster Risk Reduction, 72*, 102867.
22. Lashkari, B., and Musilek, P. 2021. A comprehensive review of blockchain consensus mechanisms. *IEEE Access, 9*, 43620–43652.
23. Lasla, N., Al-Sahan, L., Abdallah, M., and Younis, M. 2022. Green-PoW: An energy-efficient blockchain Proof-of-Work consensus algorithm. *Computer Networks, 214*, 109118.
24. Gaži, P., Kiayias, A., and Zindros, D. 2019, May. Proof-of-stake sidechains. In 2019 IEEE Symposium on Security and Privacy (SP) (pp. 139–156). IEEE.
25. Aslam, T., Maqbool, A., Akhtar, M., Mirza, A., Khan, M. A., Khan, W. Z., and Alam, S. 2022. Blockchain based enhanced ERP transaction integrity architecture and PoET consensus. *Computers, Materials & Continua, 70*(1), 1089–1109.
26. Makhdoom, I. 2020. Defense against integrity and privacy attacks in the Internet of Things (Doctoral dissertation).
27. Williams, B., and Adamson, J. 2022. PCI Compliance: Understand and implement effective PCI data security standard compliance. CRC Press.
28. Verma, R., and Dhanda, N. 2023. January. Application of Supply Chain Management in Blockchain and IoT-A Generic Use Case. In 2023 13th International Conference on Cloud Computing, Data Science & Engineering (Confluence) (pp. 405–410). IEEE.
29. Praveena Anjelin, D., and Ganesh Kumar, S. 2021. Blockchain Technology for data sharing in decentralized storage system. In Intelligent Computing and Applications: Proceedings of ICICA 2019 (pp. 369–382). Springer Singapore.
30. Pise, R., and Patil, S. 2021, February. Enhancing security of data in cloud storage using decentralized blockchain. In 2021 Third International Conference on Intelligent Communication Technologies and Virtual Mobile Networks (ICICV) (pp. 161–167). IEEE.
31. Xiao, Y., Zhang, N., Lou, W., and Hou, Y. T. 2020. A survey of distributed consensus protocols for blockchain networks. *IEEE Communications Surveys & Tutorials, 22*(2), 1432–1465.
32. Lashkari, B., and Musilek, P. 2021. A comprehensive review of blockchain consensus mechanisms. *IEEE Access, 9*, 43620–43652.
33. Hewa, T. M., Hu, Y., Liyanage, M., Kanhare, S. S., and Ylianttila, M. 2021. Survey on blockchain-based smart contracts: Technical aspects and future research. *IEEE Access, 9*, 87643–87662.
34. Zheng, Z., Xie, S., Dai, H. N., Chen, W., Chen, X., Weng, J., and Imran, M. 2020. An overview on smart contracts: Challenges, advances and platforms. *Future Generation Computer Systems, 105*, 475–491.
35. Singh, A., Parizi, R. M., Zhang, Q., Choo, K. K. R., and Dehghantanha, A. 2020. Blockchain smart contracts formalization: Approaches and challenges to address vulnerabilities. *Computers & Security, 88*, 101654.
36. Bernabe, J. B., Canovas, J. L., Hernandez-Ramos, J. L., Moreno, R. T., and Skarmeta, A. 2019. Privacy-preserving solutions for blockchain: Review and challenges. *IEEE Access, 7*, 164908–164940.

37. Amin, R., Kumar, N., Biswas, G. P., Iqbal, R., and Chang, V. 2018. A light weight authentication protocol for IoT-enabled devices in distributed Cloud Computing environment. *Future Generation Computer Systems, 78*, 1005–1019.
38. Hang, L., and Kim, D. H. 2019. Design and implementation of an integrated iot blockchain platform for sensing data integrity. *Sensors, 19*(10), 2228.
39. Sealey, N., Aijaz, A., and Holden, B. 2022. November. IOTA Tangle 2.0: Toward a Scalable, Decentralized, Smart, and Autonomous IoT Ecosystem. In 2022 International Conference on Smart Applications, Communications and Networking (SmartNets) (pp. 01–08). IEEE.
40. Sghaier Omar, A. 2020. Decentralized identity and access management framework for Internet of Things devices.
41. Archer, D. W., Bogdanov, D., Lindell, Y., Kamm, L., Nielsen, K., Pagter, J. I.,... and Wright, R. N. (2018). From keys to databases—real-world applications of secure multi-party computation. *The Computer Journal, 61*(12), 1749–1771.
42. Aceto, G., Persico, V., and Pescapé, A. 2019. A survey on information and communication technologies for industry 4.0: State-of-the-art, taxonomies, perspectives, and challenges. *IEEE Communications Surveys & Tutorials, 21*(4), 3467–3501.
43. Siddiqui, F., Beley, J., Zeadally, S., and Braught, G. 2021. Secure and lightweight communication in heterogeneous IoT environments. *Internet of Things, 14*, 100093.
44. Menon, S., and Jain, K. 2021. Blockchain Technology for transparency in agri-food supply chain: Use cases, limitations, and future directions. *IEEE Transactions on Engineering Management.*
45. Platt, M., Sedlmeir, J., Platt, D., Xu, J., Tasca, P., Vadgama, N., and Ibañez, J. I. 2021, December. The energy footprint of blockchain consensus mechanisms beyond proof-of-work. In 2021 IEEE 21st International Conference on Software Quality, Reliability and Security Companion (QRS-C) (pp. 1135–1144). IEEE.
46. Romashkova, I., Komarov, M., and Ometov, A. 2021. Demystifying Blockchain Technology for resource-constrained IoT devices: parameters, challenges and future perspective. *IEEE Access, 9*, 129264–129277.
47. Salimitari, M., Chatterjee, M., and Fallah, Y. P. 2020. A survey on consensus methods in blockchain for resource-constrained IoT networks. *Internet of Things, 11*, 100212.
48. Rana, A., Sharma, S., Nisar, K., Ibrahim, A. A. A., Dhawan, S., Chowdhry, B.,... and Goyal, N. 2022. The rise of blockchain internet of things (biot): Secured, device-to-device architecture and simulation scenarios. *Applied Sciences, 12*(15), 7694.
49. Khan, D., Jung, L. T., and Hashmani, M. A. 2021. Systematic literature review of challenges in blockchain scalability. *Applied Sciences, 11*(20), 9372.
50. Rebello, G. A. F., Camilo, G. F., Guimaraes, L. C., de Souza, L. A. C., Thomaz, G. A., and Duarte, O. C. M. 2021. A security and performance analysis of proof-based consensus protocols. *Annals of Telecommunications*, 1–21.
51. Nartey, C., Tchao, E.T., Gadze, J.D., Keelson, E., Klogo, G. S., Kommey, B., and Diawuo, K. 2021. On blockchain and IoT integration platforms: current implementation challenges and future perspectives. *Wireless Communications and Mobile Computing*, 2021, 1–25.
52. Rawat, A., Suryavanshi, R. S., Nagar, V. and Yadav, D. (2023, November). Development of Blockchain-Enabled E-Healthcare Appointment System. In International Conference on Trends in Computational and Cognitive Engineering (pp. 359–369). Singapore: Springer Nature Singapore.

Chapter 5

Transforming Vulnerability into Trust

Modernizing Government Document Dissemination with Blockchain and Machine Learning Technology

Jagdish Pimple,[1,*] *Praveen Sen*,[2] *Sweta Raut*,[3] *Shraddha Raut*[4] and *Ashish Dandekar*[5]

1. Introduction

Government agencies often face many challenges regarding disseminating critical notices, policies, regulations and related documents to the public. These challenges persist whether agencies rely on traditional paper methods or choose electronic platforms such as e-paper. Visible problems include data vulnerabilities, where sensitive information in these documents is vulnerable to data breaches and unauthorized access. In addition, there is a high risk of unauthorized changes that could lead to misinformation or confusion among the public. These challenges are exacerbated by an absence of public awareness of innovations, changes in government policy or the introduction of new regulations, which can lead to misunderstandings

[1] Department of Information Technology, St.Vincent Pallotti College of Engineering and Technology, Nagpur, India.

[2] Department of Computer Science & Business Systems, St.Vincent Pallotti College of Engineering and Technology, Nagpur, India.

[3] Department of Computer Science & Engineering, Jhulelal Institute of Technology, Nagpur, India.

[4] Department of Computer Science & Engineering, G H Raisoni College of Engineering, Amravati, India.

[5] Department of Computer Science & Engineering (Data Science), St.Vincent Pallotti College of Engineering and Technology, Nagpur.

* Corresponding author

and non-compliance. In addition, the lack of a secure and transparent document distribution process increases the risk of fraud, which ultimately undermines trust between the government and its constituents.

In response to these pressing challenges, a forward-looking solution is created that uses cutting-edge technology, especially the combination of machine learning and blockchain technology. This innovative approach aims to redefine the process of government and document distribution, providing a panacea for the problems plaguing the current system. Using the security features of blockchain technology guarantees the security and immutability of documents. When a document is stored on the blockchain, it becomes impervious to unauthorized changes, ensuring data integrity. At the same time, the public accounting inherent in blockchain technology promotes transparency in the transmission and availability of documents. Complementing this, machine learning technology offers the ability to analyze data to identify patterns, vulnerabilities and potential fraud by automating the distribution process to improve efficiency.

The advantages of this innovative approach are numerous. Document immutability ensures that documents remain unauthorized and reliable sources of information. Accessibility has improved significantly as citizens can get authentic information about government policies, announcements, regulations and government decisions with just one click, greatly reducing the risk of misinformation and misunderstanding. In addition to these advantages, the proposed solution ultimately promotes trust and accountability by ensuring the authenticity and transparency of document distribution, facilitating a symbiotic relationship between government and the public, and strengthening the basis of transparency and accountability in government operations. This approach is versatile and applications cover various scenarios. For example, it can facilitate the dissemination of policy updates and ensure that citizens have access to the most up-to-date government policies and announcements without the risk of ambiguity and fraud. Regarding legal documents, access to authentic legal documents, including regulations and government decisions, has been simplified and made more secure. It also accelerates the transmission of important information to the public, which increases efficiency and trust in the state's actions. Essentially, this innovative approach is poised to usher in a new era of secure, transparent and reliable government document distribution.

1.1 Challenges in Government Document Dissemination

The difficulties experienced by government organizations in making crucial records available to the general public are discussed, and a solution using blockchain and machine learning technology is suggested. When releasing important information to the public, whether through conventional paper-based procedures or electronic platforms, government organizations face several challenges. These difficulties include:

Data Vulnerabilities: Sensitive information is frequently contained in government documents, and there is a possibility of data vulnerabilities, such as data breaches or illegal access.

Risks of Unauthorized Alterations: It is possible to tamper with documents, resulting in unlawful changes that can have serious repercussions.

Limited Public Awareness: Public awareness of updates, changes in policies, or new ordinances may be limited, leading to misunderstandings or non-compliance with government regulations.

Heightened Risk of Fraud: The risk of fraudulent actions can increase in the absence of a clear and secure document dissemination method.

Privacy and Data Protection: Ensuring the privacy and protection of sensitive information is crucial when implementing such systems.

Interoperability: Ensuring that blockchain and machine learning systems can seamlessly integrate with existing government infrastructure and systems is a challenge.

2. Literature Survey

Concerns about the genuineness, dependability, and validity of online document authentication have been directed to various studies that have explored several potential technologies. Several investigators have concentrated on the development of document control systems based on distributed techniques. There are several blockchain-based applications in the literature that use the technology to guarantee the legitimacy, immutability and clearness of recorded information. However, little has been discovered about how self-sovereign identity technology can be used to solve present problems in document authentication systems, and this study aims to fill this gap.

Using digital signatures, text extraction, and document correlation scores, Lakmal et al. [1] introduced a novel approach to document authentication called IDStack. The three unique and separate modules that make up the IDStack procedure's modular design are the data validation, data extraction, and evaluation modules. These modules are utilized by the three user characters known as the validator, separator, and relying party. The owner or a third party with the ability to digitally sign and authenticate the machine-readable document can serve as the collector. A third-party user who can confirm the authenticity of the content and the previously signed machine-readable document is known as a verifier. A bank is an example of a trusted party—someone who has access to the document. The IDStack API is used to access the protocol. However, the suggested technique can take some time and requires multiple manual procedures. The primary issue is that each time a relying party sends a request, an IDStack API is triggered, necessitating a prior connection from each relying party. This can damage the user's privacy and make them traceable, which is troublesome. Furthermore, the topic of confirming the signer's identity is raised by the fact that IDStack's digital signature is created on self-signed certificates, which enable users to generate their own signatures.

Wavelet-atom transform-based hashing algorithms are suggested by the authors of [2] as a means of confirming the legitimacy of printed identity papers like passports and national identity cards. This article includes a digitally signed WAT-based hash that is encoded as a QR code that may be used to verify the document. Article [3] examines the use of self-sovereign identity (SSI) technology to improve document security in government operations, focusing on online document verification. SSI enables the creation of fraudulent digital identities and can replace intermediaries, improving system efficiency and privacy. The paper primarily evaluates its application for online loan processing.

A decentralized web application called DocBlock is presented by Imam et al. [4] to use the Ethereum blockchain for digital document verification. A digital document can be downloaded, reviewed, and uploaded using the suggested approach. To strengthen the document's legitimacy, the Metamask wallet must display a document that has been confirmed on the blockchain along with the file's hash linked to the organization's public key and the upload date. Because the document's hash value would have changed, the authorities would be aware of any compromise at some time. The user is responsible for keeping the IPFS (InterPlanetary File System) hash they receive when adding a document for future usage. This hash is provided just once. The system is more private because the smart contract does not keep this IPFS hash. Currently, the system searches for a file that matches an IPFS hash entered by the user and returns the original file as binary code. The paper provides a transparent and controllable system, which makes it a superior answer to previous methodologies, but it misses a crucial component: user management. A document may have little or no control over its user once it is uploaded to the blockchain.

A blockchain-based remedy for the background check process that job applicants have to go through throughout the hiring process is developed in the paper by Marella and Vijayan [5]. According to the report, a consortium blockchain would be used, in which information can only be added to the blockchain by specific publishers and validators. Every transaction (document delivery) needs to have the organizational unit's digital signature on it and be in the right format. Rather than storing the original documents on the blockchain, the system saves their hash value. Nevertheless, the suggested paradigm concentrates on organizations (publishers and verifiers) rather than users. Hyperledger Fabric [6] is used in the proof-of-concept implementation that the authors show, but the artifact is not evaluated in the article. The "S. Dhiman. et al. [7]" projected a work titled "Tesseract vs gocr a comparative study" in the year 2013 [7], in which they described things like: A process called optical character recognition is utilized to convert graphic content designs into intelligible ones. Over the years, a variety of optical character recognition software has been used in the market, each with its advantages and disadvantages. Based on varied measures or boundaries, they provided various results. However, two open-source devices, such as "Tesseract" and "GOCR," are taken into consideration as a way in this research [7]. This document [7] first provides an overview of the open-source programs "Tesseract" and "GOCR," as well as a description of how they work. This article [7] examines tools.

In 2012 [8] the authors quote, C. Patel et al. projected an article titled "Optical character recognition using the open source OCR tool tesseract: a case study" where they defined, among other things, that OCR technology was used to convert published text into mutable content. OCR is a very treasured and well-known strategy in various applications. OCR accuracy can be the subject of message processing and sharing calculations. Sometimes recovering text from an image is difficult due to different sizes, styles, orientations, complex image backgrounds, etc. This article [8] begins with an introduction to the optical character recognition strategy, the history of the Tesseract open-source OCR device, its design, and the impact of Tesseract OCR research on different types of images. This article [8] concludes with a comparative study of this device with Transym OCR devices from other companies, looking at vehicle license plates as data. From the vehicle number plate, this system [8] tried to extract the vehicle number using Transym and Tesseract and looked at these software depending on various constraints.

A framework evaluation was carried out by the authors [9] using the concepts of Privacy by Design (PbD), General Data Protection Regulation (GDPR), and SSI. Further research on the topic, focusing on the needs of banks and broadening its coverage, was done by Soltani et al. and Schlatt et al. [9]. They developed a framework to replace the eKYC process's current limitations with a fully digital procedure that made use of blockchain technology based on SSI. According to their assessment, the SSI-based design met additional stakeholder requirements and made a major contribution to an eKYC process that was more efficient. Nevertheless, the suggested structure was never put into practice and did not have a viability assessment. The author suggested a method in earlier works[11] that consists of a verification token that can be applied to blockchain-based digital asset authentication. It makes sense to utilize this function as a block to manage electronic documents. Monitoring the legitimacy of digital assets was an additional remedy. transforming the digital material into a binary file and keeping the binary file's hash on the blockchain [15]. The owner can be identified by this hash. As a result of integrity violations, this system adheres to the concept of centralization rather than decentralization, and our solution enhances current research in the areas of content extraction, distributed storage, and forgery detection.

They presented the KYC2 customer-centric framework in a paper [14]. Hyperledger Indy [14], a publicly accessible permissioned distributed ledger intended for distributed identification, served as the foundation for the framework. A financial customer onboarding scenario involving a customer and two banks is simulated by the framework. There is a dearth of current research on the application of SSI technology to document authentication. Nonetheless, some academics have looked at identifying and authentication techniques based on VC and DID. For instance, Soltani et al. [12] examined SSIs in the context of electronic know-your-customer (eKYC), focusing on the technical analysis of their solution and the client onboarding procedure. The electronic verification of a customer's identification and personally identifiable information is known as eKYC [13].

In response to the growing problem of fake documents in social media and the resulting skepticism, this paper [15] presents a versatile document authentication method. The method uses digital signatures and blockchain. The content of the document is signed using digital signature algorithms, and the signature is added to the document as a 2D barcode. This barcode can be used to obtain and verify the document and digital signature [15]. The article also presents fake detection techniques as an additional control method. In addition, signed documents are stored

Table 1: Comparison of Literature Survey.

Paper and Authors	Proposed Solution	Key Technologies and Concepts	Key Findings and Limitations
Lakmal et al. [1]	IDStack protocol for document verification	Text extraction, digital signatures, correlation score	Manual interactions and potential privacy risks due to IDStack's reliance on preconnections and frequent API triggering. Verification is self-signed, raising identity verification concerns.
F. Ahmad and L. M. Cheng [2]	Wave atom transform-based hashing for identity documents	WAT-based hash, digital signatures, QR codes	Introduces a method to verify printed identity documents using WAT-based hashing and digital signatures.
Abylay Satybaldy, Anushka Subedi and Mariusz Nowostawski [3]	Self-Sovereign Identity (SSI) technology for online document verification	Fraud-proof digital credentials, intermediaries replacement	SSI enhances document security and efficiency. Application in online loan processing is explored.
Imam et al. [4]	DocBlock for document verification using Ethereum's blockchain	Decentralized verification, Metamask wallet	Provides a decentralized solution for document verification. User control over documents may be limited.
Marella and Vijayan [5]	Blockchain-based background verification for job applicants	Consortium blockchain, hash value storage	Uses consortium blockchain to verify job applicants' background. Lacks user-focused features and lacks evaluation of the artifact.
S.Dhiman et al. [7]	Tesseract vs. gocr comparative study	Optical Character Recognition, image analysis	Compares open-source OCR tools Tesseract and GOCR based on precision and accuracy using different parameters.
C.Patel et al. [8]	Tesseract OCR case study	OCR accuracy, Tesseract performance	Investigates OCR accuracy with Tesseract on various image types and compares it to Transym OCR.
Schlatt et al. [9]	Framework based on SSI and blockchain for eKYC	SSI, GDPR, Privacy by Design (PbD)	Proposes an SSI-based framework for eKYC, Emphasizing stakeholder requirements. Lacks implementation and feasibility evaluation.
N. Nizamuddin, H. R. Hasan, and K. Salah [11]	Blockchain-based verification ID and authenticity tracking	Blockchain-based authenticity verification, decentralization	Presents solutions for blockchain-based authenticity verification and document tracking.
[14] Author	KYC2 framework using Hyperledger Indy	Hyperledger Indy, decentralized identity	Discusses a client onboarding framework using Hyperledger Indy, emphasizing decentralized identity.

Table 1: contd. ...

Table 1: contd. ...

Paper and Authors	Proposed Solution	Key Technologies and Concepts	Key Findings and Limitations
Madura Rajapashea,Muammar Adnanb [15]	Multi-format document verification using blockchain	Digital signatures, blockchain, 2D barcodes	Suggested using decentralized storage, blockchain technology, and digital signatures as part of a multiformat document verification process.
Rasha M. Abd El-Aziz [16]	Blockchain and machine learning-based document distribution	Blockchain, machine learning	Introduces a solution for secure government document distribution using blockchain and machine learning technologies.
Lakmal et al. [1]	IDStack protocol for document verification	Text extraction, digital signatures, correlation score	Manual interactions and potential privacy risks due to IDStack's reliance on preconnections and frequent API triggering. Verification is self-signed, raising identity verification concerns.
F. Ahmad and L. M. Cheng [2]	Wave atom transform-based hashing for identity documents	WAT-based hash, digital signatures, QR codes	Introduces a method to verify printed identity documents using WAT-based hashing and digital signatures.
Abylay Satybaldy, Anushka Subedi and Mariusz Nowostawski [3]	Self-Sovereign Identity (SSI) technology for online document verification	Fraud-proof digital credentials, intermediaries replacement	SSI enhances document security and efficiency. Application in online loan processing is explored.
Imam et al. [4]	DocBlock for document verification using Ethereum's blockchain	Decentralized verification, Metamask wallet	Provides a decentralized solution for document verification. User control over documents may be limited.
Marella and Vijayan [5]	Blockchain-based background verification for job applicants	Consortium blockchain, hash value storage	Uses consortium blockchain to verify job applicants' background. Lacks user-focused features and lacks evaluation of the artifact.
S. Dhiman et al. [7]	Tesseract vs. gocr comparative study	Optical Character Recognition, image analysis	Compares open-source OCR tools Tesseract and GOCR based on precision and accuracy using different parameters.
C. Patel et al. [8]	Tesseract OCR case study	OCR accuracy, Tesseract performance	Investigates OCR accuracy with Tesseract on various image types and compares it to Transym OCR.

in a storage system based on decentralized blockchain technology, which improves the overall reliability and security of the solution.

Author [16] discusses today's transition to digital document management and the urgent need for secure document authentication. It highlights the integration of Blockchain technology across multiple sectors, including government and non-government industries, to improve document security. Here, they combine Blockchain with document control processes, ensuring strong protection of digital documents sent to authorities. Verification processes, including miner verification and voluntary verification, maintain document authenticity, and strong encryption such as SHA-256 increases security. Overall, this system provides secure and intelligent document verification that improves the credibility and novelty of documents.

2.1 Comparison of Literature Survey

3. Problem Statement

Government agencies face countless challenges when releasing government documents to the public. These difficulties can take different forms depending on the nature of the documents, the distribution methods and the relevant institutions. A common problem faced by government agencies is data vulnerability, as these documents often contain sensitive information that must be secured to prevent data breaches or leaks. There is a constant risk that documents are made without authorization, especially when it comes to electronic transmission, which can lead to misinformation or confusion among the public. In addition, limited public awareness of updates, policy changes, new regulations or critical notices can lead to misunderstandings or non-compliance with government regulations. There are often concerns about the transparency of the document release process, especially when government agencies do not receive information that undermines public trust. The lack of secure and open document distribution processes can also increase the risk of fraud, including the distribution of forged documents or data manipulation.

In addition, easy access to government documents is important, but many citizens struggle to get the information they need quickly and conveniently. Some government agencies continue to use old paper-based systems, leading to inefficiencies, increased costs and environmental problems. Meeting legal and regulatory requirements is a complex challenge that agencies must navigate, and interoperability between the various systems and platforms used to distribute documents can prove problematic, especially for large government agencies. Finally, maintaining the integrity and authenticity of records, especially in digital formats, is an ongoing challenge as records must be archived, updated or reviewed over time. To address these issues, government agencies are increasingly exploring innovative solutions such as blockchain technology, machine learning and modern information systems to improve the security, transparency and efficiency of their document distribution processes while ensuring accurate information is readily available to the public.

4. Proposed Approach

The proposed method promotes the incorporation of cutting-edge technologies, in particular machine learning and blockchain technology, intending to modernize the process to address the complex issues of contemporary government document distribution.

4.1 Blockchain Technology

This solution's fundamental tenet is the implementation of blockchain technology. It provides increased security, transparency, and immutability, among other important benefits. The documents that are kept on the blockchain are protected by an impenetrable ledger that makes unauthorized changes nearly impossible. Furthermore, blockchain's public ledger function guarantees total openness in the distribution and availability of official government papers. The papers and their whole histories are made immutable, protecting them against any unwanted additions or deletions without the consent of network users, which is maybe the most crucial aspect.

4.2 Machine Learning Technology

This suggested approach relies heavily on machine learning technology to support the blockchain component. Data analysis, automation, and predictive analytics are just a few of the vital skills that machine learning offers. It can effectively analyze huge databases, find trends and potential weaknesses, and assist in the fight against fraud and security breaches. Machine learning's automated feature simplifies the distribution of documents, greatly increasing its dependability and efficiency. Predictive analytics also enables machine learning algorithms to anticipate possible problems or weaknesses in the document distribution process, enabling government agencies to take preventative measures and guarantee smoother operations. This innovative fusion of blockchain and machine learning technology intends to usher in a new era of efficient, safe, and transparent delivery of official documents.

4.3 Benefits of the Proposed Approach

By integrating these technologies, the government can achieve several benefits:

Immutable Documents: Documents warehoused on the blockchain are immutable and cannot be altered without consensus, ensuring the integrity of the information.

Accessibility: Citizens can easily access authentic information related to government policies, notices, ordinances, and government resolutions with a single click, reducing the risk of misinformation.

Trust and Accountability: The proposed solution fosters trust and accountability between the government and the public by ensuring the authenticity and transparency of document dissemination.

This approach can be applied in various scenarios, such as:

Policy Updates: Citizens can access the most up-to-date government policies and notices without ambiguity or risk of fraud.

Legal Documents: Access to authentic legal documents, including ordinances and government resolutions, is simplified and made more secure.

Public Communication: The government can disseminate information to the public in a more efficient and trustworthy manner.

5. Objectives

- **Enhanced Security:** By using the blockchain, the security of official documents is greatly improved. The risk of unauthorized access, alteration or fraud is reduced because documents are cryptographically protected on the blockchain.
- **Authentication and Verification:** Blockchain technology enables secure and efficient verification and authentication of documents. Users can trust the authenticity of documents without using intermediaries.
- **Streamlined Processes:** Automation through machine learning can improve document distribution processes and make them faster and more efficient.
- **Reduced Costs:** By removing intermediaries and automating tasks, the cost of disseminating government documents can be reduced.
- **Identity Verification:** Governments can use blockchain to securely store and verify citizen identities, reducing identity theft and fraud.
- **Supply Chain Management:** In public procurement and logistics, blockchain can be used to track and verify the authenticity of goods and reduce fraud.

6. Process Flow

7. The integration of blockchain and machine learning in securing government document distribution:

- **Document Submission:** The process begins with government agencies and individuals submitting digital documents for various purposes, such as applications, records, or requests.
- **Data Preprocessing:** Before documents are stored on the blockchain, machine learning algorithms preprocess the data. This step involves cleaning, formatting, and extracting relevant information from the documents, making them machine-readable.
- **Data Authentication:** Once the data is received from the pre-processing unit, the authenticity of the data needs to be checked with the help of ensemble learning techniques.
- **Blockchain Storage:** The authenticated documents are stored on the blockchain. Blockchain technology guarantees the security and immutability of the documents. Each document is timestamped, and a digital signature is generated to verify its authenticity.
- **Access Control:** Access to the stored documents is controlled through blockchain-based smart contracts. Machine learning models can assist in determining who is authorized to view or modify specific documents, enhancing access control.

- **Anomaly Detection:** Machine learning algorithms continuously monitor and document activities on the blockchain. They employ anomaly detection techniques to identify unusual or unauthorized document alterations or access attempts.
- **Automation:** Machine learning automation streamlines the document dissemination processes. For example, routine verification and approval tasks can be automated, reducing manual effort and improving efficiency.
- **User Authentication:** Machine learning-based user authentication methods, such as biometric recognition and two-factor authentication, ensure that only authorized personnel can access and modify documents.
- **Document Retrieval:** Users can efficiently retrieve documents through a search and retrieval system that employs machine learning algorithms to match document requests with the stored documents.
- **Continuous Improvement:** The system continually adapts and learns from new threats and challenges. Machine learning models evolve to enhance security measures and detect emerging threats effectively.

7.1 Process Flow Diagram

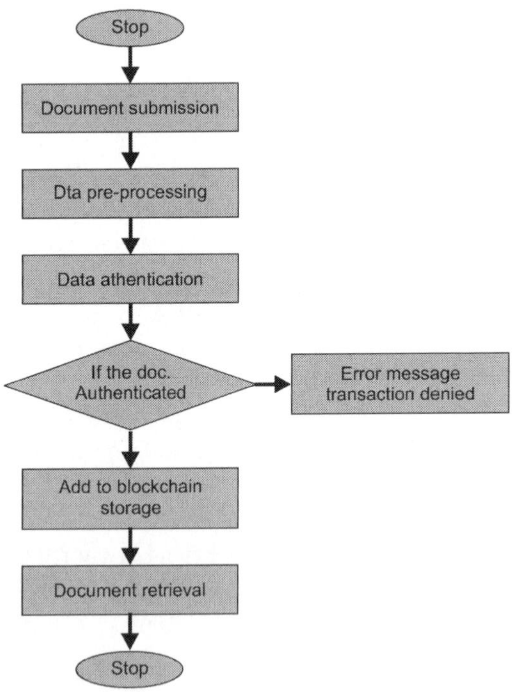

Figure 1: Process Flow of Proposed Work

7.2 System Block Diagram

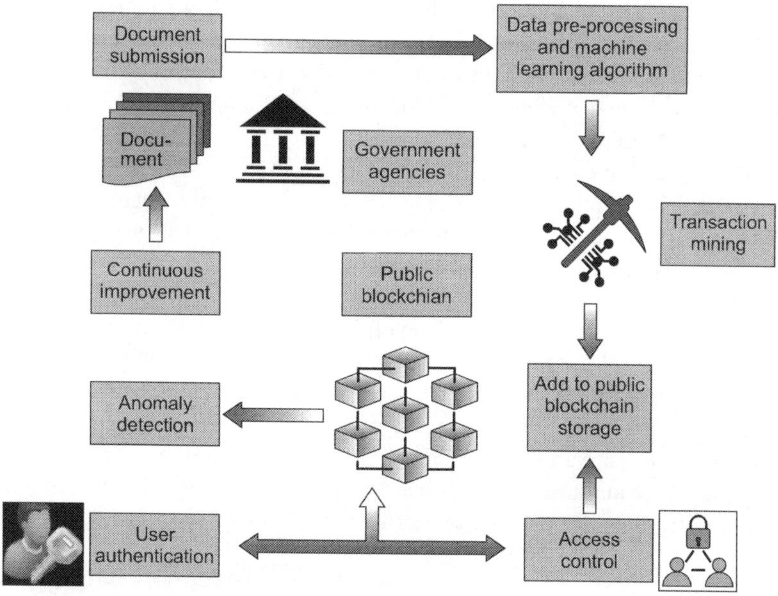

Figure 2: System Block Diagram of Proposed Work.

8. Background Study

8.1 Blockchain Technology

Blockchain technology is extremely important in today's digital environment because it offers several unique features and capabilities that will have a transformative impact on various industries. Its main priority is its unparalleled security and reliability. Blockchain's decentralized and immutable ledger records transactions using cryptographic encryption, making data virtually secure. It is invaluable in industries such as government, finance, healthcare and supply chain, where data security and trust are critical. Another important aspect of blockchain is its transparency. Blockchain transactions are usually visible to all participants, which reduces the risk of fraud and corruption. This feature is crucial in applications such as voting systems where transparency and traceability are non-negotiable. Blockchain technology also excels in efficiency. By automating complex processes and eliminating the need for intermediaries and excessive paperwork, it can significantly reduce costs and expedite operations in industries like logistics, supply chain management, and cross-border payments.

Another important aspect is decentralization. Unlike traditional centralized systems that have a solitary point of failure, the decentralized structure of blockchain makes it more flexible, less prone to hacking and less prone to downtime. This is especially important for applications such as cryptocurrencies and disaster recovery solutions.

Blockchain's immutable record-keeping capabilities ensure the accuracy and integrity of historical records in fields such as real estate, healthcare, and law. This is especially valuable for maintaining reliable historical data. Smart contract technology and innovative functionality facilitate the self-fulfillment of contracts by automating processes in areas such as legal, insurance and supply chain management.

In addition, the emergence of blockchain led to cryptocurrencies such as Bitcoin and Ethereum, which offer alternatives to traditional financial systems, enabling fast, unlimited and cost-effective transactions and expanding financial access to underserved populations. In the realm of supply chain management, blockchain is instrumental in tracking the origin and journey of products, ensuring their authenticity and quality. This traceability addresses issues like food safety and counterfeit products.

For identity management, blockchain offers a secure and decentralized approach, granting users control over their personal information and dropping the risk of data breaches and identity theft. Additionally, in sectors where regulatory compliance is paramount, such as healthcare and finance, blockchain's ability to maintain an immutable record streamlines the auditing processes and ensures compliance with regulatory requirements. Blockchain also plays a vital role in cross-border transactions, offering efficient and cost-effective solutions in a globalized world.

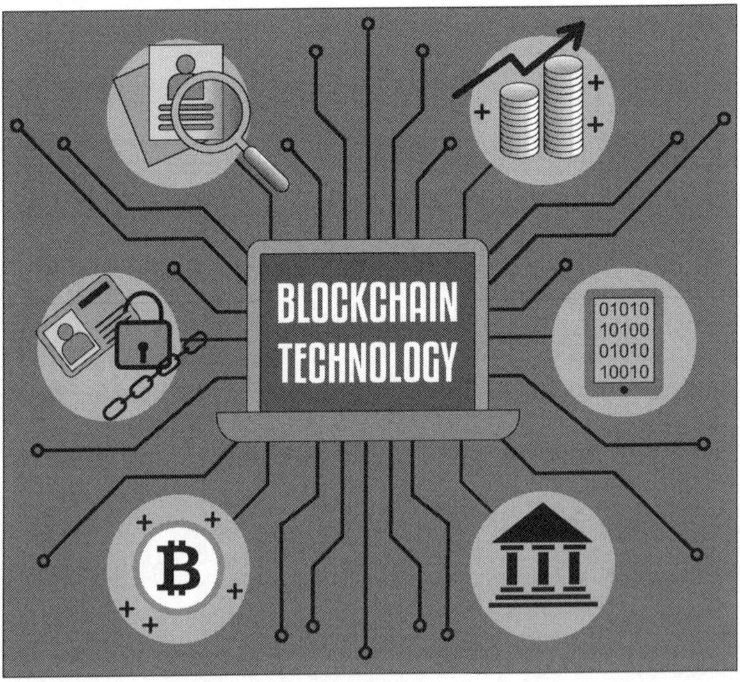

Figure 3: Revolutionizing of Blockchain Technology.
(Image-Ref-https://www.classicinformatics.com/blog/blockchain-technology-applications)

8.2 *Blockchain for Secure the Government Documents*

Protecting government records with blockchain technology requires the use of blockchain and core functionality to ensure the ultimate integrity and protection of sensitive government data. This can be achieved by creating a distributed storage system where documents are stored in an immutable ledger. Once documents are added to the blockchain, they become immune to unauthorized changes, providing additional security. Strict access control measures should be put in place to determine who can view or modify official records and restrict access to authorized personnel only. It is also extremely important to encrypt the content of the document before saving it on the blockchain, which strengthens the protection of the data against unauthorized access. Digital signatures play a dynamic role in the authenticity of documents, and every document needs a digital signature stored on the blockchain to ensure its integrity. Blockchain and timestamp functionality can record when documents are created and modified, creating an immutable audit trail that increases accountability and transparency.

The use of private or consortium blockchain can limit participation to a predefined group of authorized parties, thereby enhancing security. The implementation of smart contracts can simplify document validation processes, reduce human error and enforce predefined rules. In addition, decentralized identity management allows individuals to manage their identity and related documents, improving privacy and security. Multi-sign verification of critical documents combined with regular audits can increase the level of security. A solid legal framework for the use of blockchain in government operations is necessary, as it defines clear rights, obligations and remedies in case of disputes or violations of data security. By implementing these measures, government documents can be securely managed on the blockchain, ensuring trust, transparency, and data integrity and protecting sensitive information from unauthorized access and breaches.

9. SHA−256: Hashing

The Secure Hash Algorithm, which operates at a frequency of 256 bits, provides the proposed system with security. T o guarantee system-wide security, this suggested paper modifies the logic of the secure hash algorithm and the hashing of input characteristics. The cryptographic hash function known as SHA-256 (Secure Hash Algorithm 256-bit) is frequently used to produce hash values for data, including pictures.

1. **Read the Image:** Read the image using a Python programming language and libraries.
2. **Convert Image to Bytes:** Convert the image data into a byte array. You should convert the image into a format that can be processed by the SHA-256 algorithm.

3. **Calculate the Hash:** Use a library or built-in function to apply the SHA-256 algorithm to the byte array of the image. In Python, you can use the hashlib library for this purpose:

```python
import hashlib

# Read the image file
with open('your_image.jpg', 'rb') as file:
    image_data = file.read()

# Create an SHA-256 hash object
sha256_hash = hashlib.sha256()

# Update the hash object with the image data
sha256_hash.update(image_data)

# Get the hexadecimal representation of the hash
hash_value = sha256_hash.hexdigest()

print("SHA-256 Hash:", hash_value)
```

4. **Store or Use the Hash Value:** Use the generated hash value for the mining process and if it is mined then store the image in the block in the blockchain.

10. Consensus Mechanism

The choice of a consensus mechanism for a blockchain system that includes aggregate learning to secure documents depends on several factors, including the specific use case, the desired functionality, and the level of decentralization required. In this context, when document security is a priority and multiple machine learning models are involved, a consensus mechanism that ensures both security and efficiency is important. Two suitable options are:

10.1 Proof of Authority (PoA)

This consensus mechanism involves a predefined set of approved validators who take turns adding new blocks to the blockchain. This mechanism is very safe and efficient, making it suitable for applications where trust and security are of the greatest importance. As for the security of government documents, a coalition of trusted authorities could act as verifiers. This approach ensures that only authorized parties are involved in decisions about document security and access.

10.2 Delegated Proof of Stake (DPoS)

Its consensus mechanism contains a limited number of representatives who are chosen to confirm transactions and generate new blocks. DPoS is known for its scalability and speed. In the context of document security, a selected group of delegates could observe ensemble learning models and decisions and confirmations. This would enable efficient and quick decisions while maintaining a degree of decentralization. Both PoA and DPoS offer the advantage of scalability and efficiency while maintaining high security. They are particularly suitable for applications where document integrity and access control are critical. The choice between these mechanisms would depend on the specific requirements and governance structure of the blockchain system used to secure documents.

11. Ensemble Learning

Ensemble learning, which involves joining numerous machine learning models to improve complete performance, can be applied to improve the security of administrative documents managed on the blockchain. Group learning can be used in this context as follows:

- **Anomaly Detection:** Use multiple anomaly detection models in a cluster to detect unusual activity or unauthorized changes to documents stored on the blockchain. Combining different outlier detection algorithms can provide more reliable and accurate results, reducing the risk of false positives or false negatives.
- **Access Control:** Create a set of access control templates to define who can view or edit official documents. Each group model may use different criteria or algorithms to make access decisions, and agreement between those models may require access, adding an extra layer of security.
- **Document Content Verification:** Use ensemble learning to verify the content of documents stored on the blockchain. Different models of the unit can independently check the integrity and consistency of the document, and if any model detects anomalies, it can trigger an alarm for further investigation.
- **User Authentication:** Create a set of user authentication templates that include different authentication methods, such as biometric identification, two-factor authentication, and behavioral analytics. Access to documents can only be granted if several models in the collection successfully identify the user.
- **Predictive Security:** Combine multiple predictive security models to identify potential security threats. These models can independently analyze historical data and issue alerts when they collectively predict increased risk, enabling proactive security measures.
- **Document Retrieval and Search:** Ensemble learning can improve the efficiency and accuracy of document search and retrieval. Different models can run in parallel to search for documents requested by authorized users.
- **Continuous Improvement:** An entity can adapt and learn from new threats and vulnerabilities, improving security measures as it faces new challenges and insights.

12. Results and Discussion

The Python platform is being used experimentally to create the Blockchain-based Government document Verification Process Distribution System. This will allow candidates to utilize the system from anywhere in the world. In this case, the Kaggle dataset for document image categorization was used. Images of a variety of papers, including official government documents, are included in this dataset. Document classification and review are done with it. Once the authenticity of the data has been established by a variety of ensemble learning approaches, blockchain miners mine it and add it to the blockchain.

In this hypothetical example, four ensemble methods (Random Forest, Gradient Boosting, AdaBoost and Stacking) were applied to a multi-attribute document image classification dataset. The performance of each method is evaluated using unconventional evaluation metrics: precision, accuracy, recall, F1 rating and AUC-ROC based on the receiver operating characteristic curve (AUC-ROC). The values in the table represent the overall performance achieved by each holistic approach in a unique fitness enterprise.

- **Accuracy:** The overall accuracy of the version in making accurate predictions.
- **Precision:** The percentage of genuine high-quality predictions amongst all effective predictions.
- **ReCall**: The proportion of real wonderful predictions amongst all real nice times.
- **F1 rating:** The harmonic mean of precision and consider, representing the balance among the 2 metrics.
- **AUC-ROC:** The location below the Receiver working function Curve, a degree of the version's potential to discriminate between tremendous and negative instances.

Table 2: Comparision of Ensemble methods.

Ensemble Method	Accuracy	Precision	Recall	F1 Score	AUC-ROC
Random Forest	0.852	0.833	0.867	0.846	0.915
Gradient Boosting	0.866	0.847	0.881	0.866	0.923
AdaBoost	0.832	0.813	0.855	0.83	0.901
Stacking	0.878	0.866	0.894	0.878	0.928

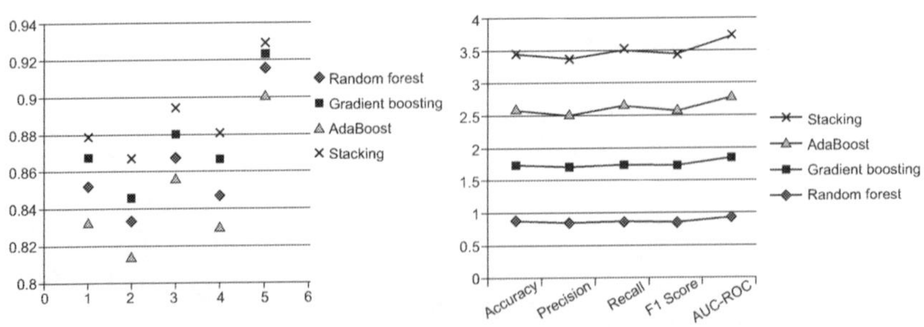

Figure 4: Comparison of Ensemble methods on Document Image Classification Dataset

13. Conclusion

Blockchain is the most motivating and progressive of technologies available, many organizations are taking advantage of Blockchain and providing safe and secure data services to users. This projected system brings together the advantages of Blockchain and implements a new strategy for government agencies to distribute government documents. The proposed solution, which uses advanced technologies such as machine learning and blockchain, offers a promising way forward. By integrating these innovations, the government can create a system that provides immutable and easily accessible records to the public. This approach not only ensures the integrity and security of government records but also makes it easier for citizens to gain real information about government policies, notifications, regulations and decisions. Citizens can access this information with one click, which in turn increases trust, transparency and accountability in government operations.

References

1. Lakmal, C., Dangalla, S., Herath, C., Wickramarathna, C., Dias, G. and Fernando, S. 2017. IDStack—The common protocol for document verification built on digital signatures. In Proceedings of the 2017 National Information Technology Conference (NITC), Colombo, Sri Lanka, 14–15 September 2017; pp. 96–99. https://doi.org/10.1109/NITC.2017.8285654.
2. Ahmad, F., and Cheng, L.M. 2019. Paper Document Authentication Using Print-Scan Resistant Image Hashing and Public-Key Cryptography, vol. 11611 LNCS, no. July. Springer International Publishing.
3. Satybaldy, A., Subedi, A., and Nowostawski, M. 2022. A Framework for Online Document Verification Using Self-Sovereign Identity Technology. *MDPI, Sensors, 22*, 8408. https://doi.org/10.3390/s22218408
4. Imam, I. T., Arafat, Y., Alam, K. S., and Shahriyar, S. A. 2021. DOC-BLOCK: A Blockchain Based Authentication System for Digital Documents. In Proceedings of the 2021 Third International Conference on Intelligent Communication Technologies and Virtual Mobile Networks (ICICV), Tirunelveli, India, 4–6 February 2021; pp. 1262–1267. https://doi.org/10.1109/ICICV50876.2021.9388428.
5. Marella, V., and Vijayan, A. 2020. Document Verification using Blockchain for Trusted CV Information. In Proceedings of the 2020 Advances in Information Systems Research (AMCIS), Virtual Conference, 10–14 August 2020; Volume 12.
6. Hyperledger Foundation. Hyperledger Fabric. 2022. Available online: https://www.hyperledger.org/use/fabric (accessed on 12 October 2022).
7. Dhiman, S., and Singh, A. 2013. "Tesseract vs gocr a comparative study," International Journal of Recent Technology and Engineering, 2(4), pp.80.
8. Patel, C., Patel, A., and Patel, D. 2012. Optical character recognition by open source OCR tool tesseract: A case study. International Journal of Computer Applications, 55(10).
9. Schlatt, V., Sedlmeir, J., Feulner, S. and Urbach, N. 2021. Designing a Framework for Digital KYC Processes Built on Blockchain-Based Self-Sovereign Identity. *Inf. Manag., 59*, 103553. https://doi.org/10.1016/j.im.2021.103553
10. Joshi, Y. S. 2014. The Future of Enterprise Printing: Securing Hardcopy Documents in the Digital Age: white paper. CIO Insight, no. July, pp. 1–12.
11. Nizamuddin, N., Hasan, H. R., and Salah, K. 2018. IPFS-blockchain-based authenticity of online publications. *Lect. Notes Comput. Sci.* (including Subser. Lect. Notes Artif. Intell. Lect. Notes Bioinformatics), vol. 10974 LNCS, no. June, pp. 199–212.
12. Soltani, R., Trang Nguyen, U., and An, A. 2018. A New Approach to Client Onboarding Using Self-Sovereign Identity and Distributed Ledger. In Proceedings of the 2018 IEEE International Conference on Internet of Things (iThings) and IEEE Green Computing and Communications

(GreenCom) and IEEE Cyber, Physical and Social Computing (CPSCom) and IEEE Smart Data (SmartData), Halifax, NS, Canada, 30 July–3 August 2018; pp. 1129–1136. https://doi.org/10.1109/Cybermatics_2018.2018.00205.

13. Parra Moyano, J. and Ross, O. 2017. KYC optimization using distributed ledger technology. *Bus. Inf. Syst. Eng., 59*, 411–423.
14. Hyperledger Foundation. 2022. Hyperledger Indy. 2022. Available online: https://www.hyperledger.org/use/hyperledger-indy (accessed on 12 October 2022).
15. Rajapashea, M., Adnanb, M., Dissanayakac, A., Guneratned, D., and Abeywardanee, K. 2020. Multi-Format Document Verification System, American Scientific Research Journal for Engineering, Technology, and Sciences (ASRJETS) ISSN (Print) 2313-4410, ISSN (Online) 2313–4402, 06 December 2020.
16. Abd El-Aziz, R.M., Aissa, A.B., Taloba A.I., and Mohamed, I.A. 2020. Experimental Analysis Of Document Verification Scheme Using Blockchain Logics Over Secured Cloud Environment. International Journal of Advanced Research in Engineering and Technology (IJARET) Volume 11, Issue 9, September 2020, pp. 709–721, Article ID: IJARET_11_09_071
17. Ganesan, N. 2017. Three foreigners jailed in Singapore for submitting fake academic certificates-Human Resources Online, Human Resources. Available at: http://www.humanresourcesonline.net/three-foreignersjailed-for-submitting-fake-academiccertificates/(Accessed: 07 February 2018).
18. Kalipa, S. 2018. Home Affairs official sold us fake papers _ IOL News, Crime and Courts IOL News. Available at: https://www.iol.co.za/news/crimecourts/home-affairs-official-sold-usfake-papers-1929048 (Accessed: 08 February 2018), 20
19. Roychoudhury, S., Bellarykar, N., and Kulkarni, V. 2016. A NLP Based Framework to Support Document Verification-as-a-Service. In Proceedings of the 2016 IEEE 20th International Enterprise Distributed Object Computing Conference (EDOC), Vienna, Austria, 5–9 September 2016; pp. 1–10. https://doi.org/10.1109/EDOC.2016.7579376
20. Lakmal, C., Dangalla, S., Herath, C., Wickramarathna, C., Dias, G., and Fernando, S. 2017. IDStack—The common protocol for document verification built on digital signatures. In Proceedings of the 2017 National Information Technology Conference (NITC), Colombo, Sri Lanka, 14–15 September 2017; pp. 96–99. https://doi.org/10.1109/NITC.2017.8285654
21. Arner, D.W., Zetzsche, D.A., Buckley, R.P. and Barberis, J.N. 2019. The identity challenge in finance: from analogue identity to digitized identification to digital KYC utilities. *Eur. Bus. Organ. Law Rev., 20*, 55–80.
22. The World Wide Web Consortium (W3C). 2022. Verifiable Credentials Data Model v1.1. 2022. Available online: https://www.w3.org/ TR/vc-data-model/ (accessed on 8 June 2022)
23. Fdhila, W., Stifter, N., Kostal, K., Saglam, C. and Sabadello, M. 2021. Methods for Decentralized Identities: Evaluation and Insights. *Int. Conf. Bus. Process. Manag., 428*, 119–135.
24. Shuaib, M., Alam, S., Alam, M.S. and Nasir, M.S. 2021. Self-sovereign identity for healthcare using blockchain. *Mater. Today Proc.*
25. Kondova, G., and Erbguth, J. 2020. Self-sovereign identity on public blockchains and the GDPR. In Proceedings of the 35th Annual ACM Symposium on Applied Computing, Brno, Czech Republic, 30 March–3 April 2020; pp. 342–345.
26. Dong, C., Wang, Z., Chen, S., and Xiang, Y. 2020. BBM: A blockchain-based model for open banking via self-sovereign identity. *Int. Conf. Blockchain, 12404*, 61–75.
27. Oliveira, S.R.M., Nascimento, M.A., and Zaiane, O.R. 2002. Digital Watermarking: Status, Limitations and Prospects. Technical Report TR 02-01, Department of Computing Science, Alberta University, Edmonton, Alberta, Canada.
28. Li, C. M., Hu, P., and Lau, W. C. 2015. Authpaper: Protecting paper-based documents and credentials using authenticated 2D barcodes. In Communications (ICC), 2015 IEEE International Conference, pp. 7400–7406, IEEE, June.
29. Ball, G. R., Stittmeyer, R., and Srihari, S. N. 2010. Writer verification in historical documents. In Proc. of SPIE Document Recognition and Retrieval, vol. 7543, pp. 1–8.
30. Chen, S., and Srihari, S. 2006. A new off-line signature verification method based on graph. In Proc. of the 18th Int. Conf. on Pattern Recognition (ICPR 06), Hong Kong, China, pp. 869–872, August.

31. Nizamuddin, N., Hasan, H. R., and Salah, K. 2018. IPFS-blockchain-based authenticity of online publications. Lect. Notes Comput. Sci. (including Subser. Lect. Notes Artif. Intell. Lect. Notes Bioinformatics), vol. 10974 LNCS, no. June, pp. 199–212.
32. Andersson, O., and Marquez, S. R. 2016. A comparison of object detection algorithms using unmanipulated testing images: Comparing SIFT, KAZE, AKAZE and OR, 0 6.
33. Hou, J. L., and Huang, C. H. 2008. A model for document validation using concurrent authentication processes. *J. Comput. Inf. Syst., 49*, 65–80.
34. Signicat. 2022. Digital Identity Solutions for the Full Customer Journey. Available online: https://www.signicat.com
35. Bolte, P. 2021. Self-Sovereign Identity: Development of an Implementation-Based Evaluation Framework for Verifiable Credential SDKs. Available online: https://bolte.cloud/assets/ssi_master_thesis.pdf

CHAPTER 6

Detection of Ovarian Cancer Using Improved Deep Learning Model

Raja Rao Budaraju[1] and *Ketan Anand*[2,*]

1. Introduction

The most common type of gynaecological cancer is commonly referred to as "ovarian cancer." Ovarian cancer is classified into four subtypes: mucous, endometrioid/clear cell, mucinous, and serous. The most common kind of ovarian cancer is serous. Only pathologists are qualified to distinguish between the four forms of cancer shown on hematologicalgical slides. Pathologists are the only medical specialists who can tell the difference using this method [1–3]. When a huge number of images must be reviewed and identified, the possibility of human error increases. Automated diagnosis, or CAD, was utilised in a variety of tests to increase pathologists' accuracy while reducing the amount of manual labour they had to undertake. The use of frameworks like CAD allows for more exact measurements, while prior information plays a less important role in the analysis. Much of our recent success in CAD innovation may be attributed to technological advancements. A rising number of studies [4–5] have used advanced medical imaging technologies such as computer-aided design. Chang et al. developed this method for analysing liver cancer based on multiphase CT images that highlight tumour characteristics. CAD is used to discriminate between benign and malignant thyroid tumours [6], a decision assistance system for cone beam computer tomography to aid in the early diagnosis of keratocystic and dental periapical cysts. This approach's purpose is to aid in the identification of these disorders. The purpose of creating this framework was to make it easier to recognise various ailments [7]. CAD systems, also known as

[1] Koneru Lakshmaiah Education Foundation, Vaddeswaram, AP, India.
[2] Department of AI&ML, Sreenidhi Institute of Science and Technology, Hyderabad, TS.
* Corresponding author

"second opinions," are a type of medical diagnostic software designed to provide an unbiased "second opinion" during the identification and diagnosis of health issues. In addition to aiding in diagnosis and treatment, CAD technology may be used to "review" patient data. This can be achieved by employing a distinct method of data analysis. This is made feasible by the utilisation of various technological applications. The data visualisation tools provide these options. CAD solutions can cut down on the time spent figuring out what a picture means while making the diagnostic process more accurate. A system for analysing breast cell histopathology images that uses a programmatically quantified support vector machine is employed. This method was used to analyse images [8]. Researchers employed computer vision and machine learning techniques like pattern recognition and image processing to detect lung nodules. The approach described follows many of the same concepts. The CAD system uses fuzzy systems and evolutionary algorithms [9], in research that combines arterial spin labelling and diffusion tensor imaging, support vector machines [10–11]. Because of the findings of this evaluation, the participants were separated into two distinct groups: those with Alzheimer's disease and those with frontotemporal dementia. A unique method of testing to detect whether a person has a brain condition was used. This method is routinely used to identify the source of a neurological condition. According to recent evaluation results, ML is widely employed in CAD because it was first utilised in biological image processing, it is possible to connect it to the concept of deep learning. You may begin right here, at the beginning. Deep learning, as opposed to task-specific algorithms, encompasses a broader range of approaches. These strategies enable the development of competence in data representation.

Deep learning algorithms such as DBNs and CNNs use raw pixel data in its unprocessed condition as input. Classic image processing systems are built around models that have been trained on object segmentation characteristics, for a more basic and appetising depiction of the original, an image might be broken into pieces. This can make navigating the image simpler. Segmenting is a technique for finding and separating specific traits and limitations within an image (such as lines, curves, etc.). This fundamental approach combines expanding regions, spreading regions, and merging regions to discover neighbouring pixels within an image. Following that, models of these features are trained to recognise them in others. Deep learning algorithms such as DBNs and CNNs use raw pixel data in its unprocessed form as input. It is faster and easier to understand because the process of computing features and segmenting objects is no longer necessary. Deep learning, a fast-expanding topic, has subsequently drawn academics from a variety of domains, including biomedical imaging [12–13]. DCNN initiation features may be utilised to segment, categorise, and display tissue histology pictures. A computerised system for diagnosing lung tumours using Deep Convolutional Neural Networks is employed [14]. Microscopy was utilised to examine these images of lung tumours, produced with the same procedure [15]. DCNNs as they are more officially termed, have been shown by Gao and colleagues to be capable of independently detecting pictures of cells collected from human epithelial-2. In practise, the structure proved to be exact as well as adaptable.

A Convolutional Neural Network is used to diagnose the kind of lung cancer present in pulmonic nodules, the purpose of this technique was to provide computer-aided decision assistance. A variety of organisations are now attempting to use CNNs in clinical imaging, with varying degrees of success. According to our findings, VGGNet has never been used to classify ovarian tumours. Wu M's pre-trained AlexNet improved from 78 percent to 84 percent after training. As the focus of our work, we employed DCNN to classify a range of ovarian cancer subtypes using histology images, both with and without augmentation. The outcomes of this study can help clinical pathologists and technologists decide whether to undertake a biopsy on a suspected tumour. AlexNet was used to build and train Deep Convolutional Neural Networks (DCNNs) that might be used to automatically detect the subtype of ovarian cancer seen in cytological pictures. The deep convolutional neural network includes seven layers in total, with five of them being convolutional layers, three being max pooling levels, and two being complete reconnect layers (DCNN). The model was then trained using two distinct sets of input data: (1) raw data from the original photographs; and (2) iterations of the same images that had been rotated and enhanced. The testing results reveal that the classification model's accuracy rose from 72.76% to 78.20% as the picture quality utilised as training data improved. The 10-fold cross-validation method was used to show that this claim is true.

The following is a list of each author's different contributions: In the first phase, one data set is analysed, the default class count for the pre-trained model was 1,000, but this was lowered to 5 to represent the scope's various subtypes. Using this method, it was feasible to reduce the node count of the pre-trained architecture from 4096 to 128. In data categorization models, network topologies such as AlexNet, Google, and VGGNet are employed. These structural characteristics can help differentiate malignant tissue from surrounding benign tissue on histology slides. The healthy and malignant tissues were close together. Because of their high degree of flexibility, multi-layer artificial neural networks are prone to overfitting the training data, which is problematic when the training dataset is small. This is especially true if only a small quantity of training data was used. This issue is more likely when the quantity of the training dataset is limited. Our primary framework was built on top of pre-trained convolutional neural networks. The ImageNet dataset was used to produce the weights for these networks, which were then used to train the models. This was done to prevent the possibility of overfitting, and it worked! Frameworks that have been trained on millions of images have been shown to extract essential visual information from images. These frameworks are suitable for the purpose. There must be more allusions to this. Some of these parts, like edges, circles, and object bulbs, maybe the building blocks for more difficult tasks involving picture recognition. According to our review of relevant published material, ovarian cancer has not been classified using VGGNet. This was discovered following an extensive assessment of the existing literature. Several papers have recently emphasised VGGNet's potential utility in cancer prediction. One of these early research projects inspired the idea for our original product.

2. Proposed Method

2.1 Dataset

The Cancer Genome Atlas Ovarian Cancer Collection (TCGA-OV) were utilised in this study [16]. Owing to the hospital's small patient population, collecting histology slides of the carcinoma subtype OC was difficult due to the absence of other types of cancer. Experimental pathologists gathered the patients' tissue and treated it so that it could be examined on slides. They prepared and coloured the cell samples using equipment and their laboratory materials. The images were captured on film with a Leica ICC50, a smaller version of the company's famed camera. This 5-megapixel camera can send video to desktop PCs as well as mobile devices.

The examination of the Cancer Genome Atlas-Ovarian Cancer database yielded pictures of carcinomas related to certain clinical circumstances. Figure 1 illustrates this concept by using a real-world event. "Total Community Genome Analysis" is the acronym for the joint initiative formed by several different organisations. Only a few additional organisations, most of which are tiny, have agreed to participate in this endeavour. After successfully logging in, anyone may access the whole repository and participate in the debate. The pictures were saved in JPG format to keep their original resolution of 220 pixels across and 220 pixels high. As a result, each pixel's size was enlarged from twice to three times its former value. The OC with malignant tumour collection contains non-carcinoma images of normal tissues and non-tumors glandular tissues for each of the four different subtypes. This website has more than simply cancer images; it also contains a variety of other information. The goal of this dataset is to use histological pictures of distinct carcinoma and non-carcinoma subtypes to automatically categorise people who have been diagnosed with oral cancer.

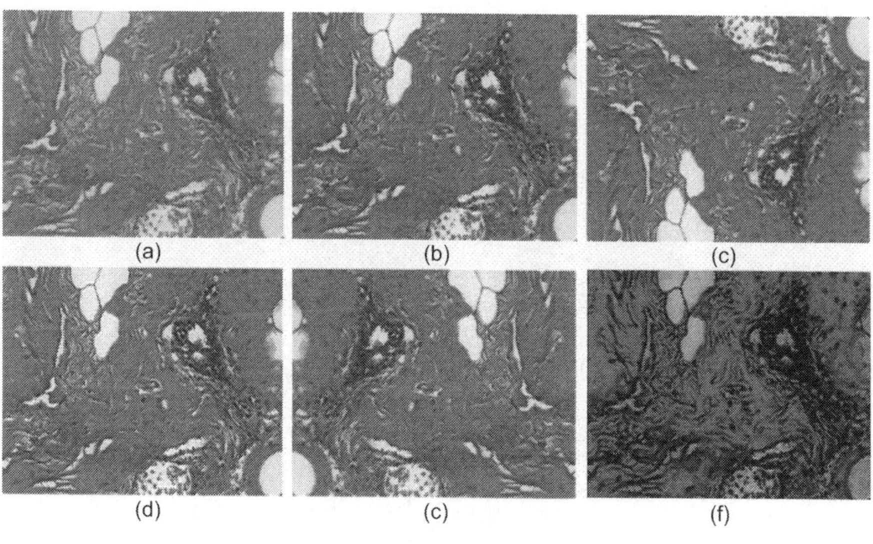

Figure 1: (a–f) Image Augmentation Techniques.

2.2 Data Augmentation

The lack of readily available training materials is a significant barrier in medical imaging. Using data augmentation, any missing data from the time of data collection will be added to this dataset, allowing it to be fully utilised. It has a significant impact on the total quantity of information in the dataset [17–18]. The images were blown up, twisted, and improved in such a way that the lecture could continue while the alterations were made to the images. A total of 29,187 images were taken when the modifications were completed. Figure 2 depicts a variety of ways that may be utilised to increase the quality of training data. If the rows and columns of each pixel are flipped, it is feasible to flip in both the horizontal and vertical directions.

Starting at zero degrees, the pictures were rotated about their axis of rotation from 0 to 3600 degrees. Table 1 offers a summary of some of the available parameter possibilities. It is critical to modify the brightness levels in the images to optimise the

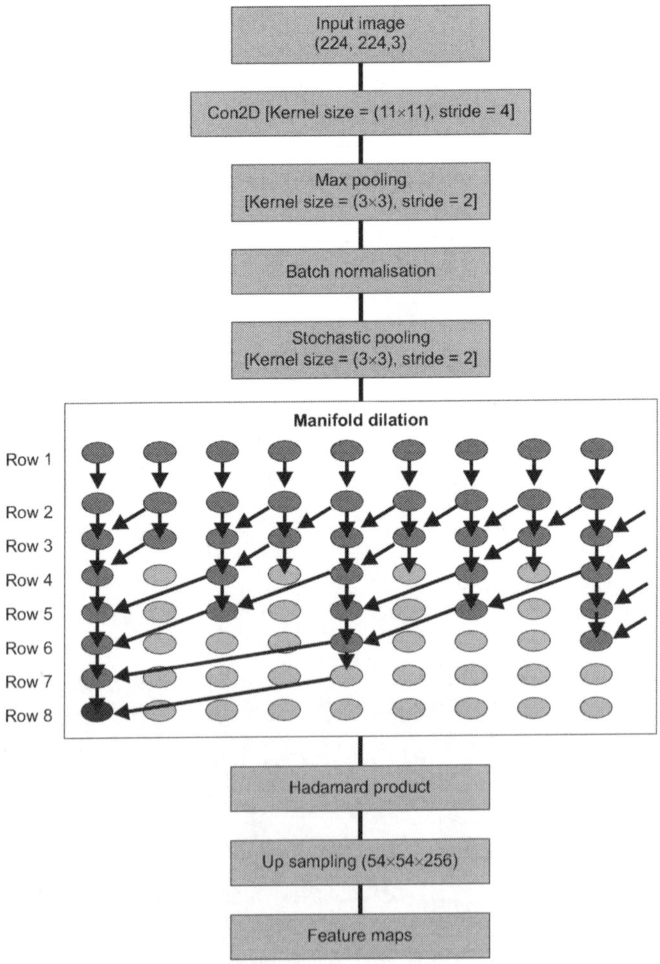

Figure 2: The Improved Model.

document validation process [19–21]. This is one of the reasons behind the current scenario. If you zoom in or out using a method called interpolation, you can get a better grasp of the model. These augmentation techniques aim to reduce instances of overfitting and underfitting in deep neural networks to improve their generalizability. The modules offered by Python make it easier to implement these tactics.

Table 2 shows how many images were captured before and following the rise in the number of shots in each category. This comparison is performed for each of the many types of images.

Table 1: Data Augmentations and Augmentation Values.

Image Parameters	Augmentation Values
Rescaling	100, 200 and 400 Percentage
Zooming	0.25, 0.50, 0.75
Rotation	30, 45, 60, 90
Shifting	3
Flipping	Horizontal and Vertical

Table 2: A Different Number of Images are Available for Every Type of Ovarian Cancer.

Type of Ovarian Cancer	Original Images	Augmented Images
Serous	286	6751
Mucinous	210	6334
Endometrioid	72	5464
Clear Cell	82	5000
Non-Cancerous	96	5638
Total	746	29,187

3. Proposed Methodology

3.1 The Improved VGG−19 Model

It is critical to be aware of the wide range of levels of complexity in a computer system; in general, CNNs are structured around these layers, with each having a distinct role. The word "filtering" is frequently used to describe this sort of function. Stacking these layers in a certain order during the last decade resulted in the development of several CNN layouts, each with its distinct appearance. This approach allows for the collection of data on a wide variety of scales. An example this is done without altering the overall quality of the image in any way. Stretching out the convolution kernel allows you to increase the size of a convolution field without introducing extra model parameters. To do this, we'll use a method called convolution kernel dilation. As the convolutions become wider, the receptive field grows, but the spatial resolution stays constant. It is not necessary to add any extra layers to prevent gridding in pre-existing neural networks with dilated convolutional layers [16]. Dilated spatial convolution units can be utilised to apply the maximum and stochastic pooling algorithms to pictures that provide considerable difficulty at the same time. Figure 2 depicts this. It's available here. To reduce the size of the feature map, use the max pooling approach. Because the pooling zone's activation is decided at random, stochastic pooling takes this activity into account [17]. Based on the current value of this variable, pooling zones are enabled. The maximum pooling approach selects the feature that occurs the most frequently in the final, filtered form of the feature map. This happens at several points throughout the pooling process. Following that, the max-pooling layer would provide a feature map that highlights the most important parts of the previous feature map. A Softmax layer can be added to a neural network before the layer that outputs the results of the network's calculations. The output layer and the softmax layer must have the same number of nodes. It is much easier to determine the pool's average and maximum size when using stochastic pooling, which is a big advantage of the approach. If you have all the required attributes, a higher value can be averaged down to a lower number until you achieve your goal average value. There is no limit to the number of times this can be repeated until the average is attained. It will continue even after the average has been determined.

The practise of max pooling, which assures that the maximum possible value is retained, exacerbates the problem. A probability map must be created for each block entry that will be used in the pooling process. This must be completed before the programme can begin. To do this, the stochastic pooling technique is employed, as discussed further below. If your random variable influences your discrete probability distribution, your discrete probability distribution may assist in keeping your random variable in equilibrium. Batch normalisation and dropouts should be utilised to avoid overfitting. The application of this product is simple and quick. Because there are no constraints on how it may be utilised, everyone can put it to good use however they see fit. Finally, "dilation" is used to expand the size of an object while retaining the same processing and memory requirements as before. A second Hadamard step of processing is then applied to both the batch-normalized characteristics and the input image. To begin, the aim is to collect features of the same size, which may be accomplished by upsampling. More research on these characteristics is recommended to better understand the differences between benign and malignant cancer cells.

To retrieve the features, multiply the feature map by the images and give it to the VGG model for further processing. The combination of the global average pool and the maximum pool yields aggregate characteristics. These improve the most important information while lowering the quantity of superfluous stuff. They can forecast cancer cells better because they have a better idea of where they are. AlexNet and other DCNNs were first defined in a blueprint published in 2012. There were six stages in all, including a softmax layer [22–24]. There were two levels of normalisation and three maximum pooling levels to choose from. It is possible to save up to 60 million unique configurations at the same time. a large amount of information. The Visual Geometry Group at Oxford University has since launched the VGG-16 network shown in Fig. 3 to replace AlexNet. As a result, the group at issue no longer has any control over what happens next. AlexNet was conceived by a group of Oxford University academics. They created it for the 2014 Image Net Challenge to increase their chances of winning the localization and classification

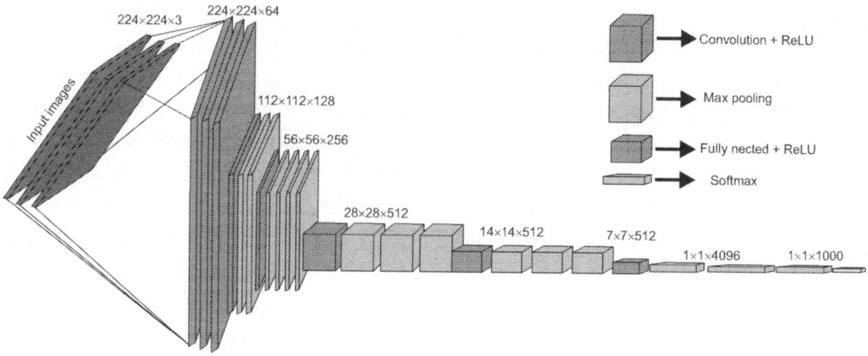

Figure 3: The VGG 16 Architecture.

project. VGG's Karen Simonyan and Andrew Zisserman created it for that purpose. When compared to prior designs, the VGG-16 has a higher level of complexity and more layers than any of the others.

Because each level in the improved VGG-16 model performs a separate function, the model has a total of sixteen levels. In its design, this approach employs thirteen layers of convolutional filters. An image with a surface area of 224 square inches on each side and a height of three inches may influence the design of the improved VGG-16. The image has been pre-trained for this shot. Because each layer contains 64 filters, the image in Fig. 2 is 224 pixels down and 64 pixels tall. This is because each layer has 64 filters. Each of these levels is made up of a single kernel and is organised in a three-by-three grid. Because the window size is two by two and the stride is two, the volume may be decreased to 112 by 112 by 64. It is feasible to achieve this aim by reducing both the game window and the game stride. The final image has 112 pixels in width, 112 pixels in height, and 128 pixels in depth. 128 filters and two more convolutions are used to achieve this aim. There are now just 56 by 56 by 128 units of volume remaining after reaching the layer with the maximum feasible pooling potential. Following that are three further convolution layers, each with 256 filter units to bring out even more information in the image. The completed picture has a dimension of 28 by 28 pixels and is made up of 256 individual pixels. The presence of a max-pool layer in the centre provides a simple distinction between the two higher-level stacks. The output on the screen is the result of processing each of these stacks with 512 filters and three convolution layers. After the last pooling layer is completed, the flattened images of a 7x7x512 volume are delivered to the FC layer to be processed. As a result of this, the FC layer is free to continue its task.

The pictures are flattened such that they are uniform in appearance on the FC layer, which has 4096 channels and a dropout of 0.5. The Softmax will be the very final layer to be put together. The remaining blocks, as shown in Fig. 2, were trained using a modified version of the improved VGG-16 model. The dropout value in this model has been drastically lowered, dropping from 0.5 to 0.2, a significant reduction. This is the current condition of the fundamental building component, which is frozen. There were 4096 nodes in the fully connected layers of the pre-trained architecture; however, in the present design, only 128 nodes are in use since some nodes have been eliminated. This is because certain nodes were redundant.

The number of classes contained in the pre-trained model is set to 1000 by default. In addition to the pre-existing category of non-cancerous disorders, this has been expanded to include five new subcategories. As a result, we should not be surprised if the overall number of categories we wish to construct corresponds to the number of subtypes we end up with. All training, validation, and testing datasets were obtained from a single original dataset [25]. Each component was given its own name, and a count of the training and validation losses was conducted at the end of each session. This total includes not only the early-stopping condition but also the accuracy of the training and validation methods. An iterative technique can be considered successful if the validation loss is less than a preset threshold.

4. Experimental Setup

Before delving into the DCNN model's performance measures and the opportunities that may be generated by modifying its hyperparameters, we'll discuss the project's general backdrop.

4.1 Implementation

The Google Collaboratory, which has a 1.569GHz Nvidia T4 GPU and 16GB of GPU RAM, was used for all testing in this study. Python frameworks like TensorFlow and Keras are used for testing and training pre-trained and created VGGNet models. These frameworks are used to test and train models. Cloud computing has made it much easier to locate GPUs.

4.2 Evaluation Metrics

In this section, we'll go through the most crucial performance indicators to monitor while working on your project. Measures such as the F1 score, and the model accuracy statistic may be useful. These performance criteria are developed using a matrix described with a focus on four major features. Some of the performance indicators used to measure staff efficiency and productivity include FP, TP, TN, and FN. There is no reason to be concerned about any of this. The abbreviation "FP" may be used to describe a false positive (FN). False positives are represented by the letters FP for benign lesions, whereas false negatives are represented by the letters TP for malignant tumours. The initials FN and TN have been assigned to malignant pictures to indicate that they belong to a distinct malignancy subtype. Examining instances where the outcome can be any two or more classes is one technique to assess the performance of deep learning systems. Let's get right to it: there's a lot of territory to cover in this discussion.

1. **Accuracy:** One method for determining forecast accuracy is the proportion of correct predictions to the total number of forecasts. This may be expressed mathematically using the following formula:

$$\textbf{Accuracy} = \frac{TP}{TP+FP+FN} \tag{1}$$

2. **Recall:** This conclusion is drawn from the precision with which positive classifications are predicted and the frequency with which such predictions are correct.

$$\textbf{Recall} = \frac{TP}{TP+FN} \tag{2}$$

3. **Precision:** The fraction of predicted events that were confirmed to have occurred.

$$\textbf{Precision} = \frac{TP}{TP+FP} \tag{3}$$

3. **F1 - Score:** This result was derived by dividing the recall/accuracy ratio by two and then adding the product to the original number.

$$\textbf{F1-Score} = \textbf{2} * \frac{Recall * Precision}{Recall + Precision} \tag{4}$$

4.3 Hyper parameter Tuning

Neural networks may form connections between inputs and outputs via a process known as self-organization, which reduces the need for external supervision. The model may have been trained using data that was not included in the testing dataset, which increases the chance of sampling noise. Because of this overfitting, there was cause for concern about the deep learning model's capacity to give reliable future estimates. We accomplished this because modifying the hyperparameters of the DCNN model was required to get the best possible degree of performance. We were able to collect correct measurements for each of our hyperparameters by following the steps outlined in this section. We made the thrilling discovery that the solution to our classification problem was a cross-entropy-based loss function. We found ourselves in this scenario when we realised it worked. To further optimise our system, we ran twenty iterations of stochastic gradient descent, often known as SGD. Before any additional training may begin, the validation accuracy must fall below a particular level. This maximum number is determined before the start of training. The validity of the validation is documented and evaluated after each epoch, and checks are also made to verify whether it has been lost. By preserving just, the model with the highest degree of accuracy and the least amount of loss, the danger of overfitting is reduced.

Several alternative learning rates were tested during deployment. Batch sizes of 32 and 64 were used to reduce the generalisation gap between training and validation loss by 0.001 and 0.0001, respectively. The 0.3 dropout helps to reduce both the amount of overfitting during training and overall overtraining. The withdrawal was beneficial to both parties involved in the circumstance. Ovarian cancer is a broad term that refers to a variety of tumours that share many clinical, histological, and molecular features. A few distinct traits separate ovarian carcinomas from other tumour forms. Ovarian carcinomas are certainly neoplasms when these criteria are present. This aspect must be carefully considered in the research of ovarian cancer. Despite this, a more clear classification based on clinicopathological and molecular markers has been developed. Currently, pathologists mostly examine the appearance of the tumour to determine which group the sickness falls into. The clinical appearance, tumour growth, and molecular genetic mutations separate type I and type II ovarian carcinomas.

5. Results

During the current round of this debate, we will go through our results and observations. According to Fig. 3, we can see the outcomes of comparing the training set to the validation set's accuracy as well as the loss per epoch for comparison. As a result, they can both exist at two separate places in space and time, making it possible for both to live at the same time. The graphs in A and B may be easily comprehended due to the evident association between the accuracy decline during training and subsequent improvement during validation. Figure 3 demonstrates this. Premature halting began in this manner in the 11th century. Our model's training accuracy climbed to 84.64 percent because of the extra dataset, and its training loss decreased to 0.021 percent. Even if only half of the errors were correctly anticipated, the model could be trained. For example, in the basic dataset, the model lost only 0.0855 points. When trained with the enriched dataset, model 2 has an accuracy of 84.64 percent, whereas model 1 has a 50% accuracy when validated without any further augmentation. A graph displaying the accuracy of the validation data is shown above the one showing the accuracy of the training data in Graph C. Based on this statistical data, we may infer that model 2 is both the most accurate and the most generalizable.

As seen in Table 3 the types of ovarian cancers, Type 1 Clear Cell, Type 2 Endometrioi, Type 3 Mucinous, Type 4 Non –Cancerous and Type 5 Serous including additional information resulted in more accurate data classification [26–27]. To determine classification accuracy, utilise the confusion matrix and the equations presented in the section on assessment metrics.

As indicated in Table 4, the inclusion of new data into our collection directly contributed to a 34–percentage point increase in overall classification accuracy. The supplemented dataset's accuracy rate was much greater than the baseline dataset's accuracy rate, which was just 50% right. The inclusion of aesthetic enhancements and tighter zooming is most likely explained by advancements in generalizability made possible by these changes. Even though both models are based on the same underlying architecture, they each use their unique method of data collection to

Table 3: Comparative A ccuracy of the Model 1 and Model 2 C lassifiers.

Ovarian Cancer Types	Classification Accuracy	
	Original Dataset	**Augmented Dataset**
Type 1	51.66%	83.95%
Type 2	65.45%	95.72%
Type 3	63.21%	84%
Type 4	59.13%	91%
Type 5	61.23%	90%

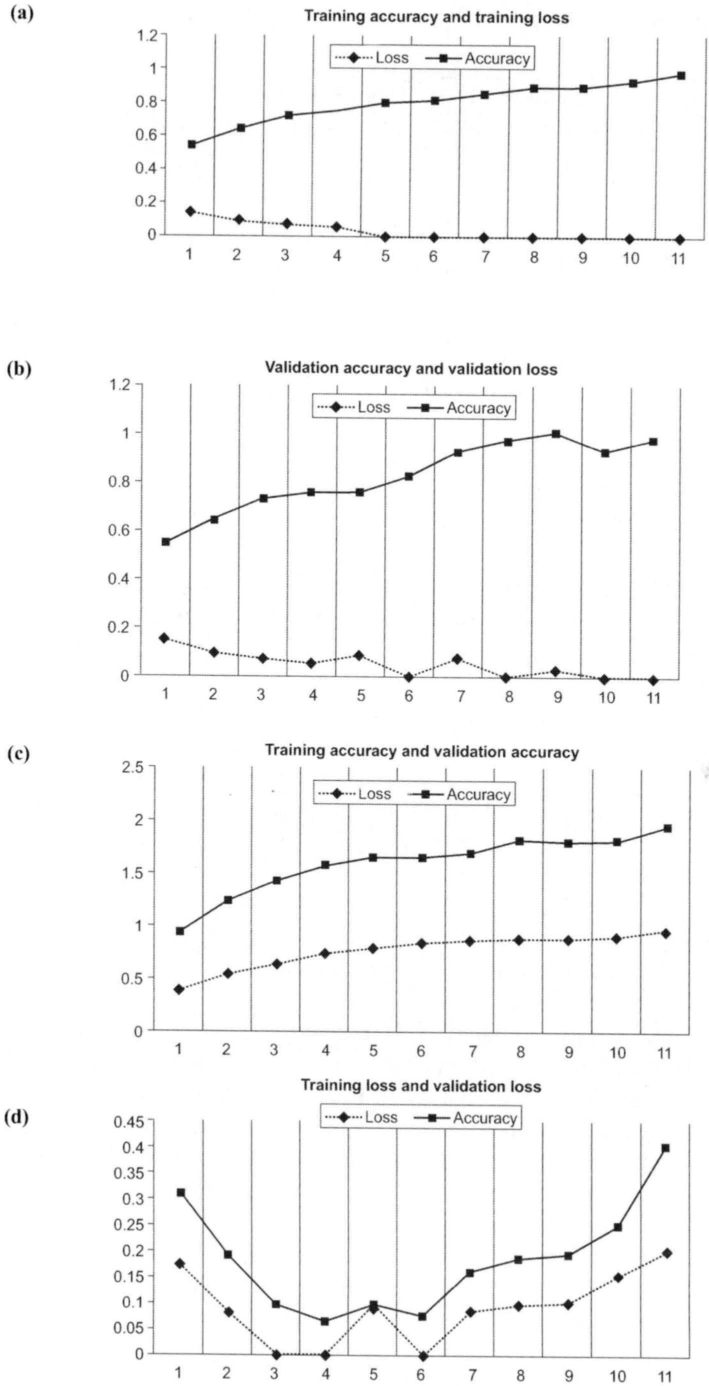

Figure 4: A Comparison of the Accuracy and Loss in the Training Dataset and the Validation Dataset, on an Epoch-by-Epoch Basis.

Table 4: Performance Evaluation Using Proposed Model: A Comparison of Their Benefits and Drawbacks.

Performance Evaluation for Improved VGG-16					
	Sub-Types	**Precision**	**Recall**	**F1- Score**	**Accuracy**
	Clear Cell	0.43	0.60	0.68	
Dataset	Endometrioid	0.55	0.63	0.74	
	Mucinous	0.50	0.45	0.67	50%
	Non-Cancerous	0.64	0.48	0.51	
	Serous	0.55	0.52	0.71	
	Clear Cell	0.80	0.88	0.94	
Augmented Dataset	Endometrioid	0.95	0.90	0.89	
	Mucinous	0.80	0.73	0.87	84.64%
	Non-Cancerous	0.94	0.66	0.85	
	Serous	0.63	0.84	0.83	

arrive at their findings. Underfitting is a major contributor to the problem, and as a result, the dataset contains too few observations to accurately describe the scenario. Likely, numerous separate factors all worked together to decide its development at the same time. It is feasible to conclude from this that the DCNN architecture may be improved by using proper picture augmentation parameters, as proven in this study.

The approaches indicated in the section on assessment metrics were used to construct a wide range of performance indicators, which were then examined in the section that followed. Table 4 shows the graphical representation of the many performance indicators that were discovered. This is seen in the diagram below. Students' performance ratings in each model 2 class and subclass have been greater than or equal to 70% of the total points available. Applying the criteria, which are all equally important, is one technique to assess a model's validity and efficacy. If you follow these guidelines, you will be able to create the most beautiful model that is even somewhat believable.

This early ending structure allowed the model to be saved at the perfect time since it could be saved at epoch 11 when its performance was at its pinnacle. This was possible because of the early termination provision. If the training dataset differs from the new dataset, we may make predictions based on our best model and the new dataset. It was compared to the preceding dataset, which served as a reference point. Figure 5 demonstrates a way of predicting and categorising a wide range of probable events.

We discovered that many images were wrongly classified due to the model's poor learning from a very small dataset of 24,742 enhanced images. This was owing to the small size of the dataset. This was due, in part, to the model being presented in this manner during prior education. Figure 6, 7 and 8 depict the decisive findings, which conform with our hypothesis.

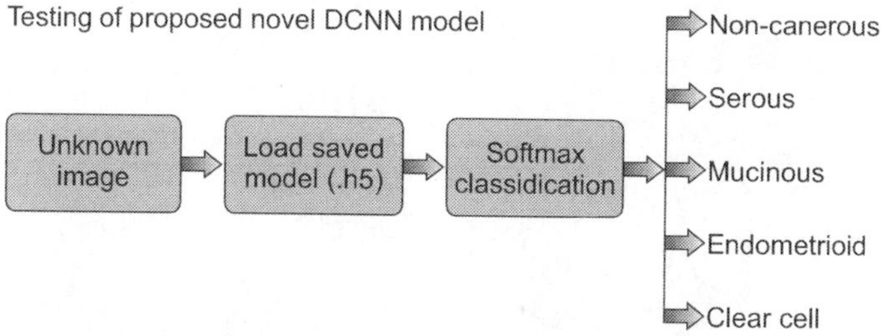

Figure 5: The P rediction and C lassification P rocess U sing an Improved VGG-16 M odel

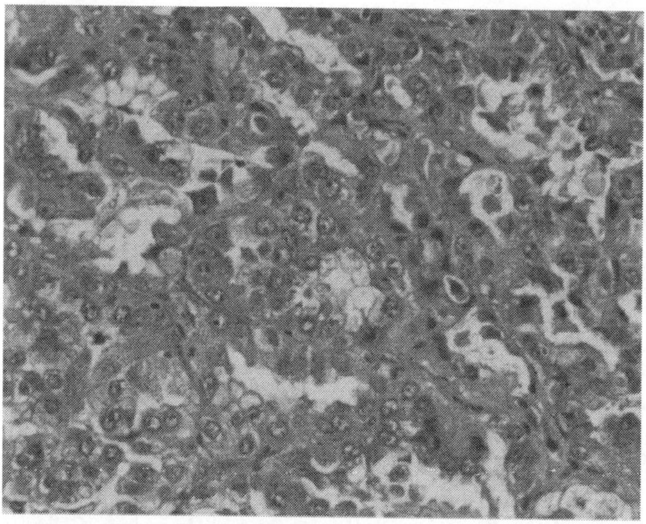

Figure 6: Representation of H&E Staining of Ovarian Clear Cell Carcinoma.

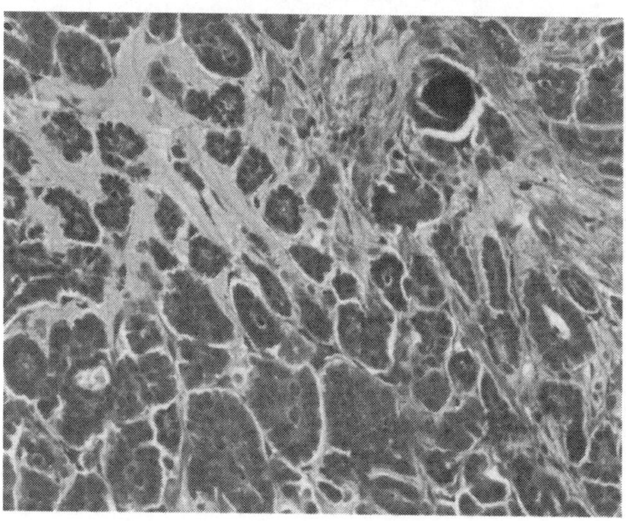

Figure 7: The Image on the Right Illustrates the Typical Pattern of H&E Staining That is Seen in Cases of Low-Grade Serous Ovarian Cancer.

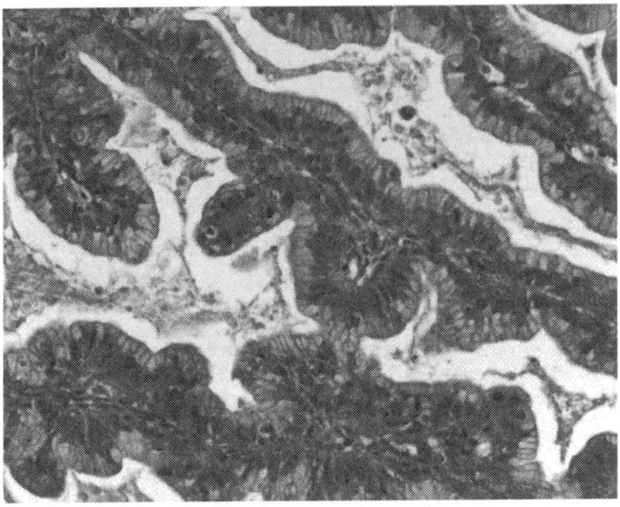

Figure 8: This is a Typical Example of H&E Staining Used to Diagnose Mucinous Ovarian Cancer.

Conclusion

To achieve the study's objectives, an ovarian cancer subtype classification based on improved VGG-16 was used. In the past, there has never been an attempt to categorize items using this novel technique. To help in the investigation, histopathological pictures were evaluated. When more images were provided, the accuracy of the categorization increased from a respectable 50% to an incredible 84.6 percent. When this research is published in a public forum and made available to the public, a pathologist will be able to properly categorise OC and its subtypes. In the next step of our research, pathologists will be able to submit images of their patients, and our researchers will be able to use these images to make diagnoses. The DCNN may be fed a variety of different pre-trained networks. This allows the DCNN's basic architecture to be fine-tuned. Certain networks are better fitted than others to handle the traffic that they encounter. A closer examination of the accuracy of various techniques will indicate how accurate they are. As a result, we'll be able to examine the accuracy of our forecasts and classifications in more depth. The two-step screening technique has achieved appropriate specificity and a distinct stage shift in early-stage ovarian cancer but not with acceptable sensitivity. This is because the procedure has two phases. This is the case when new and cutting-edge technological breakthroughs are required to give suitable sensitivity. More research is needed in both the imaging area and the field of serum biomarkers.

DATASET

The dataset will be available on request.

CONFLICTS OF INTEREST

No potential conflict of interest was reported by the authors.

FUNDING STATEMENT

There is no funding for this work.

References

1. Labidi-Galy, S. I., Treilleux, I., Goddard-Leon, S., et al. 2012. Plasmacytoid dendritic cells infiltrating ovarian cancer are associated with poor prognosis. *Oncoimmunology*, *1*(3): 380–382.
2. Tanaka, Y. O., et al. 2016. Differentiation of epithelial ovarian cancer subtypes by use of imaging and clinical data: a detailed analysis. *Cancer Imaging*, 16: 3.
3. Nair, R., and Bhagat, A. 2021. An Introduction to Clustering Algorithms in Big Data", Encyclopedia of Information Science and Technology, Fifth Edition, pp. 559–576, 2021. Available: 10.4018/978-1-7998-3479-3.ch040 [Accessed 14 June 2022].
4. Chang, C-C., Chen, H-H., Chang, Y-C., et al. 2017. Computer-aided diagnosis of liver tumors on computed tomography images. *Computer Methods and Programs in Biomedicine*, 145: 45–51.
5. Ahmad, I., Serbaya, S. H., Rizwan, A., and Mehmood, M. S. 2021. spectroscopic analysis for harnessing the quality and potential of gemstones for small and medium-sized enterprises (SMEs). Journal of Spectroscopy.
6. Almarzouki, H. Z., Alsulami, H., Rizwan, A., Basingab, M. S., Bukhari, H. and Shabaz, M. 2021. An Internet of Medical Things-Based Model for Real-Time Monitoring and Averting Stroke Sensors. Journal of Healthcare Engineering.
7. Alnuaim, A. A., Zakariah, M., Shukla, P. K., Alhadlaq, A., Hatamleh, W. A., Tarazi, H., Sureshbabu, R., and Ratna, R. 2022. Human-Computer Interaction for Recognizing Speech Emotions Using Multilayer Perceptron Classifier. Journal of Healthcare Engineering, vol. 2022, Article ID 6005446, 12 pages. https://doi.org/10.1155/2022/6005446.
8. Sathya, M., Jeyaselvi, M., Krishnasamy, L., Hazzazi, M. M., Shukla, P. K., Shukla, P. K., Nuagah, S.J. 2021. A Novel, Efficient, and Secure Anomaly Detection Technique Using DWU-ODBN for IoT-Enabled Multimedia Communication Systems. Wireless Communications and Mobile Computing, vol. 2021, Article ID 4989410, 12 pages. https://doi.org/10.1155/2021/4989410
9. Alnuaim, A. A., Zakariah, M., Alhadlaq, A., Shashidhar, C., Hatamleh, W. A., Tarazi, H., Shukla, P.K., and Ratna, R. 2022. Human-Computer Interaction with Detection of Speaker Emotions Using Convolution Neural Networks. Computational Intelligence and Neuroscience, vol. 2022, Article ID 7463091, 16 pages. https://doi.org/10.1155/2022/7463091
10. Alnuaim, A. A., Zakariah, M., Shashidhar, C., Hatamleh, W. A., Tarazi, H., Shukla, P. K., Ratna, R. 2022. Speaker Gender Recognition Based on Deep Neural Networks and ResNet50. Wireless Communications and Mobile Computing, vol. Article ID 4444388, 13 pages. https://doi.org/10.1155/2022/4444388
11. Nair, R., and Bhagat, A. 2018. An Application of Big Data Analytics in Road Transportation. Advances in Systems Analysis, Software Engineering, and High Performance Computing, pp. 39–54. Available: 10.4018/978-1-5225-3870-7.ch003 [Accessed 14 June 2022].
12. Nair, R. et al. 2022. Blockchain-Based Decentralized Cloud Solutions for Data Transfer. *Computational Intelligence and Neuroscience, vol. 2022*, pp. 1–12, 2022. Available: 10.1155/2022/8209854 [Accessed 14 June 2022].
13. Xu, Y., Jia, Z., Wang, L-B., et al. 2017. Large scale tissue histopathology image classification, segmentation, and visualization via deep convolutional activation features. BMC *Bioinformatics*, Vol. 18(1)
14. Teramoto, A., Tsukamoto, T., Kiriyama, Y., et al. 2017. Automated Classification of Lung Cancer Types from Cytological Images Using Deep Convolutional Neural Networks. *Biomed. Res. Int.*
15. Gao, Z., Wang, Zhou, L. et al. 2017. HEp-2 Cell Image Classification With Deep Convolutional Neural Networks [J]. *IEEE Journal of Biomedical and Health Informatics, 21*(2), 416–428.
16. The Cancer Genome Atlas Ovarian Cancer Collection (TCGA-OV). 2022. The Cancer Imaging Archive (TCIA) Public Access - Cancer Imaging Archive Wiki", Wiki.cancerimagingarchive.net. [Online]. Available: https://wiki.cancerimagingarchive.net/pages/viewpage.action?pageId=7569497. [Accessed: 03- Aug- 2022].

17. Kashyap, R. 2022. Breast Cancer Histopathological Image Classification Using Stochastic Dilated Residual Ghost Model. *International Journal of Information Retrieval Research, 12*(1), 1–24. Available: 10.4018/ijirr.289655 [Accessed 3 August 2022].
18. Pang, S., Yu, Z., and Orgun, M. A. 2017. A novel end-to-end classifier using domain transferred deep convolutional neural networks for biomedical images. *Computer Methods and Programs in Biomedicine, 201*, 283–293.
19. Kashyap, R. 2021. Evolution of histopathological breast cancer images classification using stochasticdilated residual ghost model. *Turkish Journal of Electrical Engineering and Computer Sciences, 29*(8), Article 12. https://doi.org/10.3906/elk-2104-40
20. Sharma H. et al. 2017. Deep convolutional neural networks for automatic classification of gastric carcinoma using whole slide images in digital histopathology. *Computer Med. Imaging Graph.*
21. Kashyap, R. 2022. Dilated residual grooming kernel model for breast cancer detection. *Pattern Recognition Letters, 159*, 157–164, 2022. Available: 10.1016/j.patrec.2022.04.037 [Accessed 17 July 2022].
22. Kashyap, R. 2021. Machine Learning for Internet of Things. *Research Anthology on Artificial Intelligence Applications in Security*, pp. 976–1002. Available: 10.4018/978-1-7998-7705-9.ch046 [Accessed 18 April 2022].
23. Krizhevsky, A., Sutskever, I., and Hinton, G. E. 2012. Image Net classification with deep convolutional neural network. In: Advances in Neural Information Processing Systems, NIPS, vol. 25, 2012, pp. 1106–1114.
24. Hinton, G. E. et al. 2012. Improving neural networks by preventing coadaption of feature detectors. ar Xiv, 1207.0580.
25. Nair, R., Gupta, S., Soni, M., Kumar Shukla, P., and Dhiman, G. 2020. An approach to minimize the energy consumption during blockchain transaction. Materials Today: Proceedings, 2020. Available: 10.1016/j.matpr.2020.10.361 [Accessed 14 June 2022].
26. Schwartz, D. et al. 2022. Ovarian cancer detection using optical coherence tomography and convolutional neural networks. *Neural Computing and Applications, 34*(11), 8977–8987.
27. Ahamad, M. M. et al. 2022. Early-stage detection of ovarian cancer based on clinical data using machine learning approaches. *Journal of Personalized Medicine, 12*(8), 1211.

CHAPTER 7

Role of AI in Supply Chain Management and Bio-Informatics Sector

Rishabh Didwania,[1,*] *Sanat Kumar Mishra,*[1] *Satvik Shukla,*[1] *Rajat Verma*[1] and *Namrata Dhanda*[2]

1. Introduction to Artificial Intelligence

AI stands at the forefront of technological advancements, starting with a fresh period of innovation and change in many different industries. At its core, AI is the study of developing computer systems that can do tasks that typically require the intelligence of humans. Development, reasoning, problem-solving, language perception, and interpretation are some of these activities. As AI continues to evolve, its applications are reshaping the landscape of various sectors, from healthcare to finance, manufacturing to transportation, opening many possibilities and efficiencies that have never happened before.

1.1 Evolution of Artificial Intelligence

Over the decades, AI has grown [1] via phases of knowledge representation, ML, and deep learning, catalyzing breakthroughs in NLP, computer vision, and reinforcement learning.

1.1.1 Historical Perspective

The history of AI began in prehistoric cultures, where humanoid entities capable of imitating human intellect were frequently depicted in myths and stories. However, the middle of the 20th century might be considered the official beginning of AI as

[1] Department of Computer Science & Engineering, Pranveer Singh Institute of Technology, Kanpur, Uttar Pradesh, India.
[2] Department of Computer Science and Engineering, Amity University Uttar Pradesh, Lucknow, India.
* Corresponding author

a scientific field. The foundation for machine intelligence was built in the 1950s by great people such as Alan Turing, who proposed the Turing Test to assess a machine's capacity for intelligence comparable to that of a person.

In the decades that followed, there were notable advancements, but there were also times when funding and interest were lower for AI as well as hopeful periods. The creation of expert systems in the 1970s and neural networks in the 1980s are two notable innovations. However, the 21st century saw an increase in the capabilities of AI, driven by developments in large data, computer power, and algorithmic sophistication.

1.1.2 Recent Advancements in Artificial Intelligence

In recent years, AI has advanced remarkably due to a combination of factors. Deep learning, a kind of machine learning, has made it possible for computers to analyze and learn from large datasets with previously unheard-of precision. Advances in NLP have made it possible for machines to understand and produce content that is similar to that of humans, transforming fields like language translation and chatbots.

The boundaries of what is possible have been further widened by the mixture of AI with other cutting-edge technologies, like quantum computing. AI is pervasive in modern life, powering recommendation systems, virtual assistants, and even driverless cars. It is no longer limited to research labs. Businesses of all sizes may now take advantage of AI capabilities for innovation and competitive advantage thanks to the democratization of tools and frameworks.

1.2 Impact of Artificial Intelligence Across Sectors

AI's impact transcends sectors, driving innovation and efficiency. AI is transforming processes, improving decision-making with data-driven insights, and causing revolutionary changes in how businesses work, ultimately changing the face of several industries, from healthcare to banking and manufacturing.

- **Healthcare:** In healthcare, AI [2] is driving monumental changes, from diagnostic tools and personalized medicine to predictive analytics for patient outcomes. Medical data is analyzed by ML algorithms to find trends and offer information that might guide treatment choices.
- **Manufacturing:** The manufacturing sector is undergoing a paradigm shift with the infusion of AI robotics and automation powered by AI. This is optimizing production processes, enhancing efficiency, and ensuring quality control. Predictive maintenance algorithms minimize downtime by anticipating equipment failures before they occur. AI-driven SCM is optimizing inventory levels and streamlining logistics.
- **Finance:** In finance, AI is revolutionizing how institutions analyze data, detect fraud, and manage risk. Algorithmic trading, robot-advisors, and fraud detection systems are some examples of AI applications in the financial sector. The industry's future is being shaped by AI's capacity to evaluate enormous capacities of financial data in the present, which facilitates quicker and better-informed execution.

- **Transportation:** The transportation sector is witnessing a transformative impact with the integration of A.I. Autonomous vehicles, guided by machine learning algorithms, and are redefining the future of mobility. Systems for managing traffic that are AI-driven maximize flow and minimize congestion. Predictive maintenance ensures the reliability of transportation infrastructure.

2. Artificial Intelligence in Supply Chain Management (SCM)

SCM is an intricate web of linked actions that starts with product creation and ends with the delivery of items to final customers. The integration of AI in SCM [3] is transforming traditional practices, offering unprecedented capabilities to enhance efficiency, transparency, and innovation throughout the supply chain.

2.1 Overview and Evolution of Artificial Intelligence in SCM

Moving in a direction to optimize different components of the supply chain, A I in SCM applies ML techniques, data analytics, and automation. Massive volumes of data are processed by A I systems, which then use the information to forecast outcomes, automate repetitive operations, and offer insightful information that helps decision-making within organizations.

The integration of AI in SCM started with basic automation and data analysis, gradually progressing to predictive analytics and optimization algorithms to enhance efficiency and decision-making in the logistics chain. Advanced technologies like ML, real-time data analytics, and AI-driven predicting models have further transformed SCM, enabling adaptive and responsive supply chains proficient in addressing dynamic market demands and uncertainties.

2.1.1 Historical Context

The historical evolution of AI in SCM can be taken back to the efforts put in the 1960s and 1970s when computer systems were first utilized to manage inventory and optimize logistics. However, limitations in computing power and data availability constrained the full potential of AI applications in SCM during this period.

2.1.2 Key Milestones

The landscape began to change in the late 20th century with the advent of more sophisticated algorithms and advancements in computing capabilities.

- **1990s:** Early adoption of expert systems for automation
- **2000s:** Rise of predictive analytics and demand forecasting
- **2010s:** Development of ML and deep learning algorithms
- **2020s:** Widespread acceptance of AI across the logistic network

The 1990s saw the emergence of optimization algorithms that could handle larger datasets, making the pathway for more effective supply or logistics chain planning. The 21st century witnessed the addition of ML and big data analytics, marking a transformative era in which A I became a cornerstone of SCM.

2.2 Revolutionizing Efficiency

By automating repetitive processes, maximizing resource allocation, and delivering real-time insights, utilizing AI in SCM has completely transformed efficiency. AI-driven predictive analytics and smart decision-making tools have streamlined operations, minimized disruptions, and elevated overall supply chain performance.

2.2.1 Predictive Analytics for Demand Forecasting

In the dynamic and ever-evolving world of SCM, accurate demand forecasting [4] is paramount for ensuring efficient operations, optimizing inventory levels, and meeting customer expectations. Conventional forecasting techniques, which frequently rely on past data and basic models of statistics, have struggled to keep pace with the increasingly complex and unpredictable nature of demand patterns.

AI is transforming SCM by introducing powerful predictive analytics tools that can examine enormous volumes of data from many sources to find trends, patterns, and anomalies, enabling more accurate and reliable demand forecasts.

2.2.1.1 Challenges in Traditional Demand Forecasting

Traditional demand forecasting [5] methods face several challenges:

- **Limited Data Sources:** Traditional methods often rely solely on historical sales data, which may not capture the full range of factors influencing demand, such as market trends, competitor activities, and external events.
- **Static Models:** Traditional models often make assumptions about demand patterns that may not hold true in real-world scenarios, leading to inaccurate forecasts.
- **Inflexibility:** Traditional models are often inflexible and unable to adjust to shifting customer behavior and market conditions.

2.2.1.2 AI–Powered Predictive Analytics for Demand Forecasting

AI-powered predictive analytics addresses the limitations of traditional methods by providing:

- **Data Integration:** Moving in a direction to provide a more complete picture of the factors influencing demand, AI algorithms can incorporate data from a variety of sources, such as previous sales data, point-of-sale data, social media sentiment, weather patterns, economic indicators, and competitor actions.
- **Dynamic Modeling:** AI systems have the capacity to learn from past data, adjust to shifting market conditions, and eventually produce more accurate projections.
- **Real-time Insights:** Real-time data analysis using AI algorithms allows companies to spot changes in demand and quickly adapt their supply networks.

2.2.1.3 Real–World Applications of AI in Demand Forecasting

AI is already being used by leading companies to revolutionize demand forecasting:

- **Walmart:** Walmart [6] uses AI to analyze customer purchase history, weather patterns, and social media trends to forecast demand for specific products at individual stores, optimizing inventory levels and reducing stockouts.
- **Amazon:** Amazon [7] uses AI to forecast demand for its vast array of products, considering factors such as product popularity, seasonal trends, and competitor offerings, ensuring efficient fulfillment and timely deliveries.
- **Coca-Cola:** Coca-Cola uses AI to forecast demand for its beverages across different regions and sales channels, including elements like weather patterns, holidays, and promotional campaigns, ensuring optimal production and distribution.

2.2.1.4 Impact o n SCM

The adoption of AI in demand forecasting is expected to have a visible impact on SCM:

- **Reduced Inventory Costs:** Industries may optimize inventory levels and cut expenditures associated with overstocking and stockouts by utilizing AI-powered demand forecasts.
- **Improved Customer Satisfaction:** Businesses can more efficiently meet consumer demand by using accurate demand projections, which can result in shorter lead times, better product availability, and more customer satisfaction.
- **Enhanced Logistic Chain Resilience:** Demand forecasting driven by AI can assist companies in anticipating and adapting to changes in demand, strengthening logistic networks' resistance to interruptions and guaranteeing business continuity.

Demand forecasting in SCM is being revolutionized by AI-powered predictive analytics, which gives companies the information they need to improve customer happiness, cut costs, and streamline processes.

2.2.2 AI—Powered Inventory Optimization

In the dynamic and ever-evolving world of SCM, inventory optimization [8] is a critical balancing act. Businesses must maintain enough inventory levels to meet the demand of the customers without incurring the high costs associated with overstocking. Traditional inventory management methods, often relying on manual calculations and static models, have struggled to keep pace with the increasingly volatile nature of supply and demand patterns.

AI is transforming SCM by introducing powerful inventory optimization [9] tools that can analyze vast amounts of data from diverse sources to identify patterns, trends, and anomalies. This data-driven thinking helps businesses to improve inventory levels with greater precision and efficiency, reducing costs, improving customer satisfaction, and enhancing supply chain resilience.

2.2.2.1 Challenges in Traditional Inventory Optimization

Traditional inventory optimization methods face several challenges:

- **Data Silos:** Traditional methods often rely on getting data via a limited number of channels, such as sales books and warehouse inventory levels, failing to capture the full picture of demand and supply dynamics.
- **Static Models:** Traditional models often use static assumptions about inventory levels, failing to consider factors such as seasonality, promotional events, and supply chain disruptions.
- **Manual Processes:** Traditional methods often involve manual calculations and decision-making, making it difficult to keep up with the real-time changes that occur in supply chains.

2.2.2.2 Revolutionizing Inventory Optimization

AI-powered inventory optimization [10] addresses the limitations of traditional methods by providing:

- **Data Collection:** AI algorithms can take data from multiple channels, including sales book, point-of-sale data, customer feedback, weather patterns, and competitor activities, providing a more comprehensive view of inventory needs.
- **Dynamic Modeling:** AI algorithms can absorb past data and adapt to varying market situations, providing more accurate inventory forecasts and optimization strategies over time.
- **Current Insights:** AI algorithms can examine data in present, enabling businesses to identify demand shifts and supply chain disruptions, making immediate adjustments to inventory levels.

2.2.2.3 Real–World Applications of AI in Inventory Optimization

AI is already being used by leading companies to revolutionize inventory optimization:

- **Adidas:** Adidas uses AI to optimize inventory levels [11] for its vast range of apparel and footwear products, considering factors such as product popularity, seasonal trends, and regional sales patterns, ensuring efficient fulfillment and timely deliveries.
- **Dell:** Dell uses AI to forecast demand for its computer components and predict potential supply chain disruptions, enabling it to optimize inventory levels and avoid stockouts, ensuring timely order fulfillment for its customers.
- **Nestlé:** Nestlé uses AI to optimize inventory levels for its diverse range of food and beverage products, considering factors such as shelf life, promotional activities, and consumer preferences, reducing waste and ensuring product availability across its retail channels.

SCM is being revolutionized by AI-powered inventory optimization, which gives companies the instruments they need to strike the difficult balance between customer happiness and cost-effectiveness.

2.3 Transparency and Traceability

Transparency and traceability are 2 things in AI systems that are increasingly vital. As AI algorithms influence decision-making in critical domains like finance and healthcare, the demand for understandable, accountable, and traceable AI models has grown. Efforts to implement ethical AI practices aim to enhance trust, allowing stakeholders to comprehend, scrutinize, and trace the decision-making processes of AI systems, thus promoting responsible deployment across various sectors.

2.3.1 Blockchain in Supply Chain Visibility

In today's interconnected world, logistics chains have become increasingly multifaceted and global, spanning multiple countries, diverse industries, and numerous intermediaries. Maintaining transparency and traceability throughout these complex networks is crucial for ensuring product authenticity, preventing counterfeiting, and safeguarding consumer trust.

Blockchain technology has emerged as a revolutionary tool for enhancing supply chain visibility [12] by providing a secure and tamper-proof digital ledger that records every step of a product's journey from origin to consumer. This distributed ledger technology empowers businesses to keep an eye on the products in present time, recognize issues, and resolve disputes efficiently.

2.3.1.1 Challenges in Traditional Supply Chain Visibility

Traditional methods of supply chain visibility often face significant challenges:

- **Limited Data Sharing:** Traditional systems often rely on siloed data, making it difficult to share information seamlessly across different parties involved in the supply chain.
- **Data Integrity:** Traditional systems may be prone to data manipulation and inaccuracies, leading to discrepancies in tracking and tracing product movements.
- **Lack of Transparency:** Traditional systems often lack transparency, making it difficult for businesses to identify and address potential issues such as counterfeiting, diversion, or unethical sourcing.

2.3.1.2 Blockchain Revolutionizing Supply Chain Visibility

Blockchain addresses [13] the limitations of traditional methods by providing:

- **Secure and Decentralized Data Sharing:** Blockchain's distributed ledger guarantees that all parties involved in the supply chain have access to a single, shared source of truth, eliminating data silos and inaccuracies.
- **Tamper-Proof Data Integrity:** Blockchain's cryptographic algorithms ensure that data cannot be altered or manipulated, providing an immutable element of that product's history.
- **Enhanced Transparency:** The transparent nature allows businesses to find products in the present time, identify issues, and resolve disputes efficiently.

2.3.1.3 Real—World Applications of Blockchain in Supply Chain Visibility

Blockchain is already being used by various industries to enhance supply chain visibility:

- **Pharmaceuticals:** Blockchain is being used to track the movement of pharmaceuticals [14] from manufacturing to distribution, ensuring product authenticity and preventing counterfeiting.
- **Food and Agriculture:** Blockchain is being used to track the origin and movement of food products [15], providing consumers with transparency into the food supply chain and ensuring food safety.
- **Luxury Goods:** Blockchain is being used to authenticate luxury goods, preventing counterfeiting and protecting brand reputation.

2.3.1.4 Impact of Blockchain on Supply Chain Management

The adoption of blockchain in supply chain visibility [16] is expected to have a significant impact on SCM:

- **Improved Product Authenticity:** Blockchain can help businesses prevent counterfeiting and ensure the authenticity of products, enhancing consumer trust and safeguarding brand reputation.
- **Enhanced Traceability:** Blockchain technology can enable real-time product traceability across the supply chain, allowing companies to quickly detect possible problems and take appropriate action.
- **Reduced Costs:** Blockchain can decrease the costs linked with product recalls, counterfeiting, and supply chain disruptions, leading to improved profitability and efficiency.

Blockchain technology is transforming supply chain visibility by giving companies a safe, transparent, and unshakable platform to follow and trace goods through their supply chain.

2.3.2 AI–Driven Data Analysis for Transparency in Supply Chain Management

Transparency in SCM has become increasingly crucial for industries to maintain consumer trust, ensure product authenticity, and uphold ethical sourcing practices. Traditional methods of gathering and analyzing supply chain data often face boundaries in terms of correctness, timeliness, and completeness, hindering the ability to achieve true transparency.

AI is transforming SCM by providing powerful data analysis tools that may find patterns, trends, and anomalies by analyzing huge sizes of data from a diversity of sources, such as supplier records, social media, and Internet of Things sensors. Businesses may discover possible problems in their supply chains, have a greater understanding of them, and make decisions that advance ethical behavior and transparency thanks to this data-driven approach.

2.3.2.1 Challenges in Traditional Supply Chain Transparency

Traditional approaches to supply chain transparency often face significant challenges:

- **Data Silos:** Traditional systems often rely on siloed data, making it difficult to aggregate and analyze information from multiple sources, leading to a fragmented view of the supply chain.

- **Limited Data Collection:** Conventional techniques could miss important information on labor practices at supplier facilities, social media sentiment, and real-time product tracking, among other things that are pertinent to supply chain transparency.
- **Inaccurate Data:** Traditional systems may be prone to data errors and inconsistencies, making it difficult to trust the information used for transparency initiatives.

2.3.2.2 AI Revolutionizing Supply Chain Transparency

AI addresses the limitations of traditional methods by providing:

- **Integrated Data Analysis:** AI algorithms can combine information from many sources, including external databases, real-time Internet of Things sensors, and internal systems, to provide a complete representation of the supply chain.
- **Advanced Analytics:** Massive data sets may be examined by AI to find minute patterns, trends, and anomalies that can point to possible problems or moral dilemmas.
- **Real-time Insights:** Because AI can analyze data in real-time, firms can proactively monitor their supply chains and spot any issues before they become more serious.

2.3.2.3 Real–World Applications of AI in Supply Chain Transparency

AI is already being used by leading companies to enhance supply chain transparency [17]:

- **Maersk:** Maersk, a global shipping and logistics company, uses AI to track shipments in real-time, providing customers with visibility into the location and status of their goods, enhancing transparency and trust.
- **Nestlé:** The multinational food and beverage giant Nestlé uses AI to assess supplier data and social media sentiment to spot any ethical issues in its supply chain. This allows the corporation to address the issues and maintain its commitment to ethical sourcing.
- **H&M:** The international fashion retailer H&M tracks the flow of raw materials and completed goods through its supply chain using AI to guarantee product authenticity and stop piracy.

2.3.2.4 Impact in SCM

The adoption of AI in supply chain transparency is expected to have a significant impact on SCM:

- **Enhanced Consumer Trust:** AI can help businesses demonstrate transparency and ethical practices throughout their supply chains, fostering consumer trust and enhancing brand reputation.
- **Improved Risk Management:** AI can recognize possible problems and hazards in supply chains, allowing companies to take preventative action to reduce risks and safeguard their brand.

- **Sustainable Sourcing:** AI may assist companies in locating and sourcing goods from moral vendors, encouraging sustainable practices and lessening the impact of their supply chains on the environment and society.

AI-driven data analysis is revolutionizing supply chain transparency, providing businesses with powerful tools to gain deeper insights into their supply chains, identify potential issues, and make informed decisions that promote ethical practices and consumer trust.

2.4 Automation and Innovation

In the dynamic and demanding world of SCM, effectiveness and automation have developed paramount practices for businesses to stay competitive and meet customer expectations. Traditional automation methods, often relying on rule-based systems and manual interventions, have struggled to keep pace with the increasingly complex and data-driven nature of supply chain operations.

2.4.1 Robotic Process Automation in SCM

RPA [18] is transforming SCM by introducing intelligent software robots that can copycat human activities to automate repetitive, rules-based tasks. A type of business process automation known as robotic process automation makes use of artificial intelligence agents or software robots. It is sometimes referred to as software robotics. These virtual assistants can work tirelessly and error-free, handling a wide range of tasks across the logistic network, from order processing and inventory management to data entry and reporting.

2.4.1.1 Challenges in Traditional Supply Chain Automation

Traditional supply chain automation methods face several challenges:

- **Limited Flexibility:** Traditional systems often lack the flexibility to adapt to changing processes and new requirements, making it difficult to automate complex and evolving workflows.
- **High Implementation Costs:** Traditional automation solutions often require significant upfront investments in software and hardware, making them less accessible to smaller businesses.
- **Limited Scalability:** Traditional systems may not scale effectively to accommodate increasing transaction volumes or expanding business operations.

2.4.1.2 RPA Revolutionizing Supply Chain Automation

RPA [19] addresses the limitations of traditional methods by providing:

- **Unmatched Flexibility:** RPA robots can learn new processes, adapt to changing requirements, and handle exceptions without requiring extensive reprogramming.
- **Lean Implementation:** Rapid and affordable implementation of RPA solutions is possible, frequently utilizing current IT infrastructure and obviating the need for substantial alterations to key systems.

- **Unparalleled Scalability:** RPA robots are easily scalable, and able to accommodate growing workloads and expanding corporate processes.

2.4.1.3 Real – World Applications of RPA in Supply Chain Automation

RPA is already being used by leading companies to revolutionize supply chain automation:

- **Amazon:** Amazon uses RPA robots to automate a wide range of tasks in its vast logistic network, including order processing, inventory management, and customer service interactions.
- **Walmart:** Walmart uses RPA robots to automate tasks such as invoice processing, supplier onboarding, and price monitoring, streamlining operations and reducing costs.
- **DHL:** DHL, a global logistics company, uses RPA robots to automate tasks such as shipment tracking, customs clearance, and documentation processing, improving efficiency and reducing errors.

2.4.1.4 Impact of RPA on SCM

The adoption of RPA in supply chain automation is anticipated to have a significant influence on SCM:

- **Reduced Costs:** RPA can automate labor-intensive tasks, leading to significant cost savings in labor and operational expenses.
- **Increased Efficiency:** RPA can streamline processes, reduce errors, and shorten cycle times, improving overall supply chain efficiency.
- **Enhanced Productivity:** By freeing up human workers to concentrate on more important and valuable duties, RPA can increase overall productivity.

RPA is revolutionizing supply chain automation, providing businesses with the tools they need to achieve unprecedented levels of efficiency, accuracy, and cost-effectiveness.

All things considered, RPA is an effective instrument that can revolutionize SCM by automating tedious operations, boosting productivity, and cutting expenses.

2.4.2 Autonomous Vehicles in Logistics

The transportation industry is on the cusp of a transformative revolution driven by the rapid advancement of autonomous vehicle (AV) [20] technology. Autonomous vehicles, or AVs for short, are automobiles that can navigate and function without the need for human interaction. AVs have enormous potential to transform logistics and SCM by improving efficiency, revolutionizing transportation, and redefining the future of goods movement.

2.4.2.1 Challenges in Traditional Logistics and Transportation

Traditional logistics and transportation methods face several challenges that AVs can address:

- **Driver Shortages:** There is a serious driver shortage in the trucking business, which makes it challenging to satisfy the increasing demand for the transportation of products. By removing the need for human drivers, AVs can help with this shortfall by guaranteeing steady and dependable employment.

- **Safety Concerns:** Truck-related traffic incidents account for a large number of injuries and fatalities. By enhancing vision, response times, and adherence to traffic laws, AVs fitted with cutting-edge sensors and AI can dramatically lower the number of accidents.
- **Inefficiency and Waste:** Traditional transportation methods often involve empty miles, detours, and inefficient route planning, leading to increased fuel consumption and wasted time. AVs can optimize routes, reduce empty miles, and adapt to real-time traffic conditions, improving overall efficiency.

2.4.2.2 AVs Revolutionizing Logistics and Transportation

AVs address the challenges of traditional logistics and transportation by offering several advantages:

- **24/7 Operations:** Due to their ability to work continuously without interruptions or rest periods, autonomous vehicles (AVs) can considerably increase vehicle utilization and improve delivery times.
- **Reduced Labor Costs:** Because AVs replace human drivers, labor expenses for wages, benefits, and training are decreased.
- **Enhanced Safety:** AVs equipped with advanced sensors and AI can operate with greater safety, reducing accidents and improving overall road safety.
- **Optimized Operations:** AVs can optimize routes, reduce empty miles, and adapt to real-time traffic conditions, improving efficiency and reducing fuel consumption.

2.4.2.3 Real—World Applications of AVs in Logistics

AVs are already being tested and deployed in various logistics applications:

- **Autonomous Trucks:** Businesses like Embark and TuSimple are creating autonomous trucks for long-distance travel to revolutionize freight transportation and lower dependency on human drivers.
- **Last-Mile Delivery:** Autonomous delivery cars for last-mile delivery are being developed by businesses like Nuro and Waymo, offering an effective and affordable alternative for deliveries in cities.
- **Warehouse Automation:** Automating processes like order fulfilment, inventory control, and material handling in warehouses with AVs boosts productivity and lowers labor expenses.

2.4.2.4 Impact of AVs on Logistics and Supply Chains

The widespread adoption of AVs in logistics is expected to have a profound impact [21] on the industry:

- **Reduced Transportation Costs:** AVs can significantly reduce transportation costs by eliminating driver salaries, reducing accidents, and optimizing routes.

- **Improved Delivery Times:** Deliveries may be made faster and more reliably because to AVs' 24/7 operation and ability to adjust to changing traffic circumstances.
- **Enhanced Logistic Network Resilience:** AVs can reduce reliance on human drivers, mitigating the impact of labor shortages and disruptions caused by driver fatigue or accidents.

AVs are poised to revolutionize logistics and SCM bringing about a paradigm shift in the transportation business. By automating tasks [22], enhancing efficiency, and reducing costs, AVs have the potential to transform the way goods are moved, making supply chains more resilient, efficient, and sustainable.

2.5 Case Studies

Within the intricate tapestry of global supply chains, the intertwining narratives of Unilever and DHL present a compelling case study ripe for exploration.

2.5.1 Unilever: Optimizing Inventory Management with AI

Unilever has experienced cost savings, better customer happiness, and a competitive advantage in the rapidly evolving consumer products industry by streamlining its inventory management procedures and utilizing AI's predictive analytics capabilities.

2.5.1.1 The Challenge: Balancing Inventory Levels

Inventory management has always been a delicate harmonizing act for dealings like Unilever. Maintaining sufficient stock levels is crucial to meet client demand and avoid stockouts, but overstocking leads to excess storage costs, product obsolescence, and financial strain.

Traditionally, inventory management relied on historical sales data, manual calculations, and human intuition, often leading to inaccurate forecasts and suboptimal inventory decisions. Unilever recognized the need for a more data-driven and predictive approach to inventory management.

2.5.1.2 AI–Powered Inventory Optimization

Unilever embraced AI as a game-changer in inventory management. o more accurately estimate future demand, the firm implemented AI-powered systems that could analyz e massive volumes of data, including past sales patterns, seasonal trends, weather predictions, and even social media sentiment.

Unilever can estimate demand more accurately because these AI algorithms can spot small trends and connections that older approaches would overlook. Unilever can maximize its inventory levels and guarantee that the correct items are available at the right time and in the right amounts by having a better understanding of future demand.

2.5.1.3 The Benefits of AI–Powered Inventory Management

The influence of AI-powered inventory management has been significant for Unilever. The company has achieved:

- **Reduced inventory costs:** By accurately predicting demand, Unilever has been able to reduce its inventory levels by up to 15%, important to substantial cost savings.
- **Better client satisfaction:** By avoiding stockouts and ensuring product availability, Unilever has enhanced customer satisfaction and loyalty.
- **Enhanced supply chain agility:** With real-time insights into demand trends, Unilever can react quickly to changes in market conditions, adapting its inventory strategies accordingly.
- **Reduced waste and obsolescence:** Accurate demand forecasting minimizes the risk of excess inventory and product obsolescence, reducing waste and associated costs.

2.5.2 DHL: Revolutionizing Warehouse Operations with Artificial Intelligence

DHL, a worldwide front-runner in logistics and transportation, is in the lead of leveraging AI to automate warehouse operations. By integrating AI into its warehouse management systems, DHL is achieving unprecedented levels of efficiency, accuracy, and productivity.

2.5.2.1 Embracing AI for Warehouse Automation

DHL's journey with AI began with a clear vision: to transform its warehouses into intelligent, autonomous environments that could handle the ever-increasing volume and complexity of shipments. The company recognized that AI could automate repetitive tasks, reduce human error, and optimize the movement of goods within warehouses.

2.5.2.2 AI–Powered Picking and Packing

One of the biggest uses of AI in DHL's warehouses is in picking and packing operations. Traditional picking and packing processes are labor-intensive and prone to errors, often leading to delays and inaccuracies.

DHL has deployed AI-powered robots that can scan product barcodes, identify items, and pick them from shelves with precision. These robots can navigate warehouse aisles autonomously, following optimized routes and minimizing travel time. Once picked, items are transferred to AI-powered packing stations where they are sorted, packaged, and labeled for shipment.

2.5.2.3 Autonomous Vehicles for Internal Logistics

To manage the transportation of items within warehouses, DHL has also implemented AV, or AGVs. These AGVs employ AI algorithms to move pallets and packages across the warehouse, greatly decreasing the need for forklifts and physical labor. The AGVs can navigate the warehouse environment safely, avoiding obstacles and coordinating their movements with other vehicles and workers. They operate 24/7, ensuring a continuous flow of goods through the warehouse.

2.5.2.4 AI—Enabled Predictive Maintenance

AI is allowing predictive maintenance, which is revolutionizing warehouse operations in addition to automating chores. AI systems can examine sensor data from technology and equipment to anticipate any problems before they happen.

By proactively scheduling maintenance based on AI predictions, DHL can diminish downtime, prevent disruptions, and extend the lifespan of its warehouse equipment. This predictive approach optimizes resource allocation and ensures the smooth operation of warehouse operations.

2.5.2.5 The Impact of AI—Powered Warehouse Automation

DHL's adoption of AI has had a profound impact on its warehouse operations:

- **Increased efficiency:** AI automation has significantly reduced labor costs, improved picking accuracy, and optimized warehouse layouts, leading to a substantial increase in overall efficiency.
- **Enhanced productivity:** The combination of AI-powered picking robots and AGVs has dramatically increased productivity, allowing DHL to handle larger volumes of shipments with fewer resources.
- **Reduced errors:** AI automation has minimized human error, leading to fewer lost or damaged items, improved customer satisfaction, and reduced claims.
- **Scalability:** AI-powered systems can be easily scaled to accommodate changing demand patterns and fluctuations in order volumes, ensuring that DHL's warehouses can adapt to future growth.

3. Artificial Intelligence in Bio—Informatics

Bioinformatics, [23] a multidisciplinary field that combines biology, computer science, and information technology, has been significantly revolutionized by the integration of A I This section will cover the whole field of AI in bio-informatics, including its history, uses in genomics, healthcare, and biological research, as well as innovations and real-world examples that highlight its revolutionary potential.

3.1 Overview and Evolution of Artificial Intelligence in Bio—Informatics

The marriage of A I and Bio-Informatics [24] has unlocked unprecedented possibilities for understanding biological data, from genetic information to complex biochemical interactions. A.I. technologies, including machine learning and data analytics, play an important role in deciphering the complexities of biological systems. This overview explores the synergies between A.I. and Bio-Informatics and sets the stage for a detailed exploration of its applications and impact.

The evolution of AI in bioinformatics [25] has progressed from basic pattern recognition to sophisticated algorithms for genomics [46], proteomics, and metabolomics analysis. ML and deep learning models have become integral, accelerating the interpretation of biological data, enabling personalized medicine, and advancing the understanding of complex biological systems.

3.1.1 Historical Context

The historical evolution of A.I. in Bio-Informatics [26] is a testament to the symbiotic relationship between technological advancements and biological research. Early efforts in the 1970s focused on algorithmic approaches for sequence alignment and structural prediction. However, it was the advent of more powerful computational tools [27] in the late 20th century that paved the way for more sophisticated AI applications in genomics and bioinformatics.

3.1.2 Key Milestones

The invention of machine learning methods for protein structure prediction, the use of neural networks in genomics, and the incorporation of deep learning for the analysis of large-scale genomic datasets are significant turning points in the history of AI in bioinformatics. The decoding of the human genome in the early 2000s marked a watershed moment, providing an unprecedented amount of data for A I applications in understanding genetic information.

3.2 Applications in Healthcare

In the healthcare industry, A I has become a potent tool for illness detection and therapy optimization [29]. To find patterns suggestive of illnesses, machine learning algorithms examine large datasets, such as molecular data, medical imaging, and patient records. This feature improves diagnostic speed and accuracy, enabling medical practitioners to customize treatment regimens for each patient and make better judgements.

3.2.1 AI–Powered Disease Diagnosis and Treatment Optimization

The power of AI is causing a significant revolution in the medical industry. Massive capacities of medical data, such as patient histories, x-rays, and genetic data, may be examined by AI algorithms, which can then be used to find trends and provide very accurate predictions. This has made it possible to diagnose diseases and optimize treatments in novel ways [30], which has improved patient outcomes and transformed the healthcare industry [31].

3.2.1.1 Challenges in Traditional Diagnosis and Treatment

Traditional diagnosis and treatment methods often face significant challenges, including:

- **Accuracy and precision:** Accurately diagnosing diseases can be challenging, especially in their early stages when symptoms may be subtle or nonspecific. Additionally, traditional methods may not always identify the underlying cause of a disease, leading to suboptimal treatment decisions.
- **Variability in treatment outcomes:** Even among people who have the same illness, therapy results might differ greatly. It is challenging to ascertain the best course of action for every patient due to this heterogeneity, which can be linked to elements including genetic variations, lifestyle choices, and comorbidities.

- **Time-consuming and resource-intensive:** Conventional techniques of diagnosis and treatment can be labor-intensive and time-consuming, frequently including several tests, consultations, and procedures. Delays in diagnosis and treatment may result from this, and patient outcomes may suffer as a result.

3.2.1.2 AI Revolutionizing Disease Diagnosis and Treatment

AI is addressing the challenges of traditional diagnosis and treatment by providing:

- **Enhanced accuracy and precision:** AI systems can discern minute patterns and provide remarkably accurate predictions by analyzing enormous volumes of data, including genetic data, medical imaging, and patient records. This may result in better treatment choices and earlier, more precise diagnoses.
- **Personalized treatment optimization:** AI can examine patient data to find genetic predispositions, lifestyle choices, and co-occurring conditions. This allows for the creation of individualized treatment regimens that are specific to the requirements and circumstances of each patient.
- **Real-time insights and decision support:** AI can offer doctors timely and more effective access to real-time insights into patient data. Improved patient outcomes, prompt therapeutic interventions, and quicker diagnosis are possible results of this.

3.2.1.3 Real−World Examples of AI in Disease Diagnosis and Treatment

In several therapeutic contexts, AI is already revolutionizing the diagnosis and treatment of diseases:

- **Medical Imaging Analysis:** When it comes to classifying abnormalities in medical images, such as X-rays, CT, and MRI scans, AI algorithms outperform human professionals in terms of correctness and precision. This might lead to the early detection of diseases including cancer and heart disease.
- **AI-powered Retinal Imaging for Diabetic Retinopathy:** As a primary cause of blindness, diabetic retinopathy is being detected early with the assistance of AI algorithms analyzing retinal pictures[37]. This may result in early detection and treatment, therefore averting visual impairment.
- **AI-assisted Mammography for Breast Cancer Detection:** The accuracy of breast cancer diagnosis is being increased by using AI algorithms to help radiologists interpret mammograms [36]. This may result in early detection and treatment, raising the likelihood of survival.
- **AI-powered Brain Scan Analysis for Neurological Disorders:** According to [38], anomalies linked to neurological illnesses including Parkinson's and Alzheimer's disease can be found in brain scans by using AI algorithms. This may result in an early diagnosis and more effective treatment strategies.
- **Genetic Risk Assessment:** AI can discover genetic variations that may predispose people to specific diseases by analyzing genomic data. Strategies for preventive and customized risk assessment can be created using this data.
- **AI-driven Genetic Counseling for Hereditary Diseases:** To identify those at risk of hereditary disorders including sickle cell anemia and Huntington's

disease, genetic data is analyzed using AI algorithms. Decisions about family planning can be informed by this knowledge and customized genetic counselling can be provided.

- **AI-powered Pharmacogenomics for Personalized Medication:** Genetic data is analyzed by AI algorithms to forecast an individual's reaction to a particular medicine. Physicians can use this info to choose the safest and most effective drugs for each patient.
- **AI-enabled Cancer Risk Assessment:** To determine a person's risk of acquiring cancer, genetic and lifestyle data are analyzed by AI algorithms. Personalized screening and preventive plans may be created with this data.
- **Predictive Analytics for Treatment Response:** AI is capable of analyzing patient data to estimate a patient's response to a given therapy. Physicians can use this information to choose the best course of action for each patient.
- **AI-powered Treatment Response Prediction for Cancer Patients:** To forecast how a patient would react to a particular cancer therapy, AI systems are examining patient data, including genetic information and tumor features. Physicians can use this information to choose the best course of action for each patient.
- **AI-enabled Prediction of Antibiotic Resistance:** To forecast antibiotic resistance, bacterial genomic data is analyzed by AI algorithms, which help in the selection of potent antibiotics and stop the spread of resistant bacteria.
- **AI-powered Mental Health Treatment Planning:** For mental health issues like anxiety and depression, AI algorithms are being utilized to evaluate patient data and forecast the chance of a certain treatment's outcome. Clinicians can use this information to create personalized treatment programmers.

3.2.1.4 Impact o n Healthcare

Healthcare is anticipated to be significantly impacted by the use of AI in illness detection and treatment:

- **Better patient outcomes:** AI is anticipated to enhance patient outcomes through the provision of more precise diagnoses, customized treatment regimens, and instantaneous insights to facilitate more informed decision-making.
- **Reduced healthcare costs:** AI is expected to reduce healthcare costs by optimizing treatment decisions, minimizing unnecessary tests and procedures, and preventing adverse events.
- **Personalized medicine:** AI is helping to develop personalized medicine, which will allow doctors to customize a patient's course of care according to their specific genetic composition, medical background, and lifestyle choices.

With the promise of earlier detection, more precise diagnoses, and personalized treatment regimens, AI is transforming the diagnosis and treatment of disease.

3.2.2 Personalized Medicine with AI: A Paradigm Shift in Healthcare

AI is enabling the creation of tailored medicine, which customizes treatment programmers for distinct patients based on their distinct genetic composition, medical history, and lifestyle characteristics. This is causing a fundamental upheaval in the healthcare industry. The identification of illnesses, the efficacy of treatments, and patient outcomes are all set to be revolutionized by this paradigm shift in healthcare.

3.2.2.1 The Promise of Personalized Medicine

Treatment choices in traditional medicine are frequently made using a "one-size-fits-all" methodology, which bases decisions on typical patient responses. Nevertheless, this method is unable to take into consideration the wide range of genetic and biological variations across people, which might result in less-than-ideal treatment results and perhaps harmful side effects.

By utilizing AI's capacity to evaluate enormous volumes of patient information, such as genetic data, medical histories, environmental exposures, and lifestyle choices, personalized medicine overcomes this constraint. This data contains subtle patterns and connections that AI systems can find, allowing healthcare professionals to customize treatment plans based on the individual needs of each patient.

3.2.2.2 Benefits of AI–Powered Personalized Medicine

The acceptance of AI in personalized medicine offers a multitude of benefits:

- **Improved Diagnostic Accuracy:** AI can analyze genetic information, medical imaging, and patient records with previously unheard-of precision, enabling earlier and more precise diagnoses. Improved treatment results and prompt action may depend on this early diagnosis.
- **Optimizing Treatment Selection:** AI can forecast a patient's response to a given treatment, which helps doctors choose the safest and most efficient drugs for each patient. With this individualized approach, side effects can be reduced and treatment effectiveness can be increased.
- **Customized Preventive Measures:** AI can recognize genetic susceptibilities, lifestyle choices, and environmental exposures that may augment a person's likelihood of contracting specific illnesses. By creating individualized preventive plans using this data, the total burden of disease can be decreased.
- **Empowering Patients:** Personalized medicine provides patients with a deeper understanding of their health and treatment options, fostering informed decision-making and promoting patient engagement in their care.

3.2.2.3 Real–World Examples in Personalized Medicine

AI is already making a significant impact on personalized medicine in various clinical settings:

- **Cancer Treatment:** To find genetic mutations and molecular traits that might inform individualized cancer treatment regimens, AI is being used to evaluate genomic data from tumors. This precision oncology approach has led to significant improvements in survival rates for certain types of cancer [32].

- **Cardiovascular Disease Management:** AI is being used to analyze patient info, including genetic info, medical history, and lifestyle factors, to predict a person's risk of emerging cardiovascular diseases [45]. This personalized risk evaluation can guide preventive measures and early intervention strategies.
- **Pharmacogenomics:** AI is being used to classify genetic variations that influence a person's response to specific medications. This information can be used to optimize drug dosing and minimize adverse reactions, improving medication safety and efficacy.
- **Mental Health Treatment:** AI is being used to examine patient data to predict treatment outcomes for mental health situations such as depression and anxiety. This info can help clinicians make informed decisions about treatment selection and personalized care plans.

3.2.2.4 Challenges and Future Directions

Despite its immense potential, personalized medicine with AI faces challenges:

- **Data Security and Privacy:** Ensuring data security and privacy for patients is essential to the moral use of personalized medicine. Robust cybersecurity safeguards and stringent data governance systems are required.
- **Addressing Healthcare disparities:** Ensuring equitable access to AI-powered personalized medicine is essential to address healthcare disparities and reduce health inequities.
- **Continuous Learning and Adaptation:** To keep their accuracy and efficacy, AI systems need to constantly learn from new data and adjust to changing medical knowledge.
- **Public Education and Awareness:** Increasing public understanding of personalized medicine and its benefits is crucial for fostering patient trust and acceptance of AI-driven healthcare solutions.

AI-powered personalized medicine holds immense promise for transforming healthcare, leading to more accurate diagnoses [33], personalized treatment plans, and improved patient outcomes. As AI technology continues to advance, it can be expected to see even more innovative applications of AI in personalized medicine, revolutionizing the way of approach to disease prevention, diagnosis [34], and treatment, leading to a healthier and more personalized healthcare future for all.

3.3 Genomics and Biological Research

In genomics and biological research, AI has become a transformative force, facilitating high-throughput data analysis, pattern recognition, and prediction of genetic variations. Machine learning algorithms, particularly deep learning models [39], are pivotal in deciphering complex genomic patterns, accelerating drug discovery [28], and unlocking new insights into the intricacies of biological processes at the molecular level.

3.3.1 AI–Powered Genetic Code Analysis

The development of AI has brought about an important change in the field of genomics, which is the study of an organism's whole genetic makeup. Researchers can now decode the complex patterns written in the genetic code and gain important insights into the biochemistry of life thanks to AI algorithms that can now analyze massive volumes of genomic data with previously unheard-of precision and speed.

3.3.1.1 Challenges in Traditional Genetic Code Analysis

Conventional techniques for analyzing genetic codes, such as manual sequencing and annotation, are frequently labor-intensive, error-prone, and time-consuming. These restrictions have slowed down the progress of genomic research and limited the comprehension of the intricate connections between proteins, genes, and biological processes.

3.3.1.2 Revolutionizing Genetic Code Analysis

By offering strong tools for automating, optimizing, and improving the precision of genomic data analysis, AI is transforming the field of genetic code analysis. Algorithms using AI can:

- Sequence large volumes of RNA and DNA with great efficiency and precision.
- Automatically identify and annotate genes, regulatory elements, and other functional genomic features.
- Analyze patterns and correlations within genomic data to identify novel genetic variants and disease-associated genes.
- Construct complex biological networks and models to simulate gene interactions and predict cellular processes.

3.3.1.3 Real–World Applications in Genetic Code Analysis

Genetic code analysis is already being significantly impacted by AI in several areas:

- **Disease Gene Discovery:** AI is being used to find genetic variations linked to complicated illnesses like neurological disorders, cancer, and heart disease. The creation of novel diagnostic instruments and individualized treatment plans depend heavily on this data.
- **Understanding Genetic Diversity:** AI is being used to analyze genetic data from diverse populations to understand human genetic variation and its impact on health [40] and disease. This research is essential for developing equitable healthcare approaches [41] and addressing genetic disparities.
- **Pharmacogenomics:** AI is being used to forecast a person's reaction to a particular drug based on that person's genetic composition. By using this information, adverse medication reactions can be reduced and personalized drug therapy can be guided.
- **Functional Genomics:** To comprehend the roles of genes and their interconnections in controlling biological processes, genomic data is analyzed using AI. Understanding the molecular causes of disease and creating new treatment targets depend on this study.

3.3.1.4 Future Directions in AI—Powered Genetic Code Analysis

The future of AI-powered genetic code analysis holds immense promise:

- **Single-Cell Genomics:** To gain an understanding of cellular heterogeneity and the emergence of disease, AI will be essential in the analysis of individual cell genomes.
- **Epigenomics:** AI will be utilized to examine DNA changes that control gene expression but do not change the DNA sequence. A better knowledge of how environmental factors affect gene expression and illness risk will be possible thanks to this research.
- **Genomics-Driven Precision Medicine:** AI will make it possible to create tailored medical strategies based on each patient's unique genetic profile, which will result in better patient outcomes and more effective therapies.

3.3.2 AI—Powered Drug Discovery: Accelerating the Path to New Therapies

The protracted and costly process of drug discovery presents a significant challenge to the pharmaceutical industry. The road from concept to market can take ten years or longer, with only a small percentage of medication candidates ultimately reaching patients. This is due to the difficulty of discovering prospective pharmacological targets and navigating clinical trials.

Drug discovery is being revolutionized by AI, which is offering strong tools to speed up the procedure, cut costs, and improve clinical trial success rates. Large-scale genomic, proteomic, and other biological data sets can be analyzed by AI algorithms to find possible drug targets, forecast interactions between drugs and targets, and improve drug design. The creation of novel treatments for a variety of ailments is being facilitated by this data-driven strategy, which is revolutionizing the drug discovery field.

3.3.2.1 Challenges in Traditional Drug Discovery

Traditional drug discovery methods are often slow, inefficient, and expensive. The challenges include:

- **Target identification:** Identifying the right biological target to modulate for therapeutic purposes is crucial for drug development. Traditional methods for target identification often rely on intuition and limited experimental data, leading to a high attrition rate.
- **Lead optimization:** Once a target is identified, researchers need to optimize drug candidates for desirable properties, such as potency, selectivity, and safety. This process is often time-consuming and involves extensive experimentation.
- **Clinical trials:** In the last phase of drug development, a drug candidate's safety and effectiveness are assessed in human volunteers. Clinical trials can cost a lot of money and take years to finish.

3.3.2.2 Revolutionizing Drug Discovery

AI is addressing the challenges of traditional drug discovery by providing:

- **Data-driven target identification:** AI systems can precisely discover possible drug targets by examining large volumes of genomic, proteomic, and other biological data. The time and expense associated with target identification are greatly decreased by this data-driven method.
- **Predictive lead optimization:** AI can optimize drug candidates by predicting their properties, such as binding affinity, selectivity, and toxicity. This predictive modeling [44] can guide researchers in selecting the most promising drug candidates and reducing the number of ineffective compounds that enter clinical trials.
- **Clinical trial prediction:** AI can evaluate patient data to forecast the chance of a clinical trial's success, enabling researchers to focus on the most promising drug ideas and save money on ones that have little chance of success.

3.3.2.3 Real–World Examples in Drug Discovery

Several companies are already using AI to accelerate drug discovery:

- **BenevolentAI:** This UK-based company uses AI to identify novel drug targets and develop new therapies for diseases such as cancer [35], neurodegenerative disorders, and rare diseases.
- **Insilico Medicine:** This US-based business drastically cuts the time and expense of medication research by designing and optimizing therapeutic candidates using AI.
- **Exscientia:** Leading UK-based AI drug discovery startup Exscientia created the patient-first CentaurAI platform, an AI system designed to produce drugs with a higher chance of success.

3.3.2.4 Impact o n Pharmaceutical Industry

The pharmaceutical sector is anticipated to be significantly impacted by the use of AI in drug discovery:

- **Reduced drug development costs:** AI is anticipated to drastically lower drug development costs by improving clinical trial success rates and optimising the R&D process.
- **Faster drug approvals:** By supplying more precise and timely data on drug safety and efficacy, AI is predicted to expedite the drug approval process.
- **Personalized medicine:** AI is making it possible to create personalised medicine, which adjusts a patient's course of care according to their genetic composition and past medical records.

AI is transforming the process of discovery and emerging new drugs. It can overcome the drawbacks of conventional approaches and expedite the distribution of novel treatments to patients.

4. Conclusion

As the confluence of A I, Blockchain, and Cryptography are examined, the transformative impact on SCM and Bio-Informatics becomes increasingly evident.

This concluding section recaps key insights, explores future trends, discusses industry implications, and offers closing thoughts on the intersection of these groundbreaking technologies.

4.1 Recapitulation of Key Points

The study of the chapters on A I in SCM, Bio-Informatics, and the synergy with Cryptography and Blockchain has highlighted the following key points:

- **AI in SCM:** A I uses technology such as blockchain to improve transparency, robotic process automation (RPA) to automate procedures, and predictive analytics to maximize efficiency. Case studies from real-world situations highlight its noticeable influence on supply chain optimization and logistics.
- **AI in Bio-Informatics:** From disease diagnosis to drug development, [42] A I is a linchpin in healthcare and genomics [43]. Its applications in medical imaging, virtual health assistants, and genetic code analysis underscore its potential to revolutionize patient care and biological research.

4.2 Future Trends and Directions

The future of A I in SCM, Bio-Informatics, and security holds promising trends:

- **Advanced Analytics:** A I will continue to evolve, leveraging advanced analytics to refine demand forecasting, personalized medicine, and genomic analysis.
- **Interoperability:** Increased interoperability between A I, blockchain, and cryptography will create seamless, secure ecosystems, fostering cross-industry collaboration.
- **Ethical Considerations:** It will be crucial to address ethical matters with algorithmic bias, security, and data privacy as AI becomes more and more popular.

4.3 Implications for the Industry

The implications of these technological integrations are far-reaching:

- **Operational Efficiency:** AI streamlines operations, reducing costs and enhancing efficiency in SCM, while accelerating drug discovery and diagnosis in healthcare.
- **Data Security:** Strong data security is provided by blockchain and encryption, which is crucial for protecting private genetic data and guaranteeing the accuracy of supply chain transactions.
- **Innovation Acceleration:** The confluence of A I, blockchain, and cryptography fosters innovation, driving advancements in healthcare, logistics, and collaborative research initiatives.

4.4 Final Summary

In the confluence of AI, Cryptography, and Blockchain, this chapter witnesses a technological symphony that harmonizes efficiency, security, and innovation across SCM and Bio-Informatics. The transformative potential is not only evident in optimized supply chains and personalized healthcare but also in the collaborative, secure sharing of sensitive genetic data for groundbreaking research.

As time goes on, it will be necessary to integrate these technologies responsibly. Maintaining a balance between innovation and ethical considerations will guarantee that blockchain, cryptography, and AI work together to create a world of unprecedented breakthroughs, transparent collaboration, and improved security in the future. The journey has only begun, but the convergence of these technologies promises to redefine possibilities, transform industries, and open the door for a time when blockchain, AI, and encryption work together to provide a foundation for wealth and progress.

References

1. Lu, Y. 2019. Artificial intelligence: A survey on evolution, models, applications and future trends. *Journal of Management Analytics, 6*(1), 1–29. https://doi.org/10.1080/23270012.2019.1570365.
2. Yadav, S. 2023. Leveraging the potential of artificial intelligence in healthcare supply chain management. *Data-Driven Technologies and Artificial Intelligence in Supply Chain*, 116–126. https://doi.org/10.1201/9781003462163-6.
3. Padmavathi, V., and Saminathan, R. 2023. Fundamentals of IoT, Artificial Intelligence, and Blockchain Approaches for Applications in Supply Chain Management. Blockchain, IoT, and AI Technologies for Supply Chain Management, 1–18. https://doi.org/10.1201/9781003264521-1.
4. Falatouri, T., Darbanian, F., Brandtner, P., and Udokwu, C. 2022. Predictive Analytics for Demand Forecasting–A Comparison of SARIMA and LSTM in Retail SCM. *Procedia Computer Science, 200*, 993–1003. https://doi.org/10.1016/j.procs.2022.01.298.
5. Ren, S., Chan, H.-L., and Siqin, T. 2019. Demand forecasting in retail operations for fashionable products: Methods, practices, and real case study. *Annals of Operations Research, 291*(1), 761–777. https://doi.org/10.1007/s10479-019-03148-8.
6. Iftikhar, R., and Khan, M. S. 2020. Social Media Big Data Analytics for Demand Forecasting. *Journal of Global Information Management, 28*(1), 103–120. https://doi.org/10.4018/jgim.2020010106.
7. Singh, B., Kumar, P., Sharma, N., and Sharma, K. P. 2020. Sales Forecast for Amazon Sales with Time Series Modeling. 2020 First International Conference on Power, Control and Computing Technologies (ICPC2T). https://doi.org/10.1109/icpc2t48082.2020.9071463.
8. Farasyn, I., Humair, S., Kahn, J. I., Neale, J. J., Rosen, O., Ruark, J., Tarlton, W., Van de Velde, W., Wegryn, G., and Willems, S. P. 2011. Inventory Optimization at Procter & Gamble: Achieving Real Benefits Through User Adoption of Inventory Tools. *Interfaces, 41*(1), 66–78. https://doi.org/10.1287/inte.1100.0546.
9. Radhakrishnan, P., Prasad, V. M., and Gopalan, M. R. 2009. Genetic Algorithm Based Inventory Optimization Analysis in Supply Chain Management. *2009 IEEE International Advance Computing Conference*. https://doi.org/10.1109/iadcc.2009.4809047.
10. El Haoud, N., and Bachiri, Z. 2019. Stochastic Artificial Intelligence benefits and Supply Chain Management inventory prediction. 2019 International Colloquium on Logistics and Supply Chain Management (LOGISTIQUA). https://doi.org/10.1109/logistiqua.2019.8907271.
11. Oosthuizen, K., Botha, E., Robertson, J., and Montecchi, M. 2020. Artificial intelligence in retail: The AI-enabled value chain. *Australasian Marketing Journal, 29*(3), 264–273. https://doi.org/10.1016/j.ausmj.2020.07.007.

12. Kaushik, A., and Jain, N. 2021. An Approach For Improving Transparency And Traceability of Industrial Supply Chain With Block chain Technology. 2021 6th International Conference on Innovative Technology in Intelligent System and Industrial Applications (CITISIA). https://doi.org/10.1109/citisia53721.2021.9719941.
13. Queiroz, M. M., Telles, R., and Bonilla, S. H. 2019. Blockchain and supply chain management integration: A systematic review of the literature. *Supply Chain Management: An International Journal, 25*(2), 241–254. https://doi.org/10.1108/scm-03-2018-0143.
14. Nguyen, A., Lamouri, S., Pellerin, R., Tamayo, S., and Lekens, B. 2021. Data analytics in pharmaceutical supply chains: State of the art, opportunities, and challenges. *International Journal of Production Research, 60*(22), 6888–6907. https://doi.org/10.1080/00207543.2021.1950937.
15. Astill, J., Dara, R. A., Campbell, M., Farber, J. M., Fraser, E. D., Sharif, S., and Yada, R. Y. 2019. Transparency in food supply chains: A review of enabling technology solutions. *Trends in Food Science & Technology, 91*, 240–247. https://doi.org/10.1016/j.tifs.2019.07.024.
16. Cole, R., Stevenson, M., and Aitken, J. 2019. Blockchain technology: Implications for operations and supply chain management. *Supply Chain Management: An International Journal, 24*(4), 469–483. https://doi.org/10.1108/scm-09-2018-0309.
17. Pournader, M., Ghaderi, H., Hassanzadegan, A., and Fahimnia, B. 2021. Artificial intelligence applications in supply chain management. *International Journal of Production Economics, 241*, 108250. https://doi.org/10.1016/j.ijpe.2021.108250.
18. Van Hoek, R., Gorm Larsen, J., and Lacity, M. 2022. Robotic process automation in Maersk procurement–applicability of action principles and research opportunities. Internation*al Journal of Physical Distribution & Logistics Management, 52*(3), 285–298. https://doi.org/10.1108/ijpdlm-09-2021-0399.
19. Khan, S., Tailor, R. K., Uygun, H., and Gujrati, R. 2022. Application of robotic process automation (RPA) for supply chain management, smart transportation and logistics. *International Journal of Health Sciences,* 11051–11063. https://doi.org/10.53730/ijhs.v6ns3.8554.
20. Zhao, J., and Lee, J. Y. 2022. Effect of Connected and Autonomous Vehicles on Supply Chain Performance. *Transportation Research Record: Journal of the Transportation Research Board, 2677*(3), 402–424. https://doi.org/10.1177/03611981221115425.
21. Ozdogru, U. 2020. Impact of exponential technologies on global supply chain management. Technology in Supply Chain Management and Logistics, 37–56. https://doi.org/10.1016/b978-0-12-815956-9.00003-x.
22. Mathew, D., Brintha, N. C., and Jappes, J. T. W. 2023. Artificial Intelligence Powered Automation for Industry 4.0. New Horizons for Industry 4.0 in Modern Business, 1–28. https://doi.org/10.1007/978-3-031-20443-2_1.
23. Ezziane, Z. 2006. Applications of artificial intelligence in bioinformatics: A review. *Expert Systems With Applications, 30*(1), 2–10. https://doi.org/10.1016/j.eswa.2005.09.042.
24. Cohen, J. 2004. Bioinformatics—an introduction for computer scientists. *ACM Computing Surveys, 36*(2), 122–158. https://doi.org/10.1145/1031120.1031122.
25. Nicolas, J. 2020. Artificial Intelligence and Bioinformatics. A Guided Tour of Artificial Intelligence Research, 209–264. https://doi.org/10.1007/978-3-030-06170-8_7.
26. Bülow, R. D., Dimitrov, D., Boor, P., and Saez-Rodriguez, J. 2021. How will artificial intelligence and bioinformatics change our understanding of IgA Nephropathy in the next decade? *Seminars in Immunopathology, 43*(5), 739–752. https://doi.org/10.1007/s00281-021-00847-y.
27. Mitra, S. 2005. Computational Intelligence in Bioinformatics. Transactions on Rough Sets III, 134–152. https://doi.org/10.1007/11427834_6.
28. Quazi, S. 2021. Role of Artificial Intelligence and Machine Learning in Bioinformatics: Drug Discovery and Drug Repurposing. https://doi.org/10.20944/preprints202105.0346.v1.
29. Ahmed, A., and Adil, K. 2023. Transforming Healthcare Systems with Artificial Intelligence: Revolutionizing Efficiency, Quality, and Patient Care. https://doi.org/10.21203/rs.3.rs-3175341/v1.
30. Alowais, S. A., Alghamdi, S. S., Alsuhebany, N., Alqahtani, T., Alshaya, A. I., Almohareb, S. N., Aldairem, A., Alrashed, M., Bin Saleh, K., Badreldin, H. A., Al Yami, M. S., Al Harbi, S., and Albekairy, A. M. 2023. Revolutionizing healthcare: The role of artificial intelligence in clinical practice. *BMC Medical Education, 23*(1). https://doi.org/10.1186/s12909-023-04698-z.

31. Davenport, T., and Kalakota, R. 2019. The potential for artificial intelligence in healthcare. *Future Healthcare Journal, 6*(2), 94–98. https://doi.org/10.7861/futurehosp.6-2-94.
32. Ko, C. 2018. Faculty Opinions recommendation of Dermatologist-level classification of skin cancer with deep neural networks. Faculty Opinions–Post-Publication Peer Review of the Biomedical Literature. https://doi.org/10.3410/f.727237185.793554281.
33. Myszczynska, M. A., Ojamies, P. N., Lacoste, A. M. B., Neil, D., Saffari, A., Mead, R., Hautbergue, G. M., Holbrook, J. D., and Ferraiuolo, L. 2020. Applications of machine learning to diagnosis and treatment of neurodegenerative diseases. *Nature Reviews Neurology, 16*(8), 440–456. https://doi.org/10.1038/s41582-020-0377-8.
34. Ahsan, M. M., Luna, S. A., and Siddique, Z. 2022. Machine-Learning-Based Disease Diagnosis: A Comprehensive Review. *Healthcare, 10*(3), 541. https://doi.org/10.3390/healthcare10030541.
35. McKinney, S. M., Sieniek, M., Godbole, V., Godwin, J., Antropova, N., Ashrafian, H., Back, T., Chesus, M., Corrado, G. S., Darzi, A., Etemadi, M., Garcia-Vicente, F., Gilbert, F. J., Halling-Brown, M., Hassabis, D., Jansen, S., Karthikesalingam, A., Kelly, C. J., King, D., Ledsam, J. R., Melnick, D., Mostofi, H., Peng, L., Reicher, J. J., Romera-Paredes, B., Sidebottom, R., Suleyman, M., Tse, D., Young, K. C., De Fauw, J., and Shetty, S. 2020. International evaluation of an AI system for breast cancer screening. *Nature, 577*(7788), 89–94. https://doi.org/10.1038/s41586-019-1799-6.
36. Kim, H.-E., Kim, H. H., Han, B.-K., Kim, K. H., Han, K., Nam, H., Lee, E. H., and Kim, E.-K. 2020. Changes in cancer detection and false-positive recall in mammography using artificial intelligence: A retrospective, multireader study. *The Lancet Digital Health, 2*(3). https://doi.org/10.1016/s2589-7500(20)30003-0.
37. Li, S., Zhao, R., and Zou, H. 2021. Artificial intelligence for diabetic retinopathy. *Chinese Medical Journal, 135*(3), 253–260. https://doi.org/10.1097/cm9.0000000000001816.
38. Becker, J., Decker, J. A., Römmele, C., Kahn, M., Messmann, H., Wehler, M., Schwarz, F., Kroencke, T., and Scheurig-Muenkler, C. 2022. Artificial Intelligence-Based Detection of Pneumonia in Chest Radiographs. *Diagnostics, 12*(6), 1465. https://doi.org/10.3390/diagnostics12061465.
39. Smith, K. P., Kang, A. D., and Kirby, J. E. 2018. Automated Interpretation of Blood Culture Gram Stains by Use of a Deep Convolutional Neural Network. *Journal of Clinical Microbiology, 56*(3). https://doi.org/10.1128/jcm.01521-17.
40. Panch, T., Szolovits, P., and Atun, R. 2018. Artificial intelligence, machine learning and health systems. *Journal of Global Health, 8*(2). https://doi.org/10.7189/jogh.08.020303.
41. Matheny, M. E., Whicher, D., and Thadaney Israni, S. 2020. Artificial Intelligence in Health Care. *JAMA, 323*(6), 509. https://doi.org/10.1001/jama.2019.21579.
42. Blanco-González, A., Cabezón, A., Seco-González, A., Conde-Torres, D., Antelo-Riveiro, P., Piñeiro, Ángel, and Garcia-Fandino, R. 2023. The Role of AI in Drug Discovery: Challenges, Opportunities, and Strategies. *Pharmaceuticals, 16*(6), 891. https://doi.org/10.3390/ph16060891.
43. Quazi, S. 2022. Artificial intelligence and machine learning in precision and genomic medicine. *Medical Oncology, 39*(8). https://doi.org/10.1007/s12032-022-01711-1.
44. Amarasingham, R., Patzer, R. E., Huesch, M., Nguyen, N. Q., and Xie, B. 2014. Implementing Electronic Health Care Predictive Analytics: Considerations And Challenges. *Health Affairs, 33*(7), 1148–1154. https://doi.org/10.1377/hlthaff.2014.0352.
45. Siontis, K. C., Noseworthy, P. A., Attia, Z. I., and Friedman, P. A. 2021. Artificial intelligence-enhanced electrocardiography in cardiovascular disease management. *Nature Reviews Cardiology, 18*(7), 465–478. https://doi.org/10.1038/s41569-020-00503-2.
46. Caudai, C., Galizia, A., Geraci, F., Le Pera, L., Morea, V., Salerno, E., Via, A., and Colombo, T. 2021. AI applications in functional genomics. *Computational and Structural Biotechnology Journal, 19*, 5762–5790. https://doi.org/10.1016/j.csbj.2021.10.009.

CHAPTER 8

Smart Contracts and Decentralized Applications

Ayush Agnihotri,[1,*] *Ashutosh Pandey,*[1] *Rajat Verma*[1] and *Namrata Dhanda*[2]

1. Introduction

The digital era has ushered in a vast landscape of opportunities for content creation and distribution. However, it has also brought forth challenges in safeguarding intellectual property rights. In this chapter, we embark on an exploration of the transformative interplay between smart contracts, decentralized applications (DApps), and Video Digital Rights Management (DRM), unraveling their collaborative potential in redefining content protection paradigms.

The advent of the digital era has revolutionized the way we consume and distribute content. Video content, in particular, has become a ubiquitous form of entertainment and information. This surge in digital video consumption has also brought to light the challenges of safeguarding intellectual property and ensuring fair compensation for content creators. Traditional Video DRM (Digital Rights Management) systems have attempted to address these challenges by employing encryption techniques and access control mechanisms [1]). However, these centralized approaches have often been criticized for their lack of transparency, vulnerability to security breaches, and inability to fully protect content from unauthorized distribution. The emergence of blockchain technology, smart contracts, and decentralized applications (DApps) presents a compelling opportunity to reimagine Video DRM and establish a more secure, transparent, and equitable ecosystem for digital content distribution.

The evolution of the digital landscape has orchestrated a seismic shift, unfurling an expansive terrain replete with opportunities for content creators and distributors alike. Within this dynamic milieu, however, lurk multifaceted challenges, casting a pervasive shadow over the sanctity of intellectual property rights. This chapter

[1] Department of Computer Science & Engineering, Pranveer Singh Institute of Technology, Kanpur, Uttar Pradesh, India.

[2] Department of Computer Science and Engineering, Amity University Uttar Pradesh, Lucknow, India.

* Corresponding author

embarks on a captivating journey, delving into the transformative interplay between smart contracts, decentralized applications (DApps), and Video Digital Rights Management (DRM). This symbiotic interweaving illuminates the uncharted potentialities that promise to redefine the entrenched paradigms of content protection.

1.1 A Transformative Metamorphosis

The proliferation of digital realms has catalyzed an era where creativity and information dissemination intersect in unprecedented ways. Content creators stand at the precipice of a digital revolution, empowered to sculpt narratives and artworks that traverse the globe with a mere click. Simultaneously, consumers traverse an expansive digital landscape, indulging in a tapestry of content spanning entertainment, education, and information. However, this digital metamorphosis is not bereft of its tribulations. Amidst this boundless expanse, the sanctity of intellectual property rights faces a labyrinth of challenges. Content creators grapple with the unyielding tide of content piracy, unauthorized distribution, and circumvention of ownership rights. These challenges underscore the imperative need for innovative frameworks that fortify content protection while fostering a conducive environment for creators to thrive.

1.2 Unveiling Collaborative

Potentials i n this landscape that is rife with challenges emerge as a convergence of technologies—smart contracts, DApps, and Video DRM—signaling a harbinger of change. Smart contracts, embodying the essence of self-executing and immutable agreements, weave a tapestry of trust and transparency within the digital realm. Their integration with Video DRM augurs a novel horizon where access rights, licensing terms, and remunerations interlace seamlessly into the fabric of digital content. Complementing these contracts are the decentralized applications, DApps, residing within the bastions of blockchain networks. These applications serve as the gateways to tamper-resistant ecosystems, where consumers and content creators converge. DApps transcend the conventional, proffering an interface that empowers creators with direct ownership and patrons with transparent access while upholding the tenets of DRM protocols.

1.3 Navigating the Perils and Promises

Amidst the promise of this transformative collaboration lies an intricate web of challenges. Scalability concerns loom large, with the demand for high-throughput content dissemination testing the thresholds of existing systems. Regulatory frameworks, yet to adapt to the burgeoning landscape, present ambiguities that necessitate refinement for harmonious integration. The horizon, however, shimmers with the gleam of potentialities. Innovations are rife, and collaborative endeavors proliferate, seeking to surmount these obstacles. The future envisions a landscape where smart contracts, DApps, and Video DRM converge seamlessly, establishing an ecosystem where creators wield authority and consumers relish an era of secure and seamless access to digital content [31].

2. Evolution of Digital Rights Management

Traditional DRM mechanisms have long been the stalwart protectors of digital content, yet they've grappled with vulnerabilities in centralized systems. The advent of blockchain technology, however, has introduced a novel paradigm for DRM, promising a decentralized, immutable, and transparent infrastructure.

2.1 Traditional DRM

Guardians of Digital Content f or decades, traditional DRM mechanisms have stood as stalwart protectors of invaluable digital content. Employing encryption, access controls, and licensing schemes, they've strived to safeguard intellectual property against unauthorized access and distribution. These systems, embedded within centralized architectures, aimed to fortify content rights, ensuring creators received their due while offering consumers-controlled access.

However, the efficacy of these traditional mechanisms encountered persistent challenges. Centralization rendered DRM systems susceptible to vulnerabilities, exposing content to piracy, illicit sharing, and unauthorized duplication. The rigid structures of these systems, which are reliant on intermediaries for validation and enforcement, often pose hindrances to seamless and transparent content distribution.

2.2 The Paradigm Shift with Blockchain Technology

The advent of blockchain technology has unfurled a novel chapter in the saga of DRM. This decentralized and immutable ledger system has emerged as a beacon of hope, promising a transformative shift from traditional DRM approaches. Blockchain's inherent characteristics, including decentralization, cryptographic security, and immutability, embody the essence of an innovative DRM paradigm [2].

Decentralization, the cornerstone of blockchain, veers away from the centralized authority prevalent in traditional DRM. The distributed architecture of blockchain networks disperses content across nodes, eliminating single points of failure and significantly fortifying security. Immutability, intrinsic to the blockchain, ensures that once recorded, data remains unalterable, mitigating tampering or unauthorized modifications. Transparency, facilitated by the transparent nature of blockchain, cultivates trust by providing a verifiable and traceable record of content transactions and ownership rights. You c an see the blockchain-based DRM system in Fig. 1 mentioned below.

Figure 1 unveils a decentralized Digital Rights Management (DRM) system powered by blockchain technology. Unlike traditional DRM, this system leverages the security and transparency of a distributed ledger to track ownership and regulate access to digital assets like music, eBooks, or software. By storing ownership records on the blockchain, creators are empowered with enhanced control, while users enjoy secure and flexible access, free from the limitations of centralized servers. This cutting-edge approach promises to revolutionize content protection and distribution, fostering a more transparent and trustworthy digital environment.

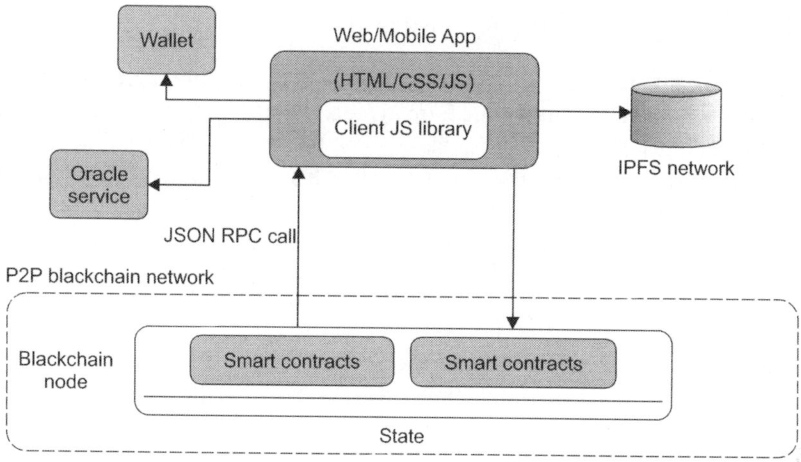

Figure 1: A Blockchain-B ased DRM System for Managing Digital Asset Ownership and Access Control.

2.3 Reimagining DRM in a Decentralized Landscape

The integration of blockchain technology into DRM heralds a paradigmatic shift, promising a decentralized, immutable, and transparent infrastructure. Smart contracts, the autonomous and self-executing codes atop blockchain networks, serve as the vanguards of this transformation. These contracts encode and enforce digital content rights, licensing terms, and royalty distributions directly within the content itself [3]. The marriage of smart contracts and DRM principles revolutionizes content protection. Access rights and usage policies, encapsulated within smart contracts, enable creators to stipulate precise terms for content access. Immutable records of transactions, stored across the decentralized blockchain network, ensure traceability and transparency. Content ownership, verifiable through blockchain's distributed ledger, fosters trust among stakeholders and alleviates concerns of unauthorized usage or infringement.

2.4 Transformative Impacts and Ongoing Innovations

The infusion of blockchain into DRM frameworks stands as a testament to innovation within the digital content landscape. Its decentralized, immutable, and transparent infrastructure addresses the vulnerabilities ingrained within centralized DRM systems. This evolution not only fortifies content protection but also promises enhanced trust, transparency, and equitable compensation for creators.

However, this transformative journey is a dynamic one, marked by ongoing innovations and iterative advancements. Challenges persist, including scalability concerns, interoperability between diverse systems, and regulatory adaptations. Collaborative efforts among industry stakeholders endeavour to surmount these obstacles, paving the way for a future where blockchain-integrated DRM fosters a secure, transparent, and thriving digital content ecosystem [30].

3. The Smart Contract Revolution

Smart contracts, epitomizing self-executing, and immutable agreements, have emerged as the linchpin in this transformative landscape. These autonomous codes operate atop blockchain networks, facilitating the creation and enforcement of contractual terms in a trustless environment. Their fusion with Video DRM principles revolutionizes content protection by encoding access rights, licensing terms, and royalty payments directly into the content itself.

3.1 The Essence of Self–Executing Agreements

Smart contracts stand as the vanguard of a transformative era, epitomizing the essence of self-executing and immutable agreements within the digital realm. These ingenious contracts, residing atop blockchain networks, herald a paradigm shift in contractual engagements by facilitating the creation and enforcement of agreements in a trustless and automated environment.

Smart contracts are the vanguards of a transformative era in contract law, representing the very essence of self-executing and immutable agreements within the digital realm. In essence, these ingenious contracts, residing atop blockchain networks, herald a monumental paradigm shift in contractual engagements, fundamentally altering the landscape of traditional agreements by fostering an environment of trustless and automated interactions.

At their core, smart contracts are self-executing agreements encoded in computer code, meticulously designed to enforce and execute the terms of a contract automatically when predetermined conditions are met. Unlike conventional contracts that rely on intermediaries or legal enforcement mechanisms, smart contracts operate autonomously within decentralized blockchain networks, cutting across geographical boundaries and eliminating the need for centralized oversight.

The transformative potential of smart contracts lies in their ability to automate and streamline contractual processes. They eliminate the dependence on intermediaries, mitigating human errors and delays inherent in traditional contract execution. By harnessing the immutability and cryptographic security of blockchain technology, smart contracts ensure that contractual terms are transparent, tamper-proof, and enforceable without the need for trust in a central authority [29].

Moreover, these contracts operate within a trustless environment, meaning that the execution of agreed-upon terms is automated and occurs only when pre-defined conditions, often referred to as "if-then" statements or clauses, are met. This automation ensures that both parties involved in the contract can rely on the predetermined code logic to enforce the terms without any need for manual intervention or third-party enforcement. The inherent characteristics of smart contracts foster a multitude of benefits. They streamline complex contractual processes, expedite transaction settlements, and reduce administrative costs by eliminating intermediaries.

Additionally, the transparency and immutability of smart contracts instil confidence among stakeholders, ensuring that the terms agreed upon are precisely executed as intended, leaving an immutable record of the transaction on the blockchain. However, as with any nascent technology, challenges exist. Smart contracts are susceptible to coding errors or vulnerabilities in the underlying blockchain infrastructure [4].

Addressing these concerns requires continuous development, auditing, and refinement of coding practices to enhance security and robustness. Nevertheless, the essence of self-executing agreements embodied in smart contracts marks a watershed moment in the evolution of contractual engagements. Their ability to operate autonomously, transparently, and securely within decentralized networks holds immense promise, redefining how agreements are conceptualized, executed, and enforced in the digital era. As this technology continues to mature and innovate, it stands poised to revolutionize not only contract law but also various industries, reshaping the very fabric of our economic and social interactions.

3.2 Underpinning Technology and Operational Autonomy

Operational autonomy defines the bedrock of smart contracts. Encoded within these autonomous codes are predefined sets of rules and conditions, meticulously crafted to execute specific actions upon fulfilment criteria. Leveraging blockchain's cryptographic security and decentralized infrastructure, these contracts operate devoid of intermediaries, seamlessly executing actions based on predetermined conditions without the need for manual intervention.

3.3 Transforming Content Protection via Fusion with Video DRM

The convergence of smart contracts with Video DRM principles ushers in a revolutionary phase in content protection. Acting as catalysts within the DRM framework, smart contracts encode, enforce, and execute contractual terms directly within the digital content. Access rights, licensing terms, and remunerations are not merely written agreements but are embedded within the very fabric of digital assets.

The convergence of smart contracts with Video DRM principles marks a seminal moment, heralding a revolutionary phase in content protection within the digital domain. Their fusion represents a catalyst that transcends the traditional paradigms of Digital Rights Management, reshaping how content access, licensing, and remuneration are encoded, enforced, and executed within the intricate framework of digital assets. Smart contracts, the linchpin of this transformative synergy, serve as the guardians of contractual integrity within the Video DRM ecosystem.

These self-executing and immutable agreements leverage the inherent capabilities of blockchain networks, revolutionizing the very essence of content protection mechanisms. Within this framework, smart contracts function as digital vaults, encapsulating access rights, licensing terms, and remunerations directly into the cryptographic fabric of digital assets. One of the pivotal transformations brought forth by smart contracts within Video DRM lies in their ability to encode contractual terms as executable code. Traditional DRM systems often relied on external servers or intermediary verification mechanisms to grant or restrict content

You can see the DRM playbook security model in t Fig. 2 mentioned below.

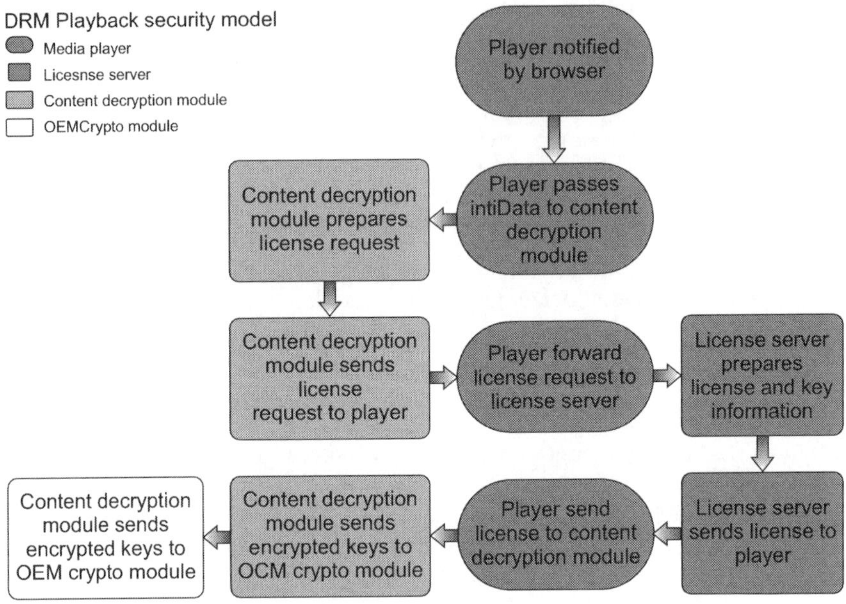

Figure 2: DRM Playbook Security Model.

Figure 2 classifies a security model called the DRM playbook, designed to protect content playback from various attacks. It utilizes layers like content decryption, OEM crypto, a license server, and a player mind for comprehensive protection. This model effectively safeguards against man-in-the-middle, reverse engineering, and side-channel attacks.

However, smart contracts embedded within the digital content itself act as gatekeepers, ensuring that access is granted only to authorized entities when pre-defined conditions encoded in the contract are met. The essence of smart contracts extends beyond the mere representation of contractual terms; they operate as autonomous entities, executing predefined actions automatically when specified conditions are fulfilled. In the context of Video DRM, this implies that licensing agreements, royalty payments, or access permissions are enforced seamlessly and transparently without the need for intermediaries, fostering a trustless environment.

Moreover, smart contracts facilitate transparent and immutable audit trails. Every transaction, alteration, or execution of contractual terms is permanently recorded on the blockchain, providing an auditable trail of all interactions related to the content. This transparent ledger ensures accountability, mitigates disputes, and fortifies the integrity of content protection measures within Video DRM. The fusion of smart contracts with Video DRM principles revolutionizes the very architecture of content protection. Access rights, licensing terms, and remunerations cease to exist as static agreements; instead, they become intrinsic attributes of the digital content, encoded within its cryptographic DNA [5]. This paradigm shift not only secures the content but also ensures that creators receive fair compensation and that consumers access content seamlessly, adhering to agreed-upon terms. However, challenges exist in ensuring the flawless integration of smart contracts within Video DRM.

Technical intricacies, interoperability concerns, and regulatory compliance require careful consideration and continuous refinement to realize the full potential of this transformative fusion.

In essence, the convergence of smart contracts with Video DRM principles signifies a monumental leap in content protection mechanisms. Their integration imbues digital assets with a self-governing and self-executing nature, transcending traditional DRM systems, and propelling content protection into an era characterized by transparency, autonomy, and trustlessness. As this fusion continues to evolve and mature, it holds the promise of reshaping not only content distribution but also the very fabric of digital ownership and intellectual property management [30].

3.4 Encoding Rights and Payments into the Content

The transformative potential of smart contracts lies in their ability to embed access rights, licensing terms, and royalty distributions directly into the content. These contracts act as immutable guardians, enabling creators to define granular access permissions and stipulate the terms of usage seamlessly. Immutable records of contractual obligations embedded within the content ensure transparent and traceable execution of payments, mitigating disputes and fortifying content protection.

3.5 Trustless and Transparent Transactions

At the core of smart contract functionality lies a trustless and transparent environment. Leveraging blockchain's distributed ledger, these contracts establish a trail of transactions that is immutable and publicly accessible. This transparency fosters trust among stakeholders by providing a verifiable record of transactions, ensuring compliance with contractual obligations and content usage rights [27].

3.6 Ongoing Evolution and Potential Innovations

The smart contract revolution within content protection continues to evolve, presenting a beacon of innovation and potential. Challenges persist, including the need for standardization, scalability concerns, and the exploration of interoperability between different blockchain networks. Collaborative efforts within the industry aim to surmount these challenges, striving for a future where smart contracts seamlessly fortify Video DRM, fostering a secure, transparent, and equitable content ecosystem.

4. Enabling Tamper—Resistant Ecosystems with DAPPS

Decentralized applications, residing on blockchain networks, synergize with smart contracts to fortify the DRM framework. DApps serve as the gateway, offering a user-friendly interface for consumers and content creators to interact within tamper-resistant ecosystems. They streamline content distribution, empower creators with direct ownership, and ensure transparent transactions while preserving the integrity of DRM protocols.

4.1 Synergy of Decentralized Applications and Smart Contracts

Decentralized applications (DApps) represent a cornerstone within the robust framework of content protection, synergizing harmoniously with smart contracts atop blockchain networks. This integration fortifies the DRM landscape, ushering in an era of resilience against tampering and unauthorized access while fostering transparent and equitable interactions.

4.2 Gateways to Tamper—Resistant Environments

DApps function as gateways, ushering consumers and content creators into tamper-resistant environments underpinned by blockchain's immutable ledger. These applications, residing on decentralized networks, offer an intuitive and user-friendly interface. They serve as portals for users to engage seamlessly with digital content, ensuring a frictionless and secure experience while upholding the integrity of DRM protocols.

Gateways to Tamper-Resistant Environments Decentralized Applications (DApps) stand as pivotal gateways, leading both consumers and content creators into a new realm of tamper-resistant environments forged by the immutable ledger of blockchain technology. Positioned within decentralized networks, these innovative applications serve as the bridge between users and the transformative potential of secure content management.

DApps offers an intuitive and user-friendly interface that seamlessly integrates with the complexities of Digital Rights Management (DRM), ensuring a frictionless and secure experience while upholding the sanctity and integrity of DRM protocols. At the heart of this paradigm shift lies the fundamental architecture of DApps, strategically designed to operate within decentralized blockchain networks. Leveraging the distributed and immutable ledger capabilities of blockchain, DApps pave the way for users to engage with digital content securely, without the need for reliance on traditional, centralized intermediaries. The role of DApps in content distribution is multifaceted. They provide a user-centric interface, offering intuitive functionalities that facilitate content discovery, consumption, and interaction within the digital realm. By residing on decentralized networks [6], DApps ensure that users can access content seamlessly while enjoying enhanced security, privacy, and transparency. One of the fundamental attributes of DApps is their ability to uphold the integrity of DRM protocols. These applications embed DRM mechanisms within their framework, ensuring that content remains protected and tamper-resistant.

Through smart integration with smart contracts, DApps enforce access rights, licensing terms, and other DRM-related functionalities seamlessly, contributing to the creation of a trustless environment. Moreover, DApps foster an environment where content creators can directly engage with their audience without the constraints imposed by centralized intermediaries. This direct interaction not only empowers creators but also fosters a more transparent and equitable relationship between creators and consumers. DApps serve as the facilitators of this direct engagement, ensuring that creators have the tools necessary to manage, distribute, and monetize their content efficiently [7]. The user-friendly nature of DApps is instrumental in driving adoption and enhancing the overall user experience within tamper-resistant environments. By offering intuitive interfaces and seamless functionalities, DApps lower the entry barriers for users, making content consumption and engagement more accessible and appealing.

However, challenges persist in the widespread adoption and integration of DApps within content distribution ecosystems. User adoption, scalability concerns, interoperability, and regulatory compliance pose hurdles that require continual refinement and innovation to overcome. In essence, DApps emerge as the gateways to usher users and content creators into tamper-resistant environments underpinned by blockchain's immutable ledger. Their intuitive interfaces, seamless functionalities, [30] and integration with DRM protocols mark them as pivotal components in reshaping the content distribution landscape, ensuring a secure, transparent, and user-centric experience for all stakeholders involved. As these gateways continue to evolve and mature, they hold the promise of revolutionizing content management and distribution paradigms.

4.3 Streamlining Content Distribution

At the crux of DApps lies their role in streamlining the labyrinthine process of content distribution. By leveraging the decentralized nature of blockchain networks, DApps facilitate direct interactions between creators and consumers. Content distribution transcends the shackles of traditional intermediaries, enabling a direct and transparent exchange of digital content.

4.4 Empowering Creators with Direct Ownership

One of the most transformative aspects of DApps lies in their ability to empower content creators with direct ownership and control. By bypassing centralized authorities, creators retain sovereignty over their intellectual property. The transparent and immutable nature of blockchain facilitates verification of ownership, instilling confidence and trust in the authenticity of content [26].

4.5 Fostering Transparent Transactions

Transparency is the bedrock of DApps' functionality within DRM frameworks. These applications leverage the blockchain's distributed ledger to meticulously record transactions. Every interaction within the ecosystem is transparent and verifiable, ensuring adherence to contractual obligations, content usage rights, and royalty distributions.

4.6 Upholding DRM Integrity Through Innovation

DApps continuously drives innovation within the DRM landscape. Their evolution seeks to fortify the integrity of DRM protocols while addressing existing challenges. Scalability, interoperability, and user experience enhancements stand as focal points for ongoing developments within DApp frameworks, promising an even more robust and user-centric content ecosystem.

Decentralized Applications (DApps) serve as catalysts for relentless innovation within the landscape of Digital Rights Management (DRM). Their evolutionary trajectory is dedicated to fortifying the integrity of DRM protocols while proactively addressing existing challenges. Key areas of focus encompass scalability enhancements, interoperability advancements, and an unwavering commitment to elevating user experiences within DApp frameworks. These collective efforts promise a future characterized by a more resilient, adaptable, and user-centric content ecosystem.

At the forefront of DApp evolution lies an unwavering dedication to bolstering the integrity of DRM protocols. The continuous innovation within DApp frameworks is not merely a pursuit of technological advancement but a strategic endeavor to fortify the very foundations of content protection mechanisms.

Scalability stands tall as a critical frontier in this ongoing journey. Efforts are underway to engineer solutions that enable DApps to accommodate growing user bases and expanding content libraries without compromising performance or network efficiency. Moreover, interoperability emerges as a pivotal focus area within the realm of DApps. The evolution of these applications aims to bridge the gaps between disparate systems, ensuring seamless communication and interaction among various content distribution platforms. Interoperability enhancements within DApps facilitate frictionless content exchange, enabling users to access and engage with diverse content offerings across multiple platforms while maintaining the integrity of DRM mechanisms [25].

Furthermore, the relentless pursuit of enhancing user experiences represents a cornerstone of innovation within DApp frameworks. User-centric design principles drive the evolution of interfaces, functionalities, and accessibility features within these applications. Streamlining content discovery, optimizing navigation, and simplifying user interactions are integral aspects of ongoing developments. DApps strives to provide users with intuitive and seamless experiences that foster engagement while upholding the security and transparency inherent in DRM protocols. However, the path to innovation within DApps is not devoid of challenges. Overcoming scalability barriers, ensuring robust interoperability across diverse platforms, and continually refining user experiences demand concerted efforts and continual refinements [8].

The dynamic nature of technological advancements necessitates a proactive approach to adapt and evolve in tandem with emerging trends and user expectations. In essence, the evolution of DApps within the DRM landscape is characterized by a relentless pursuit of innovation aimed at fortifying content protection mechanisms. Scalability, interoperability, and user-centric enhancements stand as pillars guiding the evolution of these applications. The collective endeavours within DApp frameworks promise a future where DRM protocols are more resilient, adaptable, and seamlessly integrated into user experiences, shaping a content ecosystem that prioritizes both security and user satisfaction. As these innovations continue to unfold, they hold the potential to redefine content distribution paradigms, ensuring a more robust and user-centric digital content landscape.

4.7 Collaborative Endeavours for a Secure Future

Collaborative endeavours within the realm of DApps aim to surmount challenges and push the boundaries of innovation. Efforts to address scalability concerns, enhance user experience, and foster interoperability among diverse blockchain networks seek to ensure a seamless and secure environment for creators and consumers.

5. Empowering Content Creators

The integration of smart contracts, DApps, and Video DRM stands poised to empower content creators significantly. By providing a secure, auditable, and automated environment, these technologies enable creators to retain control over their intellectual property. The transparency and automation inherent in this ecosystem foster trust among stakeholders, revolutionizing the content creation and distribution landscape.

5.1 A Paradigm of Empowerment

The convergence of smart contracts, DApps, and Video DRM heralds a monumental shift, empowering content creators with an array of tools and mechanisms that revolutionize their control over intellectual property. This symbiotic integration presents creators with a secure, auditable, and automated environment that redefines the landscape of content creation and distribution [24].

5.2 Retaining Sovereignty over Intellectual Property

At the crux of this empowerment lies the preservation of creators' sovereignty over their intellectual property. The integration of smart contracts within Video DRM enables creators to encapsulate access rights, licensing terms, and royalty distributions directly within their content. This paradigm shift from centralized control to decentralized ownership emboldens creators with unparalleled control and authority [9].

The empowerment of content creators within the realm of digital content distribution is anchored in the fundamental principle of preserving their sovereignty over intellectual property. This core tenet finds its essence in the integration of smart contracts within Video Digital Rights Management (DRM), ushering in a transformative era where creators regain unprecedented control and authority over their creative assets.

The seamless amalgamation of smart contracts with Video DRM mechanics enables creators to encapsulate access rights, licensing terms, and royalty distributions directly within their content, marking a paradigm shift from centralized control to decentralized ownership. The integration of smart contracts serves as a monumental milestone in the evolution of content ownership and control.

Traditionally, the governance and control of intellectual property within digital content were often entrusted to centralized entities or intermediaries, introducing complexities and challenges in ensuring fair compensation and secure distribution for creators. However, the advent of smart contracts disrupts this paradigm by embedding access rights, licensing agreements, and royalty distributions directly within the fabric of digital assets. At the heart of this transformation lies the encapsulation of contractual terms within the code of smart contracts. These self-executing and immutable agreements are meticulously crafted to enforce pre-defined conditions and contractual obligations without the need for intermediaries [23].

Creators can encode intricate access permissions, specify licensing terms, and automate royalty distributions within their content, ensuring that these agreements are executed autonomously and transparently. The significance of this paradigm shift transcends mere technological innovation; it represents a fundamental redefinition of ownership and control over intellectual property. By embedding contractual terms within the content itself, creators reclaim sovereignty over their creative assets. The transition from centralized control to decentralized ownership empowers creators with unparalleled authority, allowing them to assert control over how their content is accessed, utilized, and monetized. Moreover, this transition to decentralized ownership instils a sense of trust and transparency within the content ecosystem [10].

Smart contracts operate within a tamper-resistant environment, where transactions and agreements are immutable, auditable, and transparent. This transparency fosters trust among stakeholders, ensuring that creators receive fair compensation and that consumers adhere to agreed-upon usage terms. However, despite the transformative potential, challenges persist in the widespread adoption and integration of smart contracts within Video DRM frameworks. Technical complexities, standardization, and regulatory considerations require ongoing refinement and collaboration to overcome hurdles and fully realize the potential of decentralized ownership models.

In essence, the integration of smart contracts within Video DRM epitomizes a monumental shift in content ownership, placing creators at the helm of their intellectual property. This paradigm shift from centralized control to decentralized ownership empowers creators with unprecedented authority, autonomy, and transparency, redefining the very fabric of intellectual property management in the digital era. As these technologies continue to mature and evolve, they hold the promise of reshaping the relationship between creators, consumers, and content distribution platforms, ensuring a more equitable and creator-centric digital content landscape.

5.3 Secure, Auditable, and Transparent Environments

The amalgamation of smart contracts and DApps fosters environments that are inherently secure, auditable, and transparent. The immutable nature of blockchain ensures that contractual obligations, content usage rights, and royalty distributions are verifiable and traceable. This transparency alleviates concerns of unauthorized usage or infringements, fostering trust among creators, distributors, and consumers.

5.4 Automating Content Management

Automation is a cornerstone of this empowered landscape. Smart contracts automate the enforcement of contractual terms, ensuring seamless execution of access rights, licensing agreements, and royalty distributions. This automated process minimizes administrative burdens, expedites transactions, and reduces the likelihood of disputes, enabling creators to focus more on innovation and content creation.

At the core of this empowered content landscape lies the bedrock of automation, with smart contracts emerging as the vanguards orchestrating the seamless execution of contractual terms. These self-executing and immutable agreements stand as pillars of innovation, revolutionizing content management by automating the enforcement of access rights, licensing agreements, and royalty distributions within the realm of digital assets. This transformative automation mitigates administrative burdens, expedites transactions, and significantly reduces the likelihood of disputes, thereby liberating creators to channel their energies more intensely toward innovation and content creation. The advent of smart contracts within the realm of content management heralds a paradigm shift in how contractual obligations are executed.

Traditionally, managing access rights, licensing agreements, and royalty distributions entailed intricate administrative processes, human interventions, and intermediaries. However, the integration of smart contracts streamlines and automates these processes, encapsulating the essence of contractual terms within the code of digital assets. The fundamental essence of smart contracts lies in their autonomous and self-executing nature. These contracts, residing atop blockchain networks, are meticulously designed to execute predefined conditions and obligations when triggered by specified events.

Within the domain of content management, smart contracts encode access permissions, specify licensing terms, and automate royalty distributions. When predetermined conditions are met, these contracts execute the specified actions autonomously and transparently, without the need for intermediaries or manual interventions. The implications of this automation are profound. By automating the enforcement of contractual terms, smart contracts alleviate the burden of administrative tasks and reduce the likelihood of errors or disputes [11].

Creators no longer need to navigate intricate administrative processes; instead, they can rely on the inherent reliability and immutability of smart contracts to ensure the execution of contractual obligations seamlessly and transparently. Moreover, the automation facilitated by smart contracts expedites transaction settlements, enabling creators to receive timely and accurate compensation for their creative endeavors. This expediency in transactions fosters an environment conducive to innovation, allowing creators to focus more on the creative process, exploration, and ideation, rather than being entangled in administrative complexities. However, while automation through smart contracts holds immense promise, challenges persist in its widespread adoption [23].

Issues such as ensuring flawless code execution, addressing scalability concerns, and navigating regulatory landscapes require continual refinement and collaborative efforts to fully harness the potential of automated content management. In essence, the integration of smart contracts heralds an era where automation becomes the cornerstone of content management. This transformative automation not only expedites transactions and minimizes administrative burdens but also liberates creators to delve deeper into innovation and creativity. As smart contracts continue to refine and evolve, they stand poised to revolutionize not only content management but also the very essence of how creators engage with and monetize their creative assets in the digital landscape.

5.5 Nurturing Trust and Collaboration

Trust forms the bedrock of collaboration between creators, distributors, and consumers within this ecosystem. The transparency ingrained within smart contracts and DApps cultivates an environment where stakeholders engage with confidence. Creators establish direct relationships with consumers, fostering open collaboration and communication, thereby enhancing the overall quality of content and user experiences.

5.6 Revolutionizing Content Creation and Distribution

The empowerment of content creators through these integrated technologies catalyzes a revolution in the content creation and distribution landscape. Creators are no longer tethered by the constraints of centralized control. Instead, they navigate a realm where their creative process is protected, their rights are safeguarded, and their interactions with audiences are direct and transparent.

5.7 Ensuring Equitable Compensation and Recognition

Equitable compensation is a pivotal facet of this empowered landscape. The automated execution of royalty distributions via smart contracts ensures creators receive fair compensation for their work promptly. This fosters a sustainable creative economy, incentivizing creators and nurturing an environment where innovation thrives and talent is duly recognized.

5.8 Future Frontiers and Collaborative Innovations

The journey of empowerment for content creators is an ongoing evolution. Collaborative endeavors seek to address challenges such as scalability, interoperability, and regulatory adaptations. Innovations within smart contracts and DApps continue to refine user experiences, fortify security, and expand the horizons of content creation and distribution.

6. Challenges

Despite their promise, challenges persist in widespread adoption, such as scalability concerns and regulatory ambiguities. However, ongoing innovations and collaborative efforts seek to address these hurdles. The future promises a refined synergy between smart contracts, DApps, and Video DRM, fostering an ecosystem where content creators wield unprecedented control and consumers enjoy seamless, secure access to digital content [12].

6.1 Future Prospects

Persistent Challenges in Widespread Adoption Despite the transformative potential exhibited by the integration of smart contracts, DApps, and Video DRM, several hurdles impede their widespread adoption within the digital content landscape. Foremost among these challenges are concerns regarding scalability and interoperability. The burgeoning demand for high-throughput content dissemination strains existing systems, calling for scalable solutions that can accommodate a growing user base without compromising performance.

6.2 Scalability Concerns and Technological Innovations

Scalability remains a pivotal challenge as the digital content landscape continues to expand. Innovations are underway to augment existing blockchain infrastructures, seeking to enhance throughput, reduce latency, and ameliorate the limitations hampering scalability. Layer-2 scaling solutions and consensus mechanism enhancements are at the forefront of these innovations, aiming to elevate blockchain networks to meet the demands of a burgeoning user base and evolving content needs.

Scalability stands as an undeniable challenge amid the exponential expansion of the digital content landscape. As the demand for content distribution surges, the limitations of existing blockchain infrastructures come to the fore. To address these challenges, a wave of innovative solutions is underway, aiming to augment blockchain infrastructures, enhance throughput, diminish latency, and surmount the constraints that impede scalability. At the forefront of these innovations lie Layer-2 scaling solutions and enhancements in consensus mechanisms, both promising to propel blockchain networks to meet the burgeoning demands of an expanding user base and the ever-evolving needs of digital content [22]. The escalating demand for content distribution and the concurrent surge in users necessitate a radical re-evaluation of the scalability of blockchain networks. The current limitations of blockchain, primarily in terms of transaction speed, throughput, and network congestion, pose significant hurdles to accommodating the burgeoning demands of the digital content ecosystem. Recognizing these challenges, ongoing innovations are focused on augmenting the existing infrastructure to enhance its scalability.

Layer-2 scaling solutions emerge as a promising avenue in mitigating the scalability concerns plaguing blockchain networks. These solutions introduce off-chain protocols that operate in parallel to the main blockchain, alleviating congestion and enhancing scalability without compromising on the security and decentralization inherent in blockchain technology. By processing transactions off-chain and subsequently settling them on the main chain, Layer-2 solutions significantly enhance the throughput and efficiency of blockchain networks, facilitating a seamless and scalable environment for content distribution.

Furthermore, advancements in consensus mechanisms constitute another frontier in the quest for scalability. Traditional consensus mechanisms, such as Proof of Work (PoW) or Proof of Stake (PoS), have inherent limitations in their scalability potential. Innovations in consensus mechanisms, including Proof of Authority (PoA), Directed Acyclic Graphs (DAGs), and sharding, seek to revolutionize the way transactions are validated and executed within blockchain networks. These novel approaches aim to increase transaction throughput, reduce latency, and improve network efficiency, thereby addressing the scalability concerns that hinder the seamless distribution of digital content [13].

However, despite the promising strides in scalability innovations, challenges persist. Achieving consensus on a scalable solution that balances performance, security, and decentralization remains a complex task. Interoperability among different scaling solutions and achieving widespread adoption across diverse blockchain platforms also pose considerable challenges that necessitate concerted efforts and collaborative initiatives. In summary, scalability concerns stand as a pivotal challenge amid the burgeoning digital content landscape. However, ongoing innovations in Layer-2 scaling solutions and advancements in consensus mechanisms hold immense promise in overcoming these challenges.

The evolution of blockchain networks to accommodate the surging demands of content distribution signifies a transformative leap toward a scalable and efficient digital content ecosystem, ensuring a seamless and responsive environment for creators and users alike. As these technological innovations continue to refine and evolve, they hold the potential to reshape the very foundations of blockchain networks, propelling them into an era of scalable and efficient content distribution.

6.3 Regulatory Ambiguities and Adaptations

Regulatory ambiguities pose another significant obstacle to the seamless integration of smart contracts, DApps, and Video DRM into existing frameworks. The evolving nature of digital content distribution necessitates adaptations within regulatory paradigms to address novel challenges while ensuring compliance and consumer protection. Collaborative dialogues between industry stakeholders and regulatory bodies aim to pave the way for coherent and adaptable regulatory frameworks that foster innovation while safeguarding interests.

6.4 Collaborative Endeavors and Industry Synergies

Addressing these challenges necessitates collaborative efforts across industry domains. Partnerships between blockchain developers, content creators, technology innovators, and regulatory bodies are pivotal in surmounting hurdles and fostering an environment conducive to innovation and widespread adoption. Collaborative endeavors seek to bridge technological gaps, streamline regulatory compliance, and establish standards that underpin a cohesive and thriving ecosystem.

6.5 Envisioning a Refined Synergy

The future holds promise for a refined synergy between smart contracts, DApps, and Video DRM. Continuous innovations and collaborative endeavors aim to address existing challenges, forging a path toward an ecosystem where content creators wield unprecedented control while consumers relish seamless, secure access to digital content.

6.6 Technological Maturation and User–Centric Experiences

The maturation of technologies underpinning smart contracts, DApps, and Video DRM is poised to redefine user experiences within the digital content landscape. Innovations in user interfaces, enhanced interoperability between blockchain networks, and streamlined content distribution mechanisms promise a future where consumers access content effortlessly, seamlessly, and securely [14]. The evolution and maturation of the technologies underpinning smart contracts, Decentralized Applications (DApps), and Video Digital Rights Management (DRM) represent a pivotal juncture poised to redefine user experiences within the expansive and dynamic digital content landscape. These advancements, driven by ongoing innovations, encompass transformative enhancements in user interfaces, streamlined interoperability between diverse blockchain networks, and optimized content distribution mechanisms. Collectively, they promise a future where consumers can access, engage with, and enjoy digital content effortlessly, seamlessly, and securely, ushering in an era of unparalleled user-centric experiences. One of the hallmark transformations brought forth by technological maturation lies in the evolution of user interfaces within DApps and content distribution platforms. Innovations are underway to craft more intuitive, user-friendly interfaces that seamlessly integrate blockchain functionalities while remaining accessible and appealing to a diverse user base. The focus on user-centric design principles aims to simplify content discovery, optimize navigation, and enhance overall usability, ensuring that consumers can interact with digital content effortlessly and intuitively.

Enhanced interoperability between blockchain networks emerges as a fundamental enabler in reshaping user experiences within the digital content landscape. The evolution toward seamless interoperability aims to bridge the gaps between disparate blockchain platforms, fostering an environment where content can be accessed and shared across multiple networks without constraints [15].

Interoperability advancements eliminate barriers to content access, offering users a frictionless experience while navigating diverse content distribution platforms. Furthermore, the maturation of content distribution mechanisms driven by smart contracts and DApps promises a future where consumers access content securely and transparently. These advancements streamline content delivery, ensuring that users can access desired content seamlessly while adhering to predefined licensing terms and access permissions encoded within the fabric of digital assets.

The integration of smart contracts within Video DRM fundamentally alters how users interact with and consume digital content, offering a secure and transparent environment that promotes trust and compliance. However, despite the promising trajectory of technological maturation, challenges persist. Achieving widespread adoption of user-centric interfaces across diverse platforms, ensuring seamless interoperability among varied blockchain networks, and navigating regulatory landscapes require continual refinement and collaboration within the digital content ecosystem. In summary, the maturation of technologies powering smart contracts, DApps, and Video DRM heralds a transformative era in user-centric experiences within the digital content landscape.

Innovations in user interfaces, interoperability, and content distribution mechanisms promise a future where consumers interact with digital content effortlessly, seamlessly, and securely. As these technologies continue to evolve and mature, they hold the potential to redefine the very fabric of user engagement and content consumption, ensuring a more intuitive, accessible, and immersive digital content ecosystem for all stakeholders involved.

6.7 Pioneering a New Digital Frontier

The convergence of technological innovations and collaborative synergies paves the way for a new digital frontier. It envisions an ecosystem characterized by equitable access to content, fortified content protection mechanisms, and an environment that empowers creators and engenders trust among all stakeholders.

Embracing Transformation, the challenges that impede the widespread adoption of smart contracts, DApps, and Video DRM are formidable, yet not insurmountable. Collaborative efforts, technological innovations, and regulatory adaptations are propelling the industry toward a future where content distribution is seamless, creators wield unprecedented control, and consumers enjoy secure access to a diverse array of digital content [21]. The convergence of cutting-edge technological innovations and collaborative synergies within the digital content landscape heralds the dawn of a new era—a frontier marked by transformative advancements. This visionary landscape envisions an ecosystem characterized by equitable access to content, fortified content protection mechanisms, and an environment that empowers creators while fostering trust among all stakeholders involved. At the crux of this transformative landscape lies the amalgamation of technological innovations, weaving together smart contracts, Decentralized Applications (DApps), and Video Digital Rights Management (DRM). This convergence heralds a paradigm shift toward a future where the digital content ecosystem thrives on the principles of fairness, transparency, and empowerment.

This visionary future envisions equitable access to digital content for users across diverse demographics and geographies. It seeks to dismantle barriers to access by leveraging technological innovations that ensure seamless and unrestricted content consumption. Through collaborative efforts, technological advancements aim to democratize access, allowing users to explore, engage with, and derive value from an expansive array of digital content offerings. Moreover, the fortified content protection mechanisms ushered in by the convergence of these innovations represent a cornerstone of trust and security within the digital content landscape.

The integration of smart contracts within Video DRM fundamentally reshapes how content is safeguarded and distributed. By embedding access rights, licensing agreements, and royalty distributions directly within digital assets, this innovative framework ensures a secure, transparent, and tamper-resistant environment. This not only safeguards the intellectual property of creators but also engenders trust among stakeholders, promoting fair compensation and adherence to agreed-upon usage terms. Yet, the path toward this visionary frontier is not devoid of challenges. The hurdles that impede the widespread adoption of smart contracts, DApps, and Video DRM pose formidable obstacles.

However, these challenges, although significant, are not insurmountable. Collaborative efforts among industry pioneers, technological innovations aimed at scalability and interoperability, and adaptive regulatory frameworks collectively propel the industry toward a future where content distribution transcends barriers. Embracing this transformation requires a collective commitment to innovation, collaboration, and adaptability. Collaborative ecosystems, where stakeholders converge to refine and enhance these technologies, are fostering an environment conducive to unprecedented growth and evolution.

Through these concerted efforts, the industry is evolving toward a future where content distribution is seamless, creators wield unprecedented control over their intellectual property, and consumers relish secure access to a diverse array of digital content offerings. In essence, the pioneering efforts aimed at harnessing technological innovations and fostering collaborative synergies are paving the way for a new digital frontier—one characterized by inclusivity, fortified content protection mechanisms, and an empowered ecosystem that transcends boundaries. As the industry navigates these transformative waters, the collective pursuit of innovation holds the promise of reshaping the digital content landscape, propelling it toward an era defined by equitable access, empowerment, and trust among all stakeholders involved.

7. Conclusion

The convergence of smart contracts, DApps, and Video DRM heralds a new era in content protection. This collaborative synergy not only safeguards intellectual property but also redefines the relationship between content creators and consumers. As these technologies continue to mature, they will undoubtedly reshape the digital content landscape, ensuring a fair and secure environment for all stakeholders involved.

7.1 A Paradigm Shift in Content Protection

The convergence of smart contracts, DApps, and Video DRM epitomizes a watershed moment, not merely in content protection but in redefining the relationship dynamics between creators and consumers. This collaborative synergy transcends conventional paradigms, heralding an era where innovation, transparency, and fairness reign supreme within the digital content ecosystem.

Safeguarding Intellectual Property Rights At its core, this convergence stands as an impregnable fortress safeguarding the sanctity of intellectual property rights. Smart contracts, the linchpin of this transformative synergy, encode access rights, licensing terms, and royalty distributions directly within the fabric of digital assets. This immutable integration ensures creators' rights are fortified, curtailing unauthorized usage and piracy while fostering trust and transparency [16].

7.2 Reshaping the Creator–Consumer Nexus

The symbiotic integration of smart contracts and DApps engenders a shift in the relationship dynamics between creators and consumers. Creators, emboldened by direct ownership and control over their creations, forge a direct relationship with their audience. This direct engagement not only fosters greater transparency and trust but also enables creators to tailor content offerings to meet evolving consumer preferences.

Maturing Technologies and Evolutionary Trajectory, as these technologies mature, their evolutionary trajectory portends a seismic shift in the digital content landscape. Continuous innovations augment user experiences, enhance interoperability, and fortify security mechanisms. These advancements promise a future where content distribution is not just seamless but also ensures equitable compensation for creators, fostering a sustainable creative economy.

7.3 Fostering Equitable Access and Innovation

Equitable access to digital content stands as a cornerstone of this transformative landscape. The amalgamation of smart contracts and Video DRM obliterates barriers to access, ensuring that consumers enjoy secure and seamless access to diverse content offerings. This democratization of content distribution cultivates an environment where innovation thrives and talent finds recognition, irrespective of geographical barriers.

The convergence of cutting-edge technological innovations and collaborative synergies within the digital content landscape heralds the dawn of a new era—a frontier marked by transformative advancements. This visionary landscape envisions an ecosystem characterized by equitable access to content, fortified content protection mechanisms, and an environment that empowers creators while fostering trust among all stakeholders involved. At the crux of this transformative landscape lies the amalgamation of technological innovations, weaving together smart contracts, Decentralized Applications (DApps), and Video Digital Rights Management (DRM). This convergence heralds a paradigm shift toward a future where the digital content ecosystem thrives on the principles of fairness, transparency, and empowerment. This visionary future envisions equitable access to digital content for users across diverse demographics and geographies.

It seeks to dismantle barriers to access by leveraging technological innovations that ensure seamless and unrestricted content consumption. Through collaborative efforts, technological advancements aim to democratize access, allowing users to explore, engage with, and derive value from an expansive array of digital content offerings. Moreover, the fortified content protection mechanisms ushered in by the convergence of these innovations represent a cornerstone of trust and security within the digital content landscape. The integration of smart contracts within Video DRM fundamentally reshapes how content is safeguarded and distributed.

By embedding access rights, licensing agreements, and royalty distributions directly within digital assets, this innovative framework ensures a secure, transparent, and tamper-resistant environment. This not only safeguards the intellectual property of creators but also engenders trust among stakeholders, promoting fair compensation and adherence to agreed-upon usage terms. Yet, the path toward this visionary frontier is not devoid of challenges. The hurdles that impede the widespread adoption of smart contracts, DApps, and Video DRM pose formidable obstacles.

However, these challenges, although significant, are not insurmountable. Collaborative efforts among industry pioneers, technological innovations aimed at scalability and interoperability, and adaptive regulatory frameworks collectively propel the industry toward a future where content distribution transcends barriers. Embracing this transformation requires a collective commitment to innovation, collaboration, and adaptability. Collaborative ecosystems, where stakeholders converge to refine and enhance these technologies, are fostering an environment conducive to unprecedented growth and evolution. Through these concerted efforts, the industry is evolving toward a future where content distribution is seamless, creators wield unprecedented control over their intellectual property, and consumers relish secure access to a diverse array of digital content offerings.

In essence, the pioneering efforts aimed at harnessing technological innovations and fostering collaborative synergies are paving the way for a new digital frontier—one characterized by inclusivity, fortified content protection mechanisms, and an empowered ecosystem that transcends boundaries. As the industry navigates these transformative waters, the collective pursuit of innovation holds the promise of reshaping the digital content landscape, propelling it toward an era defined by equitable access, empowerment, and trust among all stakeholders involved. User Fostering Equitable Access and Innovation Equitable access to digital content stands as a cornerstone of this transformative landscape. The amalgamation of smart contracts and Video DRM obliterates barriers to access, ensuring that consumers enjoy secure and seamless access to diverse content offerings. This democratization of content distribution cultivates an environment where innovation thrives and talent finds recognition, irrespective of geographical barriers. expand ChatGPT Fostering Equitable Access and Innovation At the heart of this transformative digital landscape lies the fundamental principle of equitable access to digital content—an indispensable cornerstone shaping the future of the digital content ecosystem [17].

The seamless amalgamation of smart contracts and Video Digital Rights Management (DRM) marks a revolutionary stride in dismantling barriers to access, ensuring that consumers across diverse demographics and geographies enjoy secure, transparent, and seamless access to a myriad of content offerings. This democratization of content distribution not only engenders inclusivity but also cultivates an environment where innovation flourishes and talent finds recognition, irrespective of geographical constraints. The convergence of smart contracts and Video DRM represents a watershed moment in content distribution, heralding an era where access barriers are obliterated. Traditionally, access to premium digital content often encountered impediments due to geographical restrictions, licensing complexities, or centralized control mechanisms [20]. The integration of smart contracts within Video DRM fundamentally reshapes this landscape, embedding access rights and licensing agreements directly within digital assets. This innovative framework ensures that consumers can access a diverse array of digital content seamlessly, securely, and in adherence to predefined usage terms, transcending geographical boundaries and offering a level playing field for all users. By democratizing content distribution, this transformative landscape nurtures an environment conducive to innovation. The removal of barriers to access enables content creators, irrespective of their location, to showcase their creativity and share their content with global audiences.

This unfettered access facilitates the discovery of diverse content, propelling lesser-known creators onto the global stage and allowing innovation to thrive uninhibited by geographical constraints. Furthermore, this democratization of content distribution empowers creators and innovators by offering them a platform where talent finds recognition based on merit rather than geographical advantages.

Smart contracts and Video DRM foster an environment where content is disseminated based on its quality and relevance, empowering creators to reach audiences directly without intermediaries or gatekeepers. This empowerment not only fuels innovation but also incentivizes creators to push the boundaries of creativity, knowing that their content has the potential to resonate with audiences worldwide.

The transformative impact of equitable access to digital content extends beyond consumer empowerment; it nurtures an ecosystem where creativity, innovation, and talent converge. By transcending geographical barriers and offering a level playing field, this democratization of content distribution fosters an environment where content creators are celebrated for their ingenuity, innovation thrives, and the digital content landscape evolves into a vibrant tapestry of diverse offerings. In essence, the amalgamation of smart contracts and Video DRM to foster equitable access to digital content signifies a monumental leap toward an inclusive and innovation-driven digital content ecosystem. This transformative landscape not only empowers consumers with seamless access but also empowers creators to unleash their creativity on a global stage. As barriers to access crumble, innovation flourishes, and talent finds recognition, this visionary landscape holds the promise of reshaping the digital content ecosystem into a dynamic, inclusive, and talent-driven arena where innovation knows no bounds.

7.4 Reimagining Trust and Transparency

Trust and transparency underpin every facet of this paradigm shift. The immutable and transparent nature of blockchain, integrated within smart contracts and DApps, fosters an environment where trust is not merely an aspiration but an inherent attribute. Stakeholders engage with confidence, assured of fair compensation, transparent transactions, and adherence to contractual obligations. Collaborative Horizons and Industry Synergies Collaborative endeavors weave a tapestry of synergies among industry stakeholders. Partnerships between technology innovators, content creators, regulatory bodies, and consumers drive collective efforts to surmount challenges and harness the transformative potential of integrated technologies [19]. These collaborative horizons propel the industry toward a future marked by innovation, fairness, and inclusivity. Conclusion: Embracing Transformational Realms In essence, the convergence of smart contracts, DApps, and Video DRM propels the content protection landscape into transformational realms. It marks a technological evolution and a fundamental redefinition of the creator-consumer relationship. As these technologies evolve and integrate further, they will undoubtedly reshape the digital content landscape, establishing a fair, secure, and innovative environment for all stakeholders involved

References

1. Radziwill, N. 2018. Blockchain Revolution: How the Technology Behind Bitcoin is Changing Money, Business, and the World. *Quality Management Journal, 25*(1), 64–65. https://doi.org/10.1080/10686967.2018.1404373
2. Smart Contracts and Smart Properties Over Blockchains. 2017. *Decentralized Computing Using Blockchain Technologies and Smart Contracts*, 44–63. https://doi.org/10.4018/978-1-5225-2193-8.ch004
3. Christidis, K., and Devetsikiotis, M. 2016. Blockchains and Smart Contracts for the Internet of Things. *IEEE Access, 4*, 2292–2303. https://doi.org/10.1109/access.2016.2566339
4. Finck, M. 2018. Blockchains and Data Protection in the European Union. *European Data Protection Law Review, 4*(1), 17–35. https://doi.org/10.21552/edpl/2018/1/6
5. DAPPS 2022 Committees. 2022. *2022 IEEE International Conference on Decentralized Applications and Infrastructures (DAPPS)*. https://doi.org/10.1109/dapps55202.2022.00007
6. Gandhi, C., Shukla, N., Kaur, G., and Yadav, K. 2020. Blockchain Technology: Concept, Applications, Challenges, and Security Threats. *Blockchain Applications in IoT Ecosystem*, 77–104. https://doi.org/10.1007/978-3-030-65691-1_6
7. Drescher, D. 2017. Identifying and Protecting User Accounts. *Blockchain Basics*, 93-101. https://doi.org/10.1007/978-1-4842-2604-9_12
8. Cai, W. 2018. Decentralized Applications: The Blockchain-Empowered Software System. *IEEE Access*, 6, 53019–53033. https://doi.org/10.1109/access.2018.2870644
9. Eyal, I. 2014. Majority Is Not Enough: Bitcoin Mining Is Vulnerable. *Financial Cryptography and Data Security*, 436–454. https://doi.org/10.1007/978-3-662-45472-5_28
10. Wright, A., and De Filippi, P. 2015. Decentralized Blockchain Technology and the Rise of Lex Cryptographic. *SSRN Electronic Journal*. https://doi.org/10.2139/ssrn.2580664
11. Hewa, T., Ylianttila, M., and Liyanage, M. 2021. Survey on blockchain based smart contracts: Applications, opportunities, and challenges. *Journal of Network and Computer Applications, 177*, 102857. https://doi.org/10.1016/j.jnca.2020.102857
12. Farokhnia, S. 2023. Lazy contracts: alleviating high gas costs by secure and trustless off-chain execution of smart contracts. https://doi.org/10.14711/thesis-991013222949803412
13. Gilmour, P. M. 2022. Decentralized Blockchain Technology: Towards a Trusted and Transparent Beneficial Ownership Regime. *Financial Technology and the Law*, 185–209. https://doi.org/10.1007/978-3-030-88036-1_8
14. Hewa, T. M., Hu, Y., Liyanage, M., Kanhare, S. S., and Ylianttila, M. 2021. Survey on Blockchain-Based Smart Contracts: Technical Aspects and Future Research. *IEEE Access, 9*, 87643–87662. https://doi.org/10.1109/access.2021.3068178
15. Zou, W., Lo, D., Kochhar, P. S., Le, X.-B. D., Xia, X., Feng, Y., Chen, Z., and Xu, B. 2021. Smart Contract Development: Challenges and Opportunities. *IEEE Transactions on Software Engineering, 47*(10), 2084–2106. https://doi.org/10.1109/tse.2019.2942301
16. Khan, M., and Naz, T. 2021. Smart Contracts Based on Blockchain for Decentralized Learning Management System. *SN Computer Science, 2*(4). https://doi.org/10.1007/s42979-021-00661-1
17. Schmitz, A. J. 2022. Making Smart Contracts. *SSRN Electronic Journal*. https://doi.org/10.2139/ssrn.4188129
18. Di Angelo, M., and Salzer, G. 2020. Characterizing Types of Smart Contracts in the Ethereum Landscape. *Financial Cryptography and Data Security*, 389–404. https://doi.org/10.1007/978-3-030-54455-3_28
19. Shafagh, H., Burkhalter, L., Hithnawi, A., and Duquennoy, S. 2017. Towards Blockchain-based Auditable Storage and Sharing of IoT Data. *Proceedings of the 2017 on Cloud Computing Security Workshop*. https://doi.org/10.1145/3140649.3140656
20. Jerbi, W., Cheikhrouhou, O., Guermazi, A., Hamam, H., and Trabelsi, H. 2021. A Blockchain-based Authentication Scheme for Mobile Data Collector in IoT. *2021 International Wireless Communications and Mobile Computing (IWCMC)*. https://doi.org/10.1109/iwcmc51323.2021.9498656
21. Patel, A., and Singh, B. 2023. Implementation of Smart Contract Using Ethereum Blockchain. *Communications in Computer and Information Science*, 160–169. https://doi.org/10.1007/978-3-031-25088-0_13.

22. Christlan, C. 2010. Truth About Abs Review Uncensored. *SciVee*. https://doi.org/10.4016/18946.01
23. Taherdoost, H. 2022. A Critical Review of Blockchain Acceptance Models—Blockchain Technology Adoption Frameworks and Applications. *Computers, 11*(2), 24. https://doi.org/10.3390/computers11020024
24. Macrinici, D. 2018. Smart contract applications within blockchain technology: A systematic mapping study. *Telematics and Informatics, 35*(8), 2337–2354. https://doi.org/10.1016/j.tele.2018.10.004
25. Rouhani, S. 2019. Security, Performance, and Applications of Smart Contracts: A Systematic Survey. *IEEE Access, 7*, 50759–50779. https://doi.org/10.1109/access.2019.2911031
26. Destefanis, G. 2018. Smart contracts vulnerabilities: A call for blockchain software engineering? *2018 International Workshop on Blockchain-Oriented Software Engineering (IWBOSE)*. https://doi.org/10.1109/iwbose.2018.8327567
27. Zou, W. 2021. Smart Contract Development: Challenges and Opportunities. *IEEE Transactions on Software Engineering, 47*(10), 2084–2106. https://doi.org/10.1109/tse.2019.2942301
28. Khan, S. N. 2021. Blockchain smart contracts: Applications, challenges, and future trends. *Peer-to-Peer Networking and Applications, 14*(5), 2901–2925. https://doi.org/10.1007/s12083-021-01127-0
29. Sharma, P. 2021. Blockchain-based decentralized architecture for Cloud storage system. *Journal of Information Security and Applications, 62*, 102970. https://doi.org/10.1016/j.jisa.2021.102970
30. Khatoon, A. 2020. A Blockchain-Based Smart Contract System for Healthcare Management. *Electronics, 9*(1), 94. https://doi.org/10.3390/electronics9010094
31. Wang, S. 2019. Blockchain-Enabled Smart Contracts: Architecture, Applications, and Future Trends. *IEEE Transactions on Systems, Man, and Cybernetics: Systems, 49*(11), 2266–2277. https://doi.org/10.1109/tsmc.2019.2895123

CHAPTER 9

An Intelligent Approach Using Data Analysis Techniques for the Automobile Industry

Rajendrani Mukherjee,[1,*] *Ankur Priyadarshi*,[2] *Krishnendu Ghosh*[3] and *Aniket Samanta*[4]

1. Introduction

Data science and machine learning play key roles in predicting the future of the automobile industry. The industry has seen a remarkable application of machine learning regarding procurement, logistics, marketing, sales related issues. This research navigates the application of AI technologies in the automotive industry by performing an in-depth data analysis using a benchmark dataset. The article demonstrates how modern technologies can make the automotive industry more efficient and enhance its customer focus throughout all its operations and activities.

The subject automobile dataset was collected from the UCI machine learning repository. It possesses the specifications reminiscent of an automobile encompassing a wide range of characteristics. The attributes present in this dataset are categorical, integer bases and real in nature. Some fields in the dataset are multivariate in nature, i.e., it consists of individual measurements which are a function of more than two variables. The dataset contains information about engine type, production company, fuel type used by the automobile, engine location, number of cylinders, etc. This dataset helps in predicting the price of a car depending on the type of fuel used, manufacturer, engine type, etc.

[1] Department of CST & CSIT, University of Engineering and Management, Kolkata, India.
[2] Department of Computer Science and Engineering, Saharsa College of Engineering, India.
[3] Department of CSIT, University of Engineering and Management, Kolkata, India.
[4] Department of CST, University of Engineering and Management, Kolkata, India.
* Corresponding author

In the dataset, vehicles are initially designated with a predetermined risk factor that is specifically assigned per their price range. The symbol is then adjusted by incrementally moving it up (or down) based on the corresponding risk level, in a standardized procedure known as symbolling. According to the dataset, +3 indicates that the auto is too risky and –3 indicates that it is quite safe. So, the symboling parameter varies within the range of –3 to +3.

In this chapter, the dataset was pre-processed (deleting duplicate records, replacing NULL values with average values, discarding very few rows with junk data, etc.). To clean the dataset, replace (), fillna (), etc. functions were used. As part of the EDA (E xploratory D ata A nalysis), the research performed an analysis to find out which company is the biggest manufacturer of automobiles, which sort of fuel is most popularly used by the automobile, what kind of body trend is most popular, etc. Correlation analysis was performed to find out the relation between several continuous variables like the length, width, weight, and price of the automobile. For example, it was found that the height of the automobile doesn't show any substantial trend with the price. A regression analysis was performed on the subject dataset to build an efficient model to maximize the profit of car manufacturers.

The data analysis was represented using several data visualization techniques (box plots, scatter plots, histograms, etc.). This analysis will help to predict which automobile is worth what and it will help the customers to find automobiles specific to their choice. The car manufacturers will concentrate on the client's requirement and produce cars that are the most profit worthy.

The structure of the chapter is outlined as follows: Section 2 undertakes a comprehensive survey of the existing literature by providing detailed insights into the application of IDA (I ntelligent D ata A nalysis) across various domains. Section 3 details exploratory data analysis methodologies and the derived results graphically. Section 4 represents an analysis of correlation using maps, correlation coefficients, etc. Section 5 is dedicated to conducting regression analysis, while Section 6 focuses on performing error analysis. Finally, Section 7 concludes the work by providing insights on future scope and possibilities.

2. Literature Survey

Machine Learning has demonstrated significant success in various industries, playing a pivotal role in prediction, automation, and more. Additionally, the domain of AI has been greatly enriched by several IDA (Intelligent Data Analysis) techniques, as evidenced by the noteworthy contributions of AI [8] [9] [13] [14] [17] [22]. These advancements have provided valuable insights and empowered effective decision-making processes.

In 1997, John T. Behrens described the differences between classical data analysis and exploratory data analysis using different visualization methods [1]. ETL (Extract-Transform-Load) technology and data visualization play an important role in concluding with evidence [6]. Ehrlinger et al. conducted a comprehensive exploration of various data quality management tools, highlighting their remarkable effectiveness in addressing data quality challenges [5]. Ghosh et al. have examined several industrial datasets using different data exploration tools for exploratory analysis [2].

In 2017, V. Chang applied IDA (Intelligent Data Analysis) for weather forecasting [18]. The banking or finance sector has also benefitted from the usage of intelligent ML techniques. Priya et al. utilized a random forest for the prediction of loan Privilege for customers [3]. On the other hand, analysis of clinical data to predict and prevent diseases is also another promising avenue. Konopka et al. explored clinical study data to perform EDA (E xploratory D ata A nalysis) [4]. The authors utilized multidimensional datasets to analyze the utility of several data analysis techniques. Several researchers [7] [20] have leveraged medical data to check the implementation potential of AI techniques in curbing diseases. In 2017, Kamilaris et al. applied several intelligent algorithms in the agriculture domain and proved their usefulness [16].

The application of ML techniques in the automobile industry is an age-old practice[15] [19] [21]. Yaqi et al. clubbed random forest and nearest neighbour classifiers and detected automobile insurance frauds [10]. Another study conducted by Yibo and Wei utilized deep learning methodologies for detecting fraud [11]. Sachin S. Kamble et al. worked on automobile component manufacturing data using multi-group decision-making techniques [12].

All these related works motivated us to conduct our study in this domain. It is believed EDA (E xploratory D ata A nalysis) will make important decision making prompts and lucrative gains.

3. Exploratory Data Analysis

The automobile dataset utilized in this study was sourced exclusively from the UCI Machine Learning Repository, a renowned and reliable source of high-quality datasets. This dataset encompasses a wide range of attributes, comprising both categorical and integer/real types, providing comprehensive and diverse information for analysis and exploration. Some fields in the dataset are multivariate. The dataset contains info about engine type, production company, fuel type used by the automobile, engine location, number of cylinders, etc.

As part of the EDA (exploratory data analysis), the research performed an analysis to find out which company is the chief manufacturer of automobiles, which sort of fuel is most prevalently used by the automobile, what kind of body trend is most popular, etc. Correlation analysis was performed to find out the relation between several continuous variables like the length, width, weight, and price of the automobile.

From Fig. 1 it can be inferred that Toyota manufactured the highest number of cars. Toyota, Nissan and Mazda are the top manufacturers in terms of number of cars. From Fig. 2, it can be seen that 90% of the cars in the dataset use p etrol. The groupby () function is used to split the data into groups based on the body-style and make of the car.

From Fig. 3 it can be seen that Toyota prefers a hatchback while Nissan prefers a sedan, etc. From Fig. 4 it can be seen that Chevrolet is the brand that had the highest mileage. From Fig. 5 it can be seen that Mercedes Benz, BMW and Porsche sell the most expensive cars.

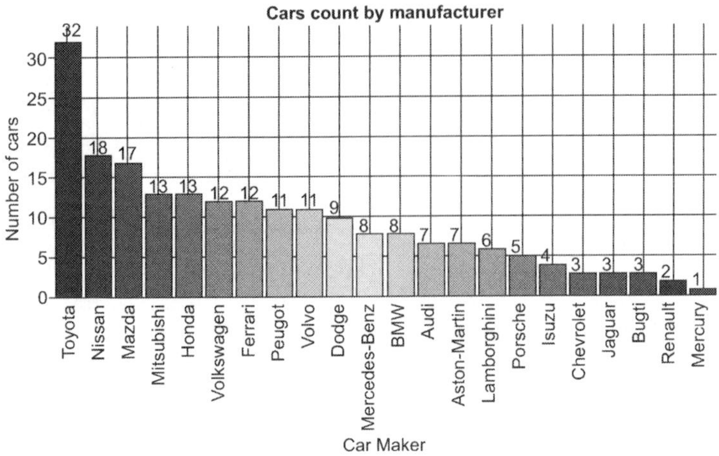

Figure 1: Analysis of Car Makers.

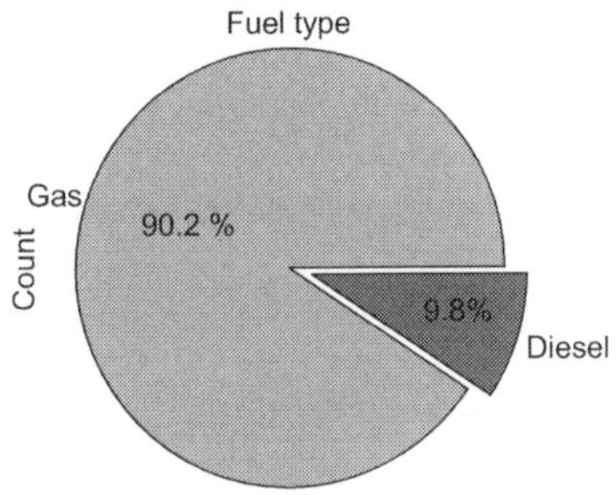

Figure 2: Analysis of Fuel Type.

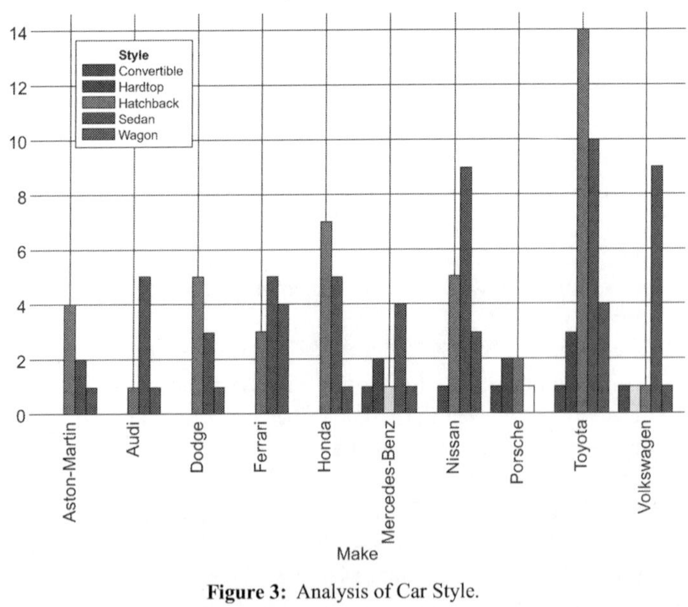

Figure 3: Analysis of Car Style.

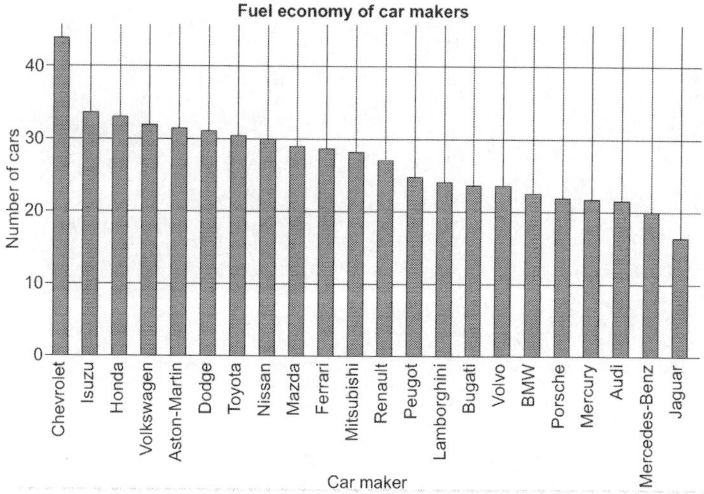

Figure 4: Analysis of Fuel E onomy or Mileage.

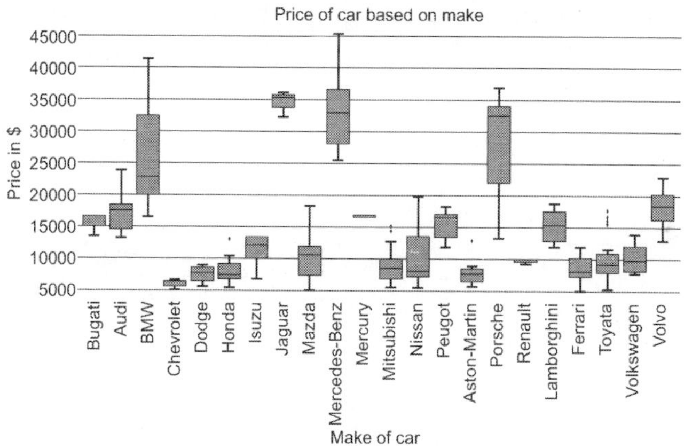

Figure 5: Analysis of Car Price Range.

4. Correlation Study

To understand correlation coefficients, a heatmap is generated and represented in Fig. 6. From the heat map it is evident that the following features, that is width, weight, engine size and horsepower have the highest correlation with the price. We can see the correlation coefficients of these four features in Table 1.

From Table 1 it can be seen that width, curb-weight, engine-size, and horsepower has a positive correlation with price. From Fig. 7 it is evident that price is negatively correlated with highway-mpg. The Corr () function was utilized to perform this correlation.

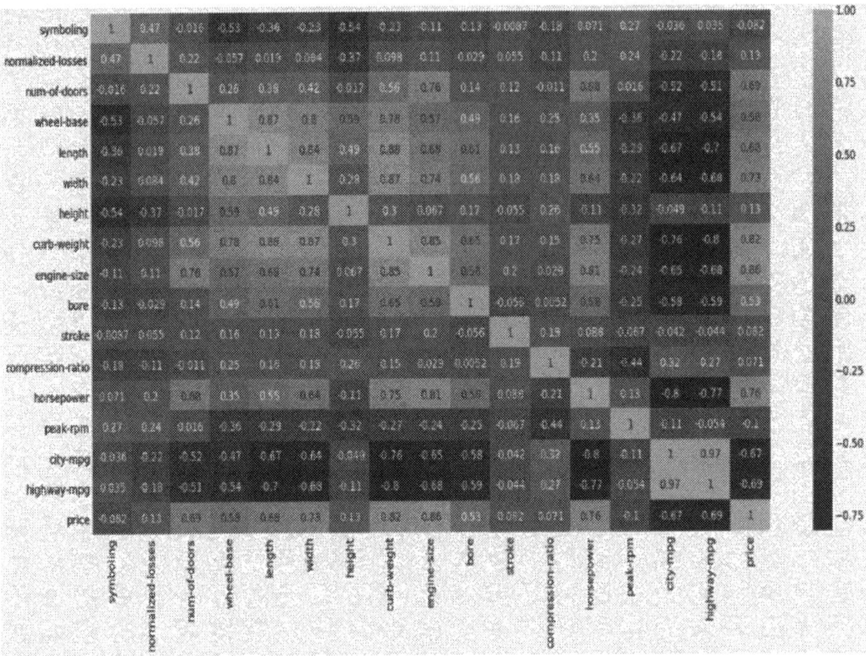

Figure 6: Representation of Correlation Coefficients.

Table 1: Correlation Coefficients.

	width	curb-weight	engine-size	horsepower	highway-mpg	Price
width	1.00	0.87	0.74	0.64	–0.68	0.73
curb-weight	0.87	1.00	0.85	0.75	–0.79	0.82
engine-size	0.74	0.85	1.00	0.81	–0.67	0.86
horsepower	0.64	0.75	0.81	1.00	–0.77	0.75
highway-mpg	–0.68	–0.8	–0.68	–0.77	1.00	–0.69
price	0.73	0.82	0.86	0.75	–0.69	1.00

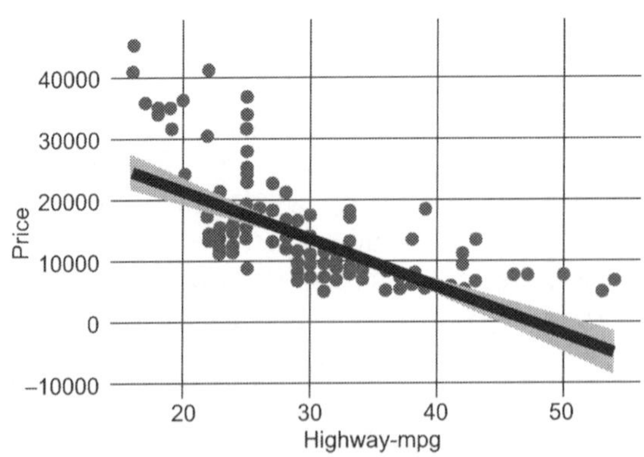

Figure 7: Representation of Correlation Coefficients.

5. Regression Analysis

A comprehensive regression analysis was conducted on the subject dataset, aiming to construct a proficient model that enhances the profitability of car manufacturers. The focal point of this analysis lies in the utilization of linear regression, where a linear function is employed to forecast the response or dependent variable based on the predictor variable. This approach holds the capability to accurately predict the outcome by establishing a direct relationship between the response and predictor variables through a linear function. The regplot () method is used to plot the data using linear regression. Here highway-mpg is the predictor variable and the price is the dependent variable.

The i ntercept value was obtained as 37470.66 and the coefficient as -789.026. The regression equation of the above model is:

$$\text{price} = 37470.66 - 789.026 * \text{highway-mpg} \tag{1}$$

Multiple Linear Regression, a widely used technique in predictive modelling, involves incorporating multiple predictors to enhance the accuracy of price prediction. In the context of this research, the analysis focused on four specific predictor variables. Based on the correlation table obtained in the previous section, it can be inferred that the following factors serve as strong indicators of price:

- Horsepower
- Curb-weight
- Engine-size
- Width

The equation of the above regression model is:

$$\begin{aligned}\text{price} = -28994.40 + (24.44 \text{ x horsepower}) + (3.36 \text{ x curb-weight}) \\ + (97.46 \text{ x engine-size}) + (283.34 \text{ x width})\end{aligned} \tag{2}$$

A comprehensive examination was undertaken to meticulously compare and contrast the distribution of the fitted values derived from the model with that of the actual values.

From Fig. 8, it is evident that the fitted values closely align with the actual values, as the two distributions exhibit a considerable overlap.

6. Error Analysis

When evaluating our models in this research study, it is essential to not only visualize the results but also employ quantitative measures to assess their accuracy. In this regard, two significant metrics, namely Mean Squared Error (MSE) and R-squared, play a crucial role in determining the model's accuracy in regression analysis. R-squared, also referred to as the coefficient of determination, measures the proximity of the data to the fitted regression line. It provides a statistical measure to evaluate how well the model represents the observed data. Similarly, the Mean Squared Error (MSE) calculates the average of the squared errors, which represents

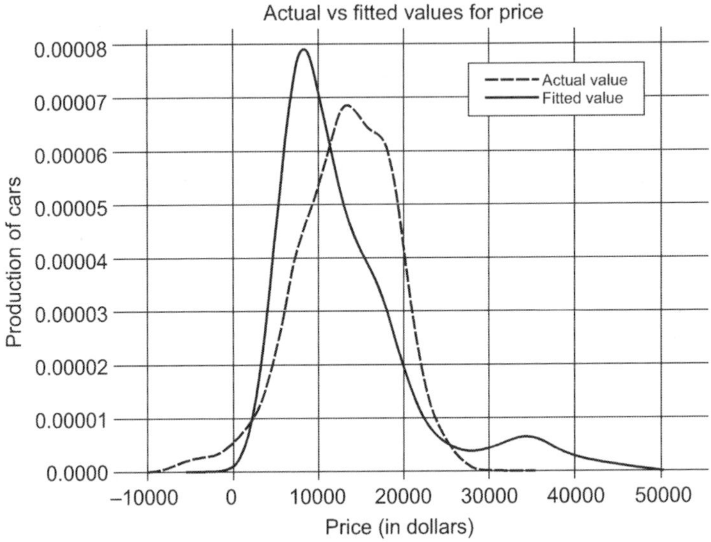

Figure 8: Representation of Actual vs Fitted Values for Price.

Table 2: Comparisons of Measure of Errors.

Model	Variable	R-Squared Value	MSE
SLR	Using Highway-mpg as a forecaster variable for price estimation and analysis.	0.47682599948512866	3.2 x10^7
MLR	Using Horsepower, Curb-weight, Engine-size, and Highway-mpg as forecaster variables, their relationship with the price of a car can be analysed.	0.7768104805018861	1.3 x10^7

the difference between the actual value (y) and the estimated value ($\hat{y}$). By utilizing these metrics, we can comprehensively evaluate the accuracy of our models and gain insights into their performance. The combination of visual analysis and quantitative measures ensures a robust evaluation process to make informed decisions based on the obtained results.

The model with the higher R-squared value is a better fit for the data while the model with the smallest MSE value is a better fit for the data. So, in this case, considering both R-squared and MSE values (as represented in Table 2) multiple linear regression model is the best fit.

7. Conclusion

Automated exploratory data analysis (autoEAD) has become an important avenue nowadays. In this research, an intelligent technique is leveraged to predict certain outcomes regarding the car manufacturing industry. As the benchmark dataset from the UCI repository is processed and executed, it was possible to gain some valuable insights such as, Toyota manufactured the highest number of cars, 90% of the cars in the dataset use petrol, Toyota prefers hatchbacks while Nissan prefers sedans, etc. As correlation analysis was done and a heatmap was generated, it was observed that width, weight, engine-size, and horsepower have a positive correlation with price while highway-mpg is negatively correlated with price. As regression analysis was performed, obtained R squared and MSE values indicated that the multiple linear regression model is the best fit. The analysed data was visualized utilizing several box plots, scatter plots, histograms, etc. Thus, this research contributes to the domain of the automobile/transportation industry with some prominent insights that are gathered by IDA (Intelligent Data Analytics). As part of the future scope, more subject datasets regarding this domain should be experimented with while different recent techniques will be explored to gain more analytical decisions for addressing economic/customer needs.

References

1. Behrens, J. T. 1997. Principles and Procedures of Exploratory Data Analysis. *Psychological Methods, 2*(2), 131–160.
2. Ghosh, A., Nashaat, M., Miller, J., Quader, S., and Marston, C. 2018. A Comprehensive Review of Tools for Exploratory Analysis of Tabular Industrial Datasets. *Visual Informatics, 2*(4), 235–253.
3. UlagaPriya, K., Pushp, S., Kalaivani, K., and Sartiha, A. 2018. Exploratory Analysis on Prediction of Loan Privilege for Customers using Random Forest. *International Journal of Engineering & Technology, 7*(2.21), 339–341.
4. Konopka, B. M., Lwow, F., Owczarz, M., and Łaczmański, L. 2018. Exploratory Data Analysis of a Clinical Study Group: Development of a Procedure for Exploring Multidimensional Data. *PLoS ONE*, pp. 1–21, DOI: 10.1371/journal.pone.0201950
5. Ehrlinger, L., Rusz, E., and Wöß, W. 2019. A Survey of Data Quality Measurement and Monitoring Tools. arXiv:1907.08138.
6. Waas, F., Wrembel, R., Freudenreich, T., Thiele, M., Koncilia, C., and Furtado, P. 2013. On-Demand ELT Architecture for Right-Time BI: Extending the Vision. *International Journal of Data Warehousing and Mining, 9*(2), 21–38, DOI:10.4018/jdwm.2013040102
7. Amar, L. A., Taha, A. A., and Mohamed, M. Y. 2020. Prediction of the final size for COVID-19 epidemic using machine learning: A case study of Egypt. *Infectious Disease Modelling*, pp. 622–634. https://doi.org/10.1016/j.idm.2020.08.008
8. Song, G., Rochas, J., El Beze, L. E., Huet, F., and Magoulès, F. 2016. K Nearest Neighbour Joins for Big Data on MapReduce: A Theoretical and Experimental Analysis. *IEEE Transactions on Knowledge and Data Engineering, 28*(9), 2376–2392. https://doi.org/10.1109/TKDE.2016.2562627
9. Judd, C. M., McClelland, G. H., and Ryan, C. S. 2017. Data Analysis: A Model Comparison Approach to Regression, ANOVA, and Beyond. 3rd edition, ISBN–978-1-315-74413-1.
10. Li, Y., Yan, C., Liu, W., Li, M. 2018. A Principle Component Analysis-based Random Forest with the Potential Nearest Neighbor Method for Automobile Insurance Fraud Identification. Applied Soft Computing, pp. 1000–1009.
11. Wang, Y., and Xu, W. 2018. Leveraging deep learning with LDA-based text analytics to detect automobile insurance fraud. *Decision Support Systems, 105*, 87–95, doi: 10.1016/j.dss.2017.11.001.
12. Kamble, S. S., Belhadi, A., Gunasekaran, A., Ganapathy, L., and Verma, S. 2021. A large multi-group decision-making technique for prioritizing the big data-driven circular economy practices in the automobile component manufacturing industry. *Technological Forecasting and Social Change, 165*, 120567, ISSN 0040-1625
13. Elliott V. 2018. Thinking about the coding process in qualitative data analysis. *Qualitative Report, 23*(11), 2850–2861. https://doi.org/10.46743/2160-3715/2018.3560
14. Mukherjee, R., and Patnaik, K. S. 2019. Introducing a Fuzzy Model for Cost Cognizant Software Test Case Prioritization. IEEE TENCON 2019: Recent Advances in Program Analysis and Software Testing (RAPAST), Page 502–507, 17–20th October, 2019, Kerala, India.
15. Hadi, W., El-Khalili, N., AlNashashibi, M., Issa, G., AlBanna, A. A. 2019. Application of data mining algorithms for improving stress prediction of automobile drivers: A case study in Jordan. *Computers in Biology and Medicine*, 103474. https://doi.org/10.1016/j.compbiomed.2019.103474

16. Kamilaris, A., Kartakoullis, A., and Prenafeta-Boldú, F. X. 2017. A review on the practice of big data analysis in agriculture", *Computers and Electronics in Agriculture*, pp. 23–37. https://doi.org/10.1016/j.compag.2017.09.037
17. Mukherjee, R., Panday, A. N., Nandy, S., and Ghosh, S. 2022. Face Mask Detector Using Convolutional Neural Networks. 5th International Conference on Innovative Computing and Communications, ICICC, New Delhi, Lecture Notes in Networks and Systems, 473, pp. 377–384, Springer, Singapore.
18. Chang V. 2017. Towards data analysis for weather cloud computing. *Knowledge-Based Systems, 127*, pp. 29–45. https://doi.org/10.1016/j.knosys.2017.03.003
19. Mukherjee, R., Datta, D., Ganguly, G., Bandopadhyay, S., Chakraborty, C., Pal, B., and Chatterjee, D. 2023. Application of Data Visualization: Realization of Car Rental System. Lecture Notes in Networks and Systems, vol. 491. Springer, Singapore. https://doi.org/10.1007/978-981-19-4193-1_4
20. Parmar, C., Barry, J. D., Hosny, A., Quackenbush, J., and Aerts, H. J. W. L. 2018. "Data Analysis Strategies in Medical Imaging. *Clinical Cancer Research, 24*(15), 3492–3499. https://doi.org/10.1158/1078-0432.ccr-18-0385
21. Pillai, R., Sivathanu, B., Mariani, M., Rana, N., Yang, B., and Dwivedi, Y. 2021. Adoption of AI-empowered industrial robots in auto component manufacturing companies. *Production Planning & Control., 33*(1): 1–17, DOI:10.1080/09537287.2021.1882689
22. Kong, D-Y., Ma, S -J., Tang, W -C., Xue, Y. 2022. The policy effect on automobile industry considering the relationship between technology, market and production: the dual-credit policy as an example. *Transportation Letters The International Journal of Transportation Research 15*(7):1–15, DOI:10.1080/19427867.2022.204028

CHAPTER 10

IoT Integration and Cryptographically Secured Solutions towards Intelligent Elevator Systems

Ankit Jain[1,*] and *Anita Shukla*[2]

1. Introduction

User authentication in smart devices has become very important in modern aspects. Authentication makes sure that only legitimate or authorized users can use these smart systems. Authentication comes under cryptographic goals [1][2]. Other goals are secrecy, data integrity and non-repudiation. Authentication can be broadly classified as entity authentication and message authentication [3][4]. In smart devices (as presented in this chapter), entity authentication is of utmost importance to make sure that only authorized users can get access to use/operate. It is very important to mention here that entity authentication for smart devices can be achieved through various possible ways as shown in below Fig. 1. The first category is something you have for example dedicated RFID cards, tags or ATMs, etc. The s econd category is something you are, that is it involves your biometric or physical attributes like a fingerprint scan, retina detection or face detection [5][6]. The third category is something you know applying the use of a PIN (Personal Identification Number), security passwords or OTP (One Time Password), etc. [7][8]. A fourth category can also be mentioned as a hybrid category which combines any two of the above-mentioned categories. So here we select something you have a category for required authentication.

[1] Department of Electronics and Communication Engineering, Pranveer Singh Institute of Technology, Kanpur, Uttar Pradesh, India.

[2] Department of BSH, Pranveer Singh Institute of Technology, Kanpur, Uttar Pradesh, India.

* Corresponding author

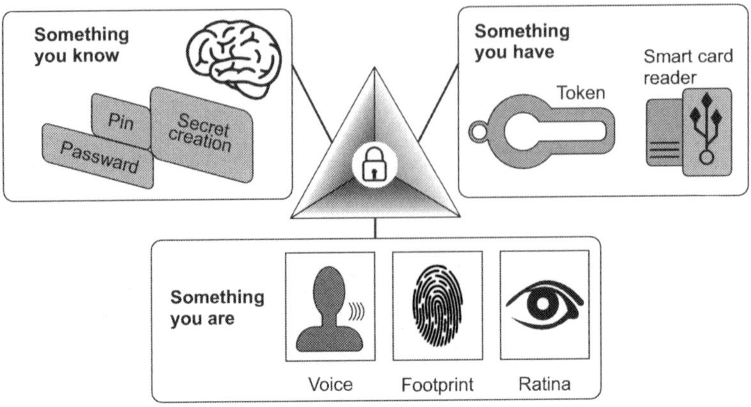

Figure 1: Showing Methods of Achieving Authentication.

Concerns about the intelligence, security, and dependability of elevators are growing among elevator users as well as manufacturers. We demonstrate this point by using a few pertinent references to show how many people have suffered from numerous flaws in recent years as a result of an over-reliance on antiquated technology.

M. McCann et al. [9], in their report, have discussed that "caught in/between' and "struck by" deaths are there because of the improper control of the elevator door. V. Zarikas et al. [10] have done a statistical survey of elevator accidents in Greece. The authors have tried to correlate various involved factors which may help to prevent future accidents. Approximately over 50% of accidents are due to electrical installation problems. N.H. Kim et al. [11] have classified various elevator failures concerning components, locations and types of failures. L.Y. Hui et al. [12] have mentioned that emergency calling through cell phones is almost impossible because of network unavailability inside elevators. So, the authors have proposed a sort messaging service (SMS) and call transmission method for emergencies to the cell phone of the person who is registered with a system called Emergency Call Management System (ECMS). J.B. Park et al. [13] have given a comparative study on safety factors related to elevator management. The authors have used reliability statistics to define the most important factors for elevator safety. R. Dhanasekaran et al. [14] have given an analytical view of elevator safety. The authors have classified elevator accidents for the age group of users and discussed some safety measurements too. N.A. Siti et al. [15] have discussed 43 types of defects related to management-maintenance issues with operational defects. Temperature-related issues and inoperative push buttons are also one of the reasons. A.A. Sakhare et al. [16] have discussed the causes of elevator failure and some advanced systems. Fire, improper usage and lock-out problems are also very dangerous and they have suggested using safety gear as a protective device. G.R. Mohite et al. [17] have discussed that some organizations like the American Society of Mechanical Engineers (ASME) have drafted some regulations regarding the safe operations of elevators. Sensor failure and faulty electronic components are the sources of serious vulnerabilities. Z. Ming et al. [18] have given an elevator monitoring system using IoT (Internet of Things). The design combines browser-server and client-server architectures. The required transmission of data can be done on a real-time basis. A. Shrestha [19] in her thesis submitted to the University of Mississippi has discussed the safety considerations for modern elevator systems along with future aspects. X. Surui [20] has discussed the involvement of human factors in elevator accidents and for that, 53 accidents have been taken as a sample and a correlation among various factors has been established using the chi-square test and ratio analysis. A. Shubham et al. [21] have reviewed the safety (as a feature) of elevator systems. Below Table 1 provides a glimpse of the consequences.

It is very important to mention here that heating effects and faulty electronic components are also the reasons for accidents. Y. Xing et al. [22] have analyzed various factors responsible for escalator/elevator accidents in metro cities of China. The authors have used the Bayesian network to find out the associated risk factors. W. Xiaolun et al. [23] have discussed vulnerabilities of elevator braking systems with a mathematical model. They have mentioned that there is a system called Beijing Elevator IoT convergence platform which monitors around eight thousand elevators. S. Lan et al. [24] have done a statistical survey of elevator accidents in China from 2002 to 2019 and they aimed to prevent future accidents by the correlation study of involved factors. Therefore, in this chapter, in contrast to the traditional embedded technology, an intelligent elevator system is presented which uses automatic sensing technology and provides various added features. The idea behind this is that any intelligent elevator system must be capable enough to cater for the needs of its users with enhanced authentication, safety and reliability features which will be very beneficial for society at large. To ensure a fully reliable and workable solution for the smart lift system, additional issues like those about people's safety inside lifts during power outages, node failure, and delayed information updates during emergencies involving high-rise buildings must be investigated and examined.

Table 1: Major Elevator Accidents in the World-Famous Elevator Collapses in the World.

S.N.	Deaths (approx)	Date (yyyy-mm-dd)	Place
1	52	1987-08-31	Welkom, South Africa
2	12	1993-06-02	North Point, Hong Kong
3	104	1995-05-10	Orkney, South Africa
4	200	2001-09-11	New York City, New York
5	12	2008-10-30	Xiapu County, China
6	11	2011-07-29	Makiivka, Ukraine
7	9	2011-08-09	Salvador de Bahia, Brazil
8	19	2012-09-13	Wuhan, China
9	10	2014-09-06	Istanbul, Turkey
10	8	2016-07-15	Longkou, China
11	11	2019-04-25	Hengshui, China

2. Literature Survey

A variety of management and monitoring strategies have been put forth for the protected operation of elevators on occasion, and matching systems have been established in response to the rapid pace of modern development and technological innovation. Here, a few pertinent studies from the previous 20 years have been chosen, and a performance comparison will be included in a later section of this work. A particular East Jakarta apartment building's traffic has been analysed by U. Komarudin et al. [25]. According to the findings, building A requires an elevator that can carry 15 passengers at a speed of 90 mpm, whereas building B also requires an elevator that can carry 15 people at a speed of 90 mpm. R.S. Karrthik et al. [26] studied an experimental elevator model that was validated by the effectiveness of the proposed control, and the outcomes were validated under various circumstances. To replicate a real-world elevator, they have built an elevator control system based on programmable logic. It is possible to increase the friction along the elevator's traction sheave. The prime mover is a servo motor, namely the Mitsubishi HJ-KS43J model, which is used for precise positioning. The SIEMENS S7 200 SMART PLC was utilised to manage complete system. The elevator system's PLC is interfaced with sensors, floor switches, hooters, and indicators to ensure optimal operation. To analyse and apply rope-less lifts using linear motors [27], concentrated on the motor drive's performance from the elevator's perspective. Based on the kinematics of an actual elevator that is required to determine certain design standards, the performance evaluation is carried out using computer simulation. Shorting out every motor winding allows for the quantitative study of free fall during a total power outage. An IOT-based lift safety monitoring system was examined by X. Pan et al. [28]. The CPU and communication module, temperature and humidity sensor module, and acceleration sensor module in this system were CC2530, DHT11, and ADXL345, respectively. The outcomes of the experiment revealed that the operation system could keep an eye on the environmental factors surrounding the elevator and achieve real-time monitoring of the elevator's operating condition. A ZigBee-based elevator's monitoring system has been studied by J. Chen et al. [29]. A microprogrammed control unit was used to operate the elevator model. The suggested system makes use of an MCU to control the operation of an elevator model. It also sends operation data to a PC monitoring system via a Zigbee information transmission module, where it is displayed together with basic data and elevator operation status information. A model of a radio frequency identification (RFID) lift has been created by M.H. Misran et al. [30]. RFID functions as an advanced security system and increases the building's security. It is appropriate for use in medical facilities. This device can restrict unwanted traffic flow during peak hours by permitting only authorized staff and patients to use the elevator through a microprogrammed control unit. Every tag has a distinct serial number that serves as the tag user's identity. The microprocessor and RFID system will interface, and the LCD display will show the outcome. An elevator monitoring system built on embedded systems and the Internet of Things was examined by S. Tu et al. [31]. When an elevator is occupied, the multi-sensor information collection offers information on vibration, acceleration, speed, driving noise, direction, and floor station, including temperature and noise information.

Additionally, the system offers technical assistance for elevator safety management. In their work, N. Cai et al. [32] addressed the fire safety regulations for elevator systems used for evacuation in extremely tall structures. The designs for elevator systems' smoke and fire protection have been thoroughly examined by the writers. Additionally, they have addressed important concerns and complications that arise while utilizing elevators in extremely tall buildings.

Using a created decision fusion system, G. Niu et al. [33] have suggested an intelligent defect diagnostic approach for induction motors for elevator traction machines. Initially, a quick introduction to the fundamentals of fusion techniques—classifier selection and decision fusion—is given. The elevator model that is operated by a voice and sensor control panel was covered by P. Cernys et al. [34] in their work. The model features speech recognition, a programmable terminal, and a logical elevator programmed that connects all of them. It is built utilising an average power controller. The Dynamic Time Warping (DTW) algorithm has been modified by them. A high-speed elevator's load simulator, which can simulate both the corresponding 1- and 3-mass systems, has been described by H.M. Ryu et al. [35]. Additionally, they have demonstrated how to simulate dynamic loads using electrical inertia, which significantly reduces the simulator system's weight and volume while facilitating simple software adjustments for the inertia value. The practicality of this simulator system has been demonstrated by experimental results, which also suggest that fault diagnosis and detection are essential for the safe operation of an elevator system. H. Hakala et al. [36] have investigated elevator traffic in terms of traffic patterns in an office building, such as the number of starts, round trips, and number of transported passengers, to realise a real-time and convenient diagnosis and satisfy the requirement of advanced maintenance of a lift system. In addition to presenting a case study on the annual energy consumption in a tall structure, they have measured and simulated traffic to investigate the energy consumption of lifts in high office buildings.

3. Proposed Method

Wireless sensor nodes connected in a multi-hop manner have been used to create a user-authenticated intelligent lift system that senses passenger traffic in real time. Additionally, an algorithm is created to prioritize and implement the lift system based on the volume of traffic. This allows all floors to connect and share traffic data, which optimizes and controls the elevator's moving and stopping. For making an intelligent and secure elevator system RFID cards, RFID Module, fire alarm and temperature sensor were integrated with the elevator system. The information on the card was written with the help of a computer. The workflow of the initialization of the card is shown in Fig. 2 below.

The user will collect the RFID card from the authority by giving his identification. On the RFID card, the authorised person will mention the floor for which the visitor is eligible and will hand it over to the visitor. The visitor will show the card to the RFID reader, it will identify the validity of the card and if the card is valid then it will record the information where it has to go. The door of the elevator will be opened. The working model flow chart has been shown in Fig. 3.

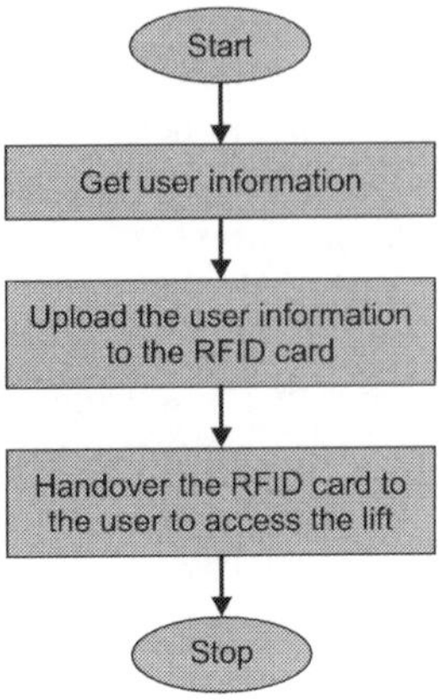

Figure 2: RFID Card Initialization Process.

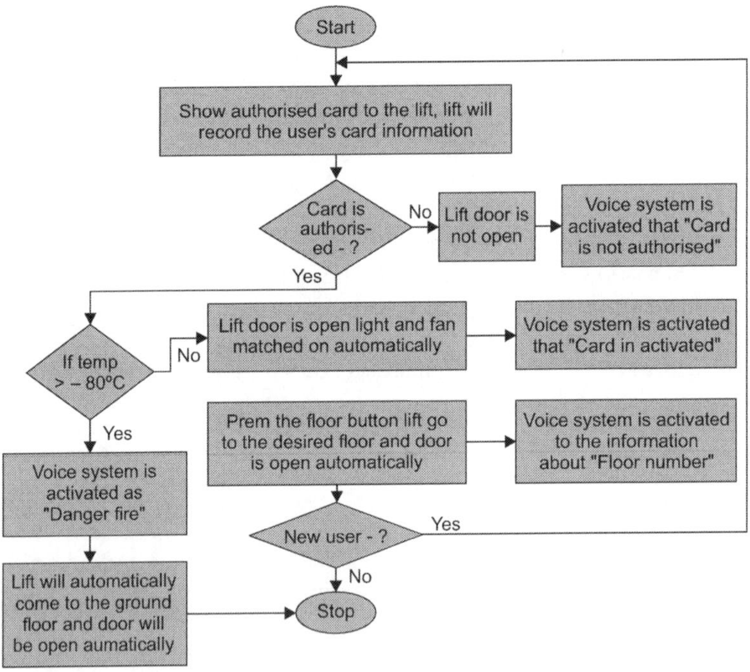

Figure 3: Flow Chart of Working Model.

The proposed system consists of the following sensors/modules and circuitry and the Block diagram of the system is depicted in Fig. 4 below.

- Arduino mega 2560
- RFID (EM18) Module
- DC motor
- L293D motor driver IC
- LM35 temperature sensor

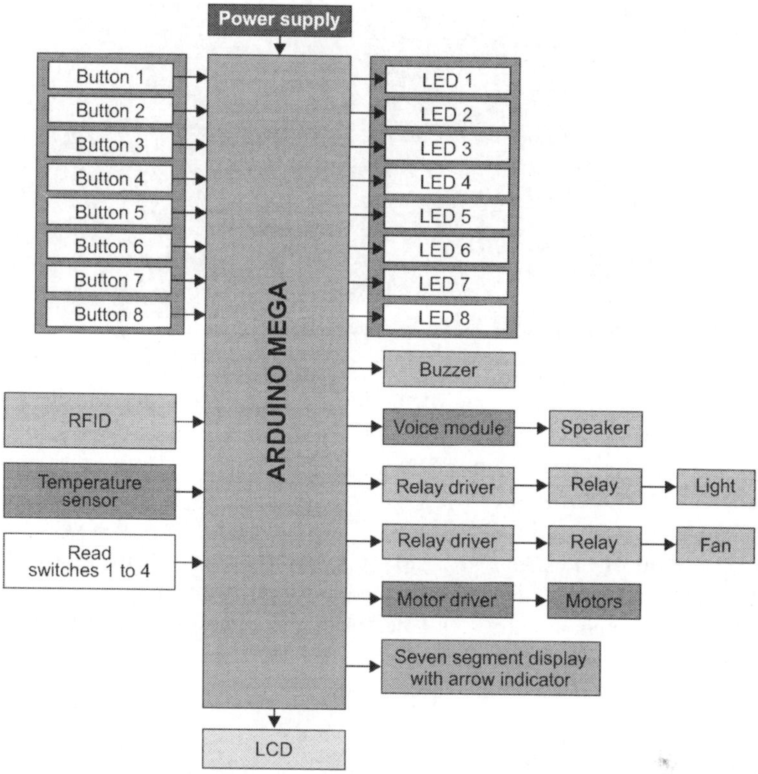

Figure 4: Block Diagram of the Proposed Elevator System. Important components used in the proposed system are given below with a brief description.

- Voice module with speaker
- ULN2003 relay driver IC
- 12V relay
- 20x4 LCD
- Push button
- Read switch
- Seven Segment Display (SSD)
- Aero Segment/Indicator Display (ASD)

- ***Arduino mega 2560*:** The Arduino Mega 2560 is a microcontroller board that is built around the ATmega 2560. 16 analogue inputs, 4 hardware serial ports (HARTs), a 16 MHz crystal oscillator, 54 digital input/output pins (15 of which are utilised as outputs for pulse width modulation, or PWM), a USB port, a power jack, an ICSP header, and a reset button are among its features. It has everything needed to keep the microcontroller running. It is easily powered by an AC-to-DC adapter or battery, to begin with, or it can be connected to a computer via a USB cable. Figure 5 below depicts an Arduino mega board in its standard form.

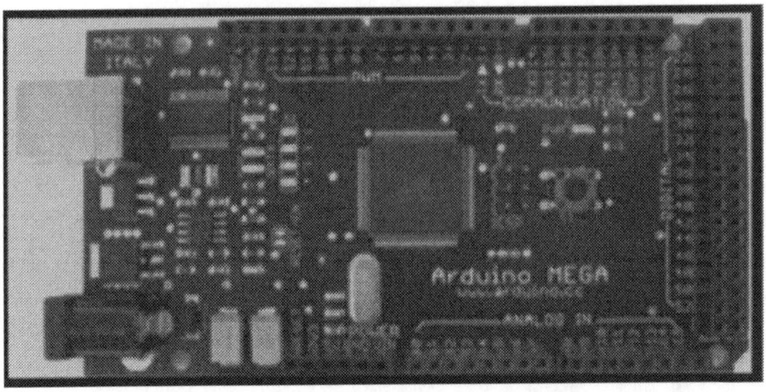

Figure 5: Showing Arduino Mega 2560 Used in the Proposed System.

- ***RFID (EM-18) module*:** One of the most popular RFID readers for reading 125 KHz tags is the EM-18. It displays several significant attributes, including affordability, low power consumption, compact form factor, and ease of use. A passive RFID tag operating at 125 KHz is energised by the 125 KHz field that the module generates through its coils. The majority of these passive RFID tags are made up of the CMOS IC EM4102, which draws sufficient power from the reader's produced field to operate. The information stored in the factory-programmed memory array will be transmitted back by the tag by altering the modulation current flowing through the coils. Figure 6 below illustrates communication between the transceiver module and the RFID tag.

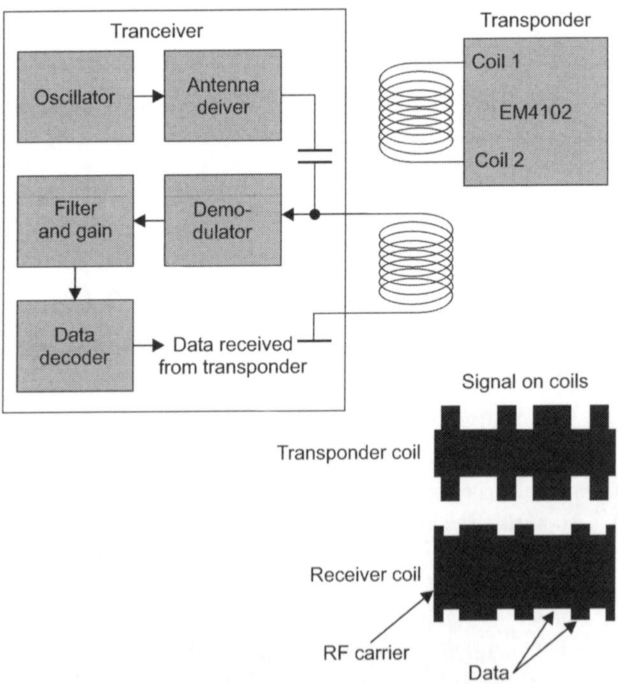

Figure 6: RFID Tag and Transceiver Module Data Transfer Process.

- ***DC motor:*** A DC motor uses mechanical energy to drive rotation by converting electrical energy from a battery or voltage source into mechanical energy. A DC power supply is connected to the motor's terminals to initiate one direction of rotation in the motor. The direction of rotation is reversed if the power supply's polarity is switched. Revolutions per minute (RPM) is the unit of measurement for DC motor speed. The DC motor accelerates with increasing supply voltage. But we are unable to use the supply voltage above its rated level. A DC motor's speed is determined by its load. A DC motor runs at its fastest when there is no load. The speed drops as the load increases. The DC motor might be harmed by overloading because of the considerable heat produced by the high current consumption. A DC motor is shown in Fig. 7 below and the circuit diagram of the elevator is shown just after in Fig. 8.

Implementation of the hardware circuit (step-by-step approach): As we already mention the Arduino mega 2560 works as the heart of this elevator module. The implementation strategy is discussed in the following steps below:

- The proposed elevator system has one dedicated control panel on which 8 switches have been placed with 20x4 LCD display. Switch numbers 1 to 4 are used to operate the floor number (i.e., floor 0 to floor 3), switches number 5 and 6 are used to start and stop the elevator respectively, switch numbers 7 and 8 are used to switch 'on' and 'off' the light and fan manually.
- The elevator control panel also has 8 LEDs corresponding to 8 switches. When any of the switches is pressed by the user then the corresponding LED glows and also indicates the corresponding button. All switches are placed in a pull up configuration.
- When an authentic user scans his or her dedicated RFID card on the EM-18 RFID module then the user's name will be displayed on the LCD with a welcome message.
- To save energy consumption, the light and fans of the elevator will be 'on' only when the authentic user enters the elevator, i.e. a user who has a valid dedicated RFID card. The l ight and fan placed in the elevator are connected with two relays. To operate these relays, we use a BC547 transistor as a relay driver. (User can on/off the light and fan by manual switch also).

Figure 7: Showing the DC motor Used in the Proposed System.

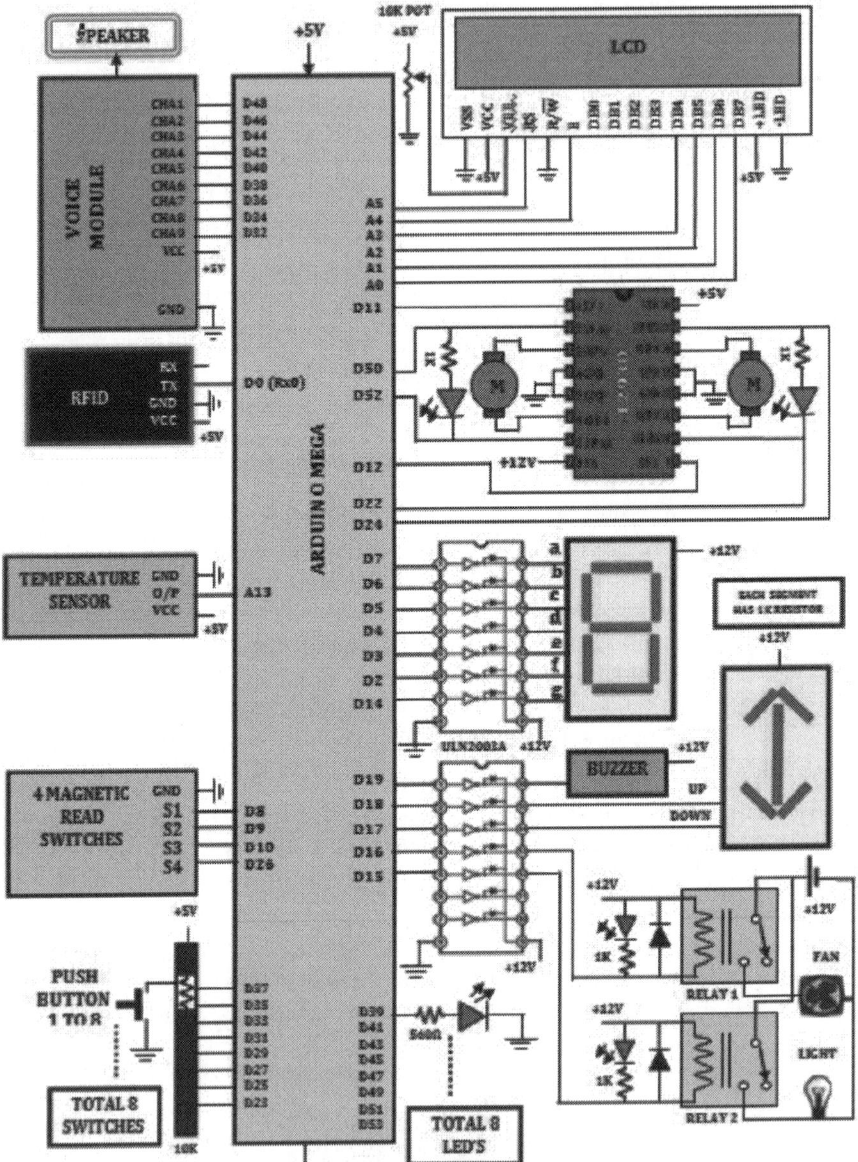

Figure 8: Circuit Diagram of the Elevator System.

- The genuine user will then be prompted by the system to press the floor button within ten seconds. If the user does not press any floor button within this interval, then the system will prompt the user to scan their RFID card again.
- If a valid RFID card has been detected through an authentic user and when the user presses the floor button then the DC motor gets activated and the elevator starts to move to the desired floor where a gate opens with voice output. The voice system announces the floor number with a welcome message.

- Two 12V DC motors are used for the demo purpose. One DC motor is used to operate the elevator and the other is used to open/close the gate of the elevator. Since the microcontroller will not be able to provide the desired current to operate the DC motor that is why the L293D motor driver module is used to operate the motor.
- To detect the floor number, four magnetic read switches are placed on four floors. One magnet is placed on the compartment of the elevator, whenever the elevator moves, the magnet will come into contact with the magnetic read switch which is placed at every floor and the signal will be generated which will used to detect the floor number.
- Like other elevator systems, here we also use Seven Segment Display (SSD) and Aero Segment to show the floor number and direction of movement of the elevator on every floor. We use 12V common anode SSDS and ASD for which we use two ULN2003 driver ICs.
- For fire safety, we use the LM35 temperature sensor. If the temperature goes beyond the threshold value (this threshold value can be adjusted in coding), a buzzer beeps automatically and the elevator directly comes to the ground floor automatically with a voice alert.

4. Analysis, Advantages and Implementation R esults

- **User authentication:** The proposed elevator system provides user authentication (something you have) uniquely. The user needs to show his or her dedicated RFID card to use the elevator system and unauthorized users would not be able to use the system because their RFID details are not stored in the system. There are plenty of benefits to using a dedicated RFID card for user authentication because text-based passwords have many associated vulnerabilities [37–39]. A brief overview of security attacks on text-based passwords is shown in below Table 2. Various password-cracking tools, their characteristics and corresponding web addresses are shown below in T **able 3** for reference.

Table 2: Showing Various Attacks on Passwords with a Brief Description.

SN	Attack type	Description
1	Brute force	Trying all possible combinations
2	Dictionary attack	Words in the dictionary are checked as possible passwords
3	Spyware attack	A specific application is installed on the targeted computer
4	Shoulder-surfing attack	Password can be seen by another person (very common in crowded areas)
5	Guessing attack	The i ntruder tries possible keywords such as passwords like name, mobile number, account number, etc.
6	Phishing	The intruder sends a fraudulent message to capture the user's password and other details
7	MITM (Man-in-the Middle Attack)	The i ntruder can be in the middle of the user and the other party and pretends to be genuine (to both parties)

Thus, it is obvious that all the above-mentioned attacks (Table 2) do not apply to the proposed system and all the password -cracking tools (Table 3) are also not feasible on the proposed system and it guarantees authenticated access to the elevator system.

- **Multilayer authentication:** The proposed system can be enhanced by introducing a GSM module or IoT technology. By pressing the get OTP button placed on the lift panel, the user will get the OTP (something you know) in his cell phone. The OTP will be entered through the hex key pad provided on the elevator panel then the elevator will move to the desired floor. The i ntroduction of the GSM module or Node MCU (For IoT) will not affect the total cost of the proposed system much and it will make the system more secure and reliable.
- **Energy saving:** The proposed system provides an optimized energy consumption feature as a value addition. Usually in existing elevator systems, the light and fan remains switched on even when nobody is using them [40]. In the proposed system, the light and fan turn on only after successful user authentication. Light and fan remain off in case of unsuccessful authentication and it makes sure that electricity is consumed only when an authentic user is using the system hence optimized energy consumption is achieved. This value-added feature makes it a cost cost-efficient and intelligent elevator system too.

Table 3: Showing Various Password-racking Tools with their Respective Web Addresses.

SN	Tool name	Characterstics	Web address
1	Aircrack-ng	Very famous wi-fi password-cracking tool	http://www.aircrack-ng.org/
2	John the Ripper	Can detect the hashing used in a password	http://www.openwall.com/john/
3	Rainbow Crack	Make rainbow tables for performing the attack	http://project-rainbowcrack.com/
4	Cain and Abel	Can launch brute force and dictionary attacks	http://www.oxid.it/cain.html
5	L0phtCrack	Famous for cracking W indows passwords	http://www.l0phtcrack.com/
6	Ophcrack	Forms LM hashes and rainbow tables	http://ophcrack.sourceforge.net/
7	Crack	Used for UNIX based systems	http://www.crypticide.com/alecm/software/crack/c50-faq.html
8	Hashcat	Very fast password breaking-can launch brute force, dictionary, hybrid and rule-based attacks	https://www.hashcat.net/
9	SAMInside	Rainbow tables, dictionary and hybrid attacks	http://www.insidepro.com/
10	DaveGrohl	dictionary attacks and incremental attacks	https://github.com/octomagon/davegrohl
11	Ncrack	Used for breaking password-based authentication	https://nmap.org/ncrack/
12	THC Hydra	Used for breaking password -based authentication	https://www.thc.org/thc-hydra/

- **Enhanced safety:** An additional fire safety system is incorporated into the proposed system safety. Usually in existing elevator systems, there is no protection against fire [41] but in the proposed system, the elevator automatically comes to the ground floor in case of fire with a buzzer alarm and voice alert. The same is applicable even when the inside temperature goes beyond the predefined threshold. It is worth mentioning here that many accidents have been reported earlier because the elevator got stuck on the top floor (or higher floors) in the case of fire and rescue operations and it became difficult to rescue people hence lots of casualties were reported. Since the proposed elevator system comes to the ground floor with voice alerts in the case of fire, the required rescue operation will be easy and it can save many lives [42][46].
- **Implementation results:** Original pictures for different parts and modules of the elevator have been shown in Fig. 9 (a, b, c and d).

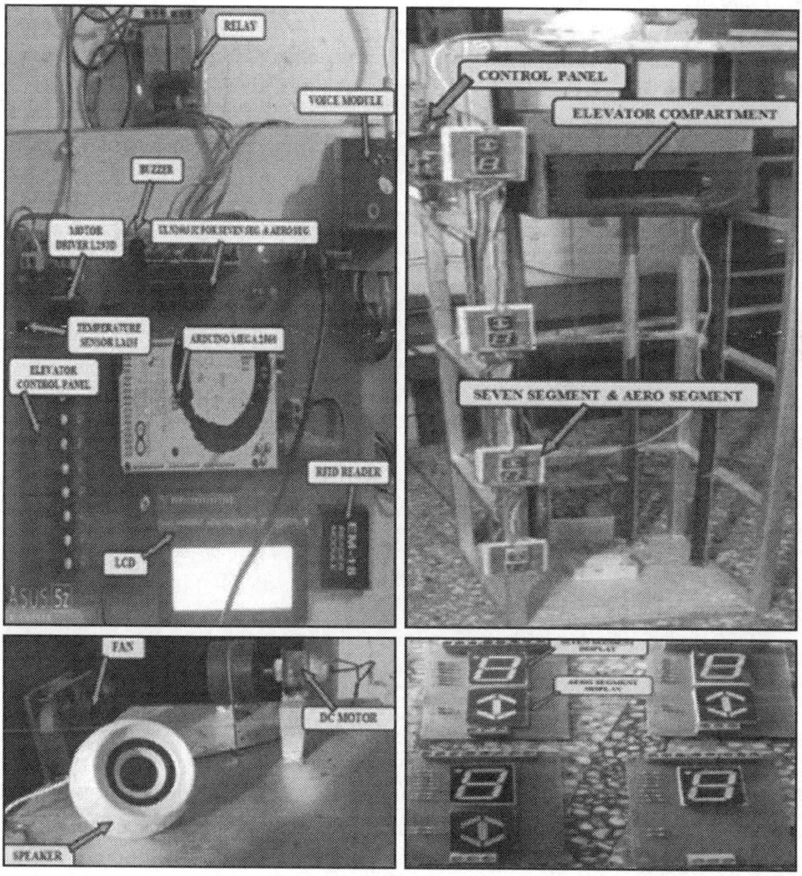

Figure 9: (a) Control panel of elevator model (b) Elevator model (c). Top of the elevator (d). SSD & ASD.

- **Cost analysis:** The proposed elevator system is a very cost-effective solution with many value-added features like user authentication, fire safety energy savings, etc. The total prototype cost is even less than 7600 INR (Indian Rupees) as shown in Table 4 below:

Table 4: Showing Cost Analysis of the Proposed System.

S.N.	Component(S)	Quantity	Rate (INR)	Amount (INR)
1	Arduino Mega	1	800	800
2	RFID module	1	400	400
3	RFID tags	5	20	100
4	16x2 LCD	1	180	180
5	LM35 Temperature sensor	1	60	60
6	Magnetic read switch	4	120	480
7	Push button	8	2	16
8	DC motor 12V/20RPM	2	180	360
9	Motor driver IC L293d	1	60	60
10	Voice module with SD card	1	800	800
11	Speaker 5W	1	300	300
12	Relay driver IC ULN2003A	2	20	40
13	Fan 12V	1	80	80
14	Light 12V	1	30	30
15	Seven segment display	4	40	160
16	Aero Indicator display	4	40	160
17	Buzzer 12V	1	12	12
18	Relay 12V	2	12	24
19	LED's	-----	10	10
20	Passive components (Resistance, capacitor 1000μF/25V)	-----	10	10
21	Lift structure	-----	2000	2000
22	DC Socket	1	8	8
23	Other miscellaneous components/items	----	1500	1500
Total				7590

5. Performance Comparison

Before moving to the performance comparison Table 5, the technical gist of every paper is given briefly for the better understanding of the readers of this paper

- **Udin Komarudin et al. [25]:** This study includes traffic analysis for an apartment complex in East Jakarta. The authors have examined an apartment block in East Jakarta that is made up of two towers, Tower A and B, each measuring 87.3 metres in height and having a total of 25 storeys (24 stories plus a basement) and one story on top. Building A has 727 occupants, whereas building B has 748 occupants. A traffic study is necessary to decide the best elevator type, and the Indonesian National Standard, SNI 03-6573-2001, was the research method applied. As per the obtained results, building A requires an elevator that can accommodate 15 people (1000 kg) at a speed of 90 mpm

Table 5: Showing the Performance Comparison of the Proposed System with Existing Methods.

SN	Papers	Technology used	Important components used	Uniqueness in Proposed system
1	U. Komarudin et al. [25]	The sort of elevator that is best for the building is determined by using elevator traffic analysis. This study employs the Indonesian National Standard, SNI 03-6573-2001, as its methodology.	Elevator capacity is simulated (No hardware)	1. In our system we which uses using RFID system (something you have) to operate the elevator.
2	RS Karrthik et al. [26]	Programmable logic based elevator control system is constructed to simulate an actual elevator which uses a method for a traction-based elevator system that doesn't need a counterweight.	Servo motor (Mitsubishi's HJ-KS43J), SIEMENS S7 200 SMART PLC, floor switches, indicator	
3	A. So et al. [27]	Analysis of the application of ropeless elevators using linear motors focusing It is used to increase the performance of the motor drive	Mathematical modeling and Simulation based (No hardware)	2. The elevator is highly secured 3. Energy saving module is incorporated.
4	X. Pan et al. [28]	Communication is done with the help of IoT, Fault tree analysis is used to analyze the reliability of the system	CC2530 processor, DHT11, ADXL345	4. Voice output
5	J. Chen et al. [29]	Communication is done with the help of Zigbee	PIC series microcontroller, Stepper motor	
6	M.H. Misran et al. [30]	Communication is done with the help of RFID	PIC series microcontroller, LCD, relay, buzzer, push buttons	5. We can extend this system to IoT as well
7	S. Tu et al [31]	Communication is done with the help of IoT	ARM processor, Touch screen, LCD, vibration sensor, accelerometer, switches	
8	N. Cai et al. [32]	No communication	Fire and smoke sensor	

Table 5: contd. ...

... *Table 5: contd.*

SN	Papers	Technology used	Important components used	Uniqueness in Proposed system
9	G. Niu et al. [33]	Decision fusion system is used to design the system	Induction motor, Elevator traction machine	
10	P. Cernys et al. [34]	Voice control system	Voice module, DC motors, pulley mechanism, frequency convertor circuit	
11	H.M Ryu et al. [35]	Electrical inertia is used in dynamic load simulation techniques to significantly reduce the weight and volume of the simulator system.	Simulation based (No hardware)	
12	H. Hakala et al. [36]	Energy consumption of an elevator in tall office buildings is studied by measurement and traffic simulation	Simulation based (No hardware)	
13	Proposed system	A highly secure RFID based elevator security system is used. We can extent this system to communicate with IoT technology as well.	Arduino Mega, RFID module, 16x2 LCD, Seven segment display, Voice module, Node MCU, Relay module, ULN 2003A relay driver IC, DC motor, Regulator IC 7805 with filter capacitor 1000μF/25V	

and can accommodate a total of 3 units. Similarly, building B also requires an elevator that can accommodate 15 people (1000 kg) at a speed of 90 mpm and can accommodate a total of 3 units. The sole basis for the work is simulation.

- **RS Karrthik et al. [26]:** The authors have presented the experimental model which is verified by the functionality of the proposed control and obtained results were verified under different conditions. To replicate a real-world lift, they have built an elevator control system based on programmable logic. The elevator car is balanced based on the diverting pulley arrangement, and a counterweightless system was thus built up. This technique can be used to increase friction along the lift's traction sheave. The prime mover is a servo motor, namely the Mitsubishi HJ-KS43J model, which is used for precise positioning. The SIEMENS S7 200 SMART PLC was utilised to manage the complete system. The elevator system's PLC is interfaced with sensors, floor switches, hooters, and indicators to ensure optimal operation.
- **A. So et al. [27]:** This study examines the use of linear motors in rope-less elevator applications, with a particular emphasis on the motor drive's performance as seen from the lift's perspective. With an emphasis on the overall performance of the elevator system, the authors have created a fundamental mathematical framework to analyse and design the controllers and drive of a multi-car lift system employing linear permanent magnet synchronous motors. Utilising computer simulation and the required kinematics of an actual lift to meet certain design requirements, the performance is assessed. By short-circuiting every motor winding, the free fall during a total power outage is statistically investigated. By short -circuiting every motor winding, the free fall during a total power outage is statistically investigated. Additionally, they have researched various techniques for achieving a low, fairly constant descent speed and have analytically determined values that are confirmed by simulation.
- **X Pan et al. [28]:** The Internet of Things (IoT)-based elevator safety monitoring system design is presented in this research. The system's processor and communication module is the CC2530, while the temperature and humidity sensors are DHT11, and the accelerometer sensor module is ADXL345. The power supply module incorporates the chip LM2576S to obtain power from both 3V and 3.3V sources. The reliability of the system designed in this research is analysed using the fault tree analysis method. Additionally, the system was tested in a residential area, where the average relative errors of velocity, air humidity, and temperature were determined to be 1.36%, 1.40%, and 0.47%, respectively. The test results show that this user-friendly system can achieve real-time monitoring of the elevator 's operating status and can monitor the environmental parameters of the elevator with high precision.
- **J. Chen et al. [29]:** The authors have examined the Zigbee-based elevator monitoring system, which uses MCU to regulate lift functions. Additionally, information about elevator operations is transmitted by a Zigbee information transmission module that is linked to a PC monitoring system. The Zigbee module and MCU must communicate to convert the transmission signal's voltage value.

- **M.H. Misran et al. [30]:** A security and safety system based on the RFID application has been created by the authors. The suggested solution is appropriate for use in hospitals and improves building security. During busy hours, this technology can reduce traffic congestion by granting access to authorised users only. The microprocessor and RFID system are integrated, and an LCD display shows the outcomes.
- **S. Tu et al. [31]:** This study provides an Internet of Things-based embedded system-based elevator monitoring system. The WinCE operating system is adapted to the ARM9 master control chip S3C2410 in the proposed system to enable centralized monitoring and automatic remote failure alarm for elevator operation. The elevator parameter monitoring terminals and the remote control centre establish a WAN link. The embedded operating system is changed to an ARM embedded system and C++ is employed. The CS8900A high-performance industrial Ethernet controller provides a fast, dependable, and high real-time performance for network connectivity. The system is effective with high promotion value, according to the results.
- **N. Cai et al. [32]:** The standards for fire safety on elevator systems used for evacuation in extremely tall buildings are the main topic of this essay. This work examines potential elevator system designs for smoke and fire protection during evacuation since ultra- tall buildings do not have enough fire safety measures in place for evacuation. Two main safety concerns have been highlighted as hazards: the first is the spread of smoke via the lobby and elevator shafts, and the second is the need for construction that is fire-resistant in case of large fires.
- **G. Niu et al. [33]:** This work uses a built decision fusion system to propose an intelligent problem detection approach for lift traction machines using induction motors. The writers have presented a created decision fusion system and talked about the fundamentals of fusion techniques. The steps of feature calculation, classification, classifier selection, and decision fusion are given in the decision fusion system. Using this technique, the defect diagnostics of a lift motor produced excellent results. It is discovered that the decision fusion technology used in current monitoring is excellent and reasonably priced.
- **P. Cernys et al. [34]:** The elevator model operated by a voice and sensor control panel is analysed in this study. The model connects them all with a logical lift programme, programmable terminal, and speech recognition, among other things. Automation, voice signal recognition, and control technology specialisations can be taught to students using this model as an instructional tool. A version of the widely recognised DTW (Dynamic Time Warping) technique is employed. Eight Lithuanian words (floor number, go, stop) and two phrases (hello and goodbye) make up the set of voice commands. It was discovered that eight voice commands had 100% recognition accuracy. People with disabilities can benefit greatly from voice-activated technology.
- **H.M. Ryu et al. [35]:** The simulator that the authors discussed can electrically provide a dynamic load for high-speed elevator s. By adjusting the torque of a DC machine, the suggested method can replicate the load characteristics of an elevator system. In comparison to the traditional system, the new one

is significantly lighter, smaller, and more adaptable. In the typical approach, the entire elevator system's inertia is implemented through the employment of mechanical flywheels with a big radius. In this study, the dynamic load simulation methods utilising electrical inertia have been adopted, resulting in a significant reduction in the volume and weight of the simulator system, as well as easy software adjustment of the inertia value.

- **H. Hakala et al. [36]:** In this work, the energy consumption of elevators in tall office buildings is studied by measurements and traffic simulations. Elevator load and travel distributions are used in a method that computes energy usage. Measurements conducted in an office block with a single occupant corroborated the modelling results. It was also examined to compare the percentages of savings made by various driving systems, pieces of equipment, and control systems. An annual case study of a tall building's energy use has been provided by the writers. This case study claims that for the most cutting-edge and exacting office buildings, the MX machinery with TMS9900 GATM control system offers a true green elevator product.

6. Conclusion and Future Scope

This paper presents, a user-authenticated intelligent elevator system. The proposed system provides entity authentication (something you have) with RFID cards. The need for user authenticated and intelligent elevator systems are highly desired now days in hospital buildings emergencies etc. Many elevator accidents have been reported in past years due to various loopholes in elevator systems. The proposed systems not only provide user authentication but are also equipped with various safety features, timer mechanisms and energy- saving capabilities. The proposed system is cost- efficient and better than previously presented systems in various aspects such as improved authentication, and enhanced fire security with voice output. The future work of the proposed system seems very promising as other enhanced features like multilayer authentication and inclusion of IoT concept can also be added to make it more useful, user friendly and customized for plenty of applications.

References

1. Menezes, A. J., Oorschot, P. C. V., and Vanstone, S. A. 2001.Handbook of applied cryptography, 5th edition, CRC Press Inc., USA, ISBN: 9780849385230.
2. Stallings, W. 2005. Cryptography and network security, principles and practices, seventh edition, Prentice Hall, ISBN-13:978-0134444284, ISBN-10:0134444280.
3. Anand, D., Khemchandani, V., Sabharawal, M., Cheikhrouhou, O., and Fredj, O. B. 2021.Lightweight technical implementation of single sign-on authentication and key agreement mechanism for multiserver architecture-based systems, Security and communication networks, volume 2021, article id 9940183, 1–15, DOI: https://doi.org/10.1155/2021/9940183
4. Yang, M., Wei, S., Jiang, R., Ali, F., and Yang, B. 2021. Single-message-based cooperative authentication scheme for intelligent transportation systems, Computers & electrical engineering, volume 96, part B, DOI: https://doi.org/10.1016/j.compeleceng.2021.107390
5. Thomas, P. A., and Mathew, K. P. 2021.A broad review on non-intrusive active user authentication in biometrics, Journal of ambient intelligence and humanized computing, 1–22, DOI: https://doi.org/10.1007/s12652-021-03301-x

6. Xu, H., Yin, X., Zhu, F., and Li, P. 2021. An enhanced secure authentication scheme with one more tag for RFID systems. *IEEE Sensors Journal, 21*(15), 17189–17199, DOI: 10.1109/JSEN.2021.3077575
7. Yıldırım, M., and Mackie, I. 2019. Encouraging users to improve password security and memorability. *International Journal of Information Security, 18*, 741–759, DOI: https://doi.org/10.1007/s10207-019-00429-y
8. Chang, B., Li, T., Wang, Q., Zhu, W. T., and Deng, R. H. 2018. Making a good thing better: enhancing password/PIN-based user authentication with smartwatch, Cybersecurity, volume 1, article number 7, 2018, 1–13, DOI: https://doi.org/10.1186/s42400-018-0009-4
9. McCann, M., and Zaleski, N. 2006. Deaths and injuries involving elevators and escalators—a report of the center to protect workers' rights, 1–27, available at https://www.elcosh.org/record/document/1232/d000397.pdf
10. Zarikas, V., Loupis, M., Papanikolaou, N., and Kyritsi, C. 2013. Statistical survey of elevator accidents in Greece, *Safety Science, 59,* 2013, 93–103, DOI: https://doi.org/10.1016/j.ssci.2013.05.001
11. Kim, N. H., and Jeong, B. H. 2014. Classification of elevator failure using the analysis of failure case. *Journal of the Korea Safety Management & Science, 16*(3), 209–218, DOI: https://doi.org/10.12812/ksms.2014.16.3.209
12. Hui, L. Y., and Gi, K. H. 2014. Development an elevator emergency call transmission system. *International Journal of Control and Automation, 7*(8), 311–316, DOI: http://dx.doi.org/10.14257/ijca.2014.7.8.28
13. Park, J. B., and Shin, S. J. 2014. The comparative study on safety factors at elevator management system operation. *The Journal of the Institute of Internet, Broadcasting and Communication, 14*(6), 159–162, DOI: http://dx.doi.org/10.7236/JIIBC.2014.14.6.159
14. Dhanasekaran, R., Vinothkanna, R., Reddy, S., Kumar, N., and Kalyan, B. A. 2016. A review of escalator accidents and preclusion. *GRD Journal for Engineering, 1*(12), 132–136, available at https://www.grdjournals.com/uploads/article/GRDJE/V01/I12/0073/GRDJEV01I120073.pdf
15. Siti, N. A., Asmone, A. S., and Chew, M. Y. L. 2018. An assessment of maintainability of elevator system to improve facilities management knowledge-base. *IOP Conference Series: Earth and Environmental Science, 117*, 1–8, DOI: 10.1088/1755-1315/117/1/012025
16. Sakhare, A. A., Patil, P. D., Pawar, N. R., Mohite, G. R., and Rashinkar, S. R. 2018. A review on new advanced safety systems for residential elevators. *Asian Journal of Convergence in Technology, 4*(3), 1–5, available at https://asianssr.org/index.php/ajct/article/view/833/670
17. Mohite, G. R., Patil, P. D., Pawar, N. R., Sakhare, A. A., and Rashinkar, S. R. 2018. A review paper on major & minor causes of accidents in elevators in metro city. *Journal for Research, 4*(10), 7–11, available at https://www.academia.edu/38124722/A_REVIEW_PAPER_ON_MAJOR_AND_MINOR_CAUSES_OF_ACCIDENTS_IN_ELEVATORS_IN_METRO_CITY_J4RV4I10005
18. Ming, Z., Han, S., Zhang, Z., and Xia, S. 2019. Elevator safety monitoring system based on internet of things, *14*, number 8, 2018, 121–133, DOI: https://doi.org/10.3991/ijoe.v14i08.9179
19. Shrestha, A. 2019.0 Safety considerations for the design of modern elevator systems, University of Mississippi, honors theses, 1–64, available at https://egrove.olemiss.edu/hon_thesis/1195?utm_source=egrove.olemiss.edu%2Fhon_thesis%2F1195&utm_medium=PDF&utm_campaign=PDFCoverPages
20. Surui, X. 2019. Analysis of human factors of elevator accidents based on HFACS. *China Safety Science Journal, 29*(7), 70–75, DOI: 10.16265/j.cnki.issn1003-3033.2019.07.012
21. Shubham, A., Gauri, C., Akshay, T., and Patil, P. D. 2020. A review paper on new safety system in elevator. *International Journal of Advanced Research in Science & Technology, 7*(1), 138–145, DOI: XX.072020/IJARST
22. Xing, Y., Chen, S., Zhu, S., and Lu, J. 2020. Analysis factors that influence escalator-related injuries in metro stations based on bayesian networks: a case study in China. *International Journal of Environmental Research and Public Health, 17*(2), 1–21, DOI: 10.3390/ijerph17020481
23. W. Xiaolun, X. Zhang, Research on the risk points of elevator braking system, *International Conference on Artificial Intelligence and Computer Engineering,* 2021, 28–31, DOI: 10.1109/ICAICE51518.2020.00011
24. Lan, S., Jiang, S., Qiu, J., Wan, Z., Chen, L., Li, G., and Alam, J. 2021. Statistical analysis of typical elevator accidents in China from 2002 to 2019, *6*(2), 193–208, DOI: https://doi.org/10.2478/amns.2021.2.00003

25. Komarudin, U., Suryaman, N. N., and Farida, I. 2021. Optimization of passenger elevator design for apartment building, *Turkish Journal of Computer and Mathematics Education, 12*(8), 1537–1545, DOI:
26. Karrthik, R. S., Prabhu, S. R., Megavarthini, K. K., Nandakumar, K., and Mathesh, B. 2020. *IOP Conference Series: Materials Science and Engineering, 764*, 1–9, DOI: 10.1088/1757-899X/764/1/012051
27. So, A., and Chan, W. L. 2018. A study of linear PMSM driven ropeless elevators, *Building Services Engineering Research and Technology, 40*, 93–108, DOI: https://doi.org/10.1177/0143624418797604
28. Pan, X. 2016. The design and reliability analysis of elevator monitoring system based on the internet of things, *International Journal of Smart Home, 10*(12), 183–192, DOI: http://dx.doi.org/10.14257/ijsh.2016.10.12.17
29. Chen, J.W., Yang, H.J., and Fan, W.L. 2014. Research of elevator remote monitoring system based on zigbee technology, Advanced materials research, volume 1049-1050, 1222–1226, DOI: https://doi.org/10.4028/www.scientific.net/AMR.1049-1050.1222
30. Misran, M. H., Said, M. A. M., Othman, M. A., Sulaiman, H. A., Aziz, M. Z. A. A., Mustaffa, I., Rahim, R. A., and Zainudin, M. N. S. 2014. Design and development of RFID based elevator, *International Symposium on Technology Management and Emerging Technologies*, 252–257, DOI: 10.1109/ISTMET.2014.6936515
31. Tu, S. L., Wu, Z. Y., and Qian, B. 2013. Research of the elevator monitoring system based on the internet of things. *Applied Mechanics and Materials*, 423–426, 2380–2385, DOI: https://doi.org/10.4028/www.scientific.net/AMM.423-426.2380
32. Cai, N., and Chow, W. K. 2011.Fire safety requirements on lift system for evacuation in supertall buildings. *International Journal on Engineering Performance-based Fire Codes, 10*(2), 17–23, available at https://moam.info/fire-safety-requirements-on-lift-system-for-evacuation-in-supertall-_5a1855841723dd12c27f85e4.html
33. Niu, G., Lee, S. S., Yang, B. S., and Lee, S. J. 2008. Decision fusion system for fault diagnosis of elevator traction machine. *Journal of Mechanical Science and Technology, 22*, 85–95, DOI: https://doi.org/10.1007/s12206-007-1010-0
34. Cernys, P., Kubilius, V., Macerauskas, V., and Ratkevicius, K. 2003. Intelligent control of the lift model. *Second IEEE International Workshop on Intelligent Data Acquisition and Advanced Computing Systems: Technology and Applications*, 428–431, DOI: 10.1109/IDAACS.2003.1249600
35. Ryu, H. M., Kim, S. J. Sul, S. K., Kwon, T. S., Kim, K. S., Shim, Y. S., and Seok, K. R. 2002. Dynamic load simulator for high-speed elevator system, *Proceedings of the Power Conversion Conference*, 885–889, DOI: 10.1109/PCC.2002.997638
36. Hakala, H., Siikonen, M. L., Tyni, T., and Ylinen, J. 2001. Energy-efficient elevators for tall buildings, Proceedings of 6th world congress at CTBUH 559–574, available at https://global.ctbuh.org/resources/papers/download/1684-energy-efficient-elevators-for-tall-buildings.pdf
37. Wash, R., and Rader, E. 2021. Prioritizing security over usability: strategies for how people choose passwords. *Journal of Cybersecurity, 7*(1), 1–17, DOI: https://doi.org/10.1093/cybsec/tyab012
38. Ali, M., Baloch, A., Waheed, A., Zareei, M., Manzoor, R., Sajid, H., and Alanazi, F. 2021. A simple and secure reformation-based password scheme, *IEEE Access, 9*, 11655–11674, DOI: 10.1109/ACCESS.2020.3049052
39. Chanda, K. 2016. Password security: an analysis of password strengths and vulnerabilities, *International Journal of Computer Network and Information Security, 8*(7), 23–30, DOI: 10.5815/ijcnis.2016.07.04
40. Passenger elevators. n.d. Mitsubishi electric, available at https://www.mitsubishielectric.com/elevator /products/basic/elevators/nexiez_mr/pdf/product_guide.pdf
41. Cai, N., and Chow, W. K. 2019. Numerical studies on fire hazards of elevator evacuation in supertall buildings. *Indoor and Built Environment, 28*(2), 247–263, DOI: https://doi.org/10.1177/1420326X17751593
42. Ronchi, E., and Nilsson, D. 2013. Fire evacuation in high-rise buildings: a review of human behaviour and modelling research. *Fire Science Reviews, 2*(7), 1–21, DOI: https://doi.org/10.1186/2193-0414-2-7
43. Kodmany, K. A. 2015. Tall buildings and elevators: a review of recent technological advances. *Buildings*, *5*(3), 2015, 1070–1104, DOI: https://doi.org/10.3390/buildings5031070

44. Chen, Y., Wang, C., Yap, J. B. H., Li, H., Hu, H. S., Chen, C. C., and Lai, K. K. 2020. Fire evacuation process using both elevators and staircases for aging people: simulation case study on personnel distribution in high-rise nursing home. *Discrete Dynamics in Nature and Society,* 2020, article id 5365126, 2020, 1–14, DOI: https://doi.org/10.1155/2020/5365126
45. Minegishi, Y. 2021. Occupant evacuation elevators as a measure for crowd management and evacuation for people with mobility impairments in high-rise buildings. *Japan Architectural Review, 4*(2), 2021, 391–402, DOI: 10.1002/2475-8876.12219
46. Komarudin, U., Suryaman, N. N,, and Farida, I. 2021. Optimization of passenger elevator design for apartment building. *Turkish Journal of Computer and Mathematics Education, 12*(8), 2021, 1537–1545, DOI: https://doi.org/10.17762/turcomat.v12i8.3192

CHAPTER 11

Artificial Intelligence (AI) enabled Digital Twin for Smart Manufacturing

Anupriya Bharti,[1] *Bramah Hazela,*[2,*] *Pallavi Asthana*[1] and *Shikha Singh*[1]

1. Introduction

In 2011, the concept of Industry 4.0 was introduced at the Hannover Fair. The Fourth Industrial Revolution or Industry 4.0 introduced a rapid change in the manufacturing processes and became an immediate focus for Germany and other European countries. The integration of various advanced technologies such as IoT, AI, Digital Twin, Cyber-Physical Systems, Cloud Computing and Data Analytics have greatly evolved the manufacturing process and the industry from traditional methods to a highly intelligent and connected environment capable of making data-driven decisions and advanced automation.

According to [1], I ndustry 4.0 is an integration of the nine pillars in Fig. 1 which are as follows:

i. ***IoT (Internet of Things)***
   This is the most important pillar of Industry 4.0 that enables the connection of physical devices and facilitates the communication and data exchange among the systems.

ii. **Cloud Computing**
   Manufacturing requires scalability and accessibility to storage and other services that are provided by cloud computing. It is a powerful tool that allows the sharing of storage and other resources along with excellent computing abilities without the need for any local infrastructures for it.

[1] Amity School of Engineering & Technology, Lucknow Amity University Uttar Pradesh, India.

[2] Department of Computer Science & Engineering, Amity School of Engineering & Technology Lucknow, Amity University Uttar Pradesh, India.

* Corresponding author

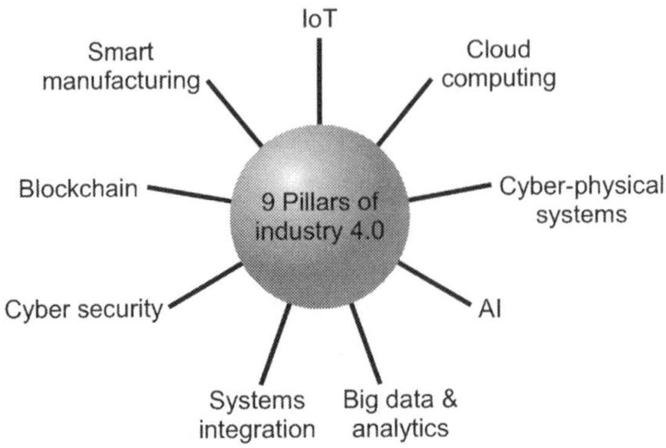

Figure 1: The Nine Pillars of Industry 4.0.

iii. **Cyber-Physical Systems**
CPS is heavy machinery that is required to enhance automation by integrating physical processes and precise control. It is a technology where physical processes meet the digital technologies's computing abilities, facilitating a seamless interaction between the physical and digital worlds.

iv. **Artificial Intelligence**
AI is a fundamental requirement that empowers machines and applications with the ability to learn and adapt to make reasonable decisions. It helps to improve the efficiency and capability of the processes.

v. **Big D ata and Analytics**
With the growing industry and need for detailed data for various processes big data is an essential technology that can help to deal with the enormous amount of data generated by intelligent technologies. It can help to collect, organize and retrieve data in no time from the huge pile of data collected from numerous sources for processes like maintenance, analytics, security and many more.

vi. **Systems Integration**
This simply involves connecting and coordinating the diverse systems and technologies to work together seamlessly. This is an essential part of smart manufacturing which helps the systems to be aware of other processes and exchange data among themselves in real-time.

vii. **Cyber Security**
This is the other most important part of Industry 4.0. It is an essential technology that ensures security from external cyber-attacks and protection from internal threats.

viii. **Blockchain**
A revolutionary technology that enables decentralized systems and enhanced security for transparent and tamper-proof records of any kind of transaction. It is an advanced technology that can be used to establish trust in diverse technologies in the future.

xi. **Smart Manufacturing**
Smart manufacturing is the use of enabling and smart technologies in the processes of manufacturing to optimize processes and respond to dynamic environments.

1.1 Smart Manufacturing and its Components

Smart manufacturing has a huge impact on industrial processes by introducing advanced technologies to enhance efficiency, flexibility, and response mechanisms across the entire manufacturing chain. The main approach of smart manufacturing is to combine advanced technologies to develop more interconnected intelligent systems for better adaptability, optimized production, and decision-making mechanisms. Smart manufacturing greatly redefines traditional manufacturing practices creating a new era of intelligent and interconnected network of data-driven industrial processes.

Smart manufacturing consists of the following described main components for the collection of data and decision-making processes:

- **IoT:**
  IoT in smart manufacturing provides a way for communication and collaboration among the different devices in the smart manufacturing environment. It is an essential component of the smart manufacturing ecosystem. It allows devices to comprehend quick decisions while engaging in real-time communication. IoT in the industry involves the integration of sensors, actuators, and smart devices into manufacturing processes to enable real-time communication and data collection.
- **AI and ML:**
  The revolutionary technologies that are used in smart manufacturing for the automation of decision-making and better responsive models that use analytics, pattern recognition and prediction in smart manufacturing systems.
- **Digital Twin technology:**
  Digital Twin technology is the main component of the smart manufacturing system. This technology creates virtual replicas of physical systems used in manufacturing practices to provide a real-time digital counterpart for monitoring and analysis of that system. It greatly helps in performance optimization and predictive maintenance by enabling the simulation of that physical system.
- **Advanced Robotics and Automation:**
  Robotics and automation technologies involve the deployment of intelligent robotic systems capable of doing repetitive tasks precisely and more efficiently than being done manually. It leads to fewer errors and increased efficiency in material handling, assembly, and quality control processes.
- **Cloud computing:**
  Cloud computing provides scalable and centralized resources for storage, processing and analysis of big data generated in smart manufacturing. It enhances collaboration and deployment of advanced data analytics.

- **Cyber-Physical Systems:**
  CPS involves combining physical processes with digital technologies creating interconnected systems to adapt to the changing conditions. It improves real-time monitoring, coordination, and control over manufacturing processes.

1.2 Digital Twin Technology in Smart Manufacturing

Digital Twin is an important component and foundation of smart manufacturing. Through the years the digital twin has been given many definitions (see Table 1).

The digital twins allow much easier manufacturing system environment designing, process analysis, and optimizing for better results. As illustrated in Figure 2, digital twin technology due to its unique applications is a very valuable asset to a smart manufacturing industry. The evolutionary system adapts throughout time digitally or in conjunction with a cyber-physical system. A large quantity of data is collected and analysed in digital twin technology, therefore,

Table 1: Definitions of 'Digital Twins'.

Author	Definitions
NASA 2012 [2]	"A Digital Twin is an integrated Multiphysics, multiscale, probabilistic simulation of an as-built vehicle or system that uses the best available physical models, sensor updates, fleet history, etc., to mirror the life of its corresponding flying twin."
Y. Chen 2017 [3]	"A digital twin is a computerized model of a physical device or system that represents all functional features and links with the working elements"
Z. Liu, N. Meyendorf, and N. Mrad 2018 [4]	"The digital twin is actually a living model of the physical asset or system, which continually adapts to operational changes based on the collected online data and information, and can forecast the future of the corresponding physical counterpart"
Y. Zheng, S. Yang, and H. Cheng 2018 [5]	"A Digital Twin is a set of virtual information that fully describes a potential or actual physical production from the micro atomic level to the macro geometrical level"
R. Vrabîc, J. A. Erkoyuncu, P. Butala, and R. Roy 2018 [6]	"A digital twin is a digital representation of a physical item or assembly using integrated simulations and service data. The digital representation holds information from multiple sources across the product life cycle. This information is continuously updated and is visualised in a variety of ways to predict current and future conditions, in both design and operational environments, to enhance decision making"
A.Madni, C. Madni, and S. Lucero 2019 [7]	"A Digital Twin is a virtual instance of a physical system (twin) that is continually updated with the latter's performance, maintenance, and health status data throughout the physical system's life cycle"
Tao, Fei, Meng Zhang, and Andrew Yeh Chris Nee 2019 [8]	"Digital twin is a virtual representation of a physical object or system that interacts with each other throughout the processes in its lifecycle."
T. Bergs, S. Gierlings, T. Auerbach, A. Klink, D. Schraknepper, T. Augspurger 2021 [9]	"A digital twin is an automatic two-way connection between real-world and virtual simulation in digital twin spaces"

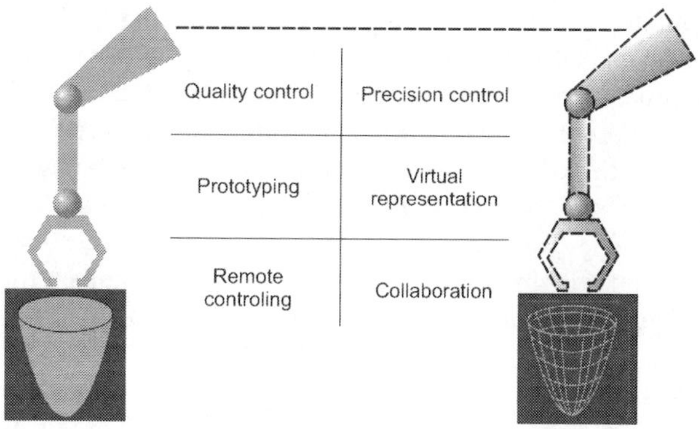

Figure 2: Digital Twin Technology.

AI-enabled analytics and machine learning models are powerful tools to support long-term and large-scale requirements. According to [10] it has three main parts:

i. Physical system in real space
ii. Virtual replica of the physical system in virtual space
iii. Connection between virtual and real space systems

The Digital twin is a virtual representation of a physical object or system that interacts with each other throughout the processes in its lifecycle [8]. It provides intelligence, optimization and prediction capabilities for better monitoring and control. The digital twins have a great potential to enhance the new generation of smart manufacturing and [12] industrial economic growth or F-factor which are productivity and quality in the manufacturing processes.

The Digital Twin technology [11] can give real-time status of the physical assets about their performance and feedback. Digital twin coupled with the power of AI result in a powerful tool capable of predicting any fault or failure and analysing big data. It introduces a connected infrastructure of devices and applications that can communicate with each other as well as share feedback to the AI or machine learning models.

1.3 AI—enabled Digital Twins

Artificial Intelligence is a revolutionary technology that is designed to automate the process of learning and application of knowledge. AI-enabled digital twins are equipped with mechanisms to learn and understand correlations in the smart manufacturing environment. The aggregation of data with the goal of optimize and analyzing processes makes it an excellent tool to implement in a smart environment. Previously, many studies have been implemented on AI digital twins.

According to [12], the AI-enabled DTs have better chances of completing a complex task and high adaptability in a changing environment. AI can be used to create user behaviour or customer preference models from historical data or feedback. It greatly enhances the beauty of the digital twin technology by improving the efficiency of data analysis and prediction algorithms. Model granularity based on a complex environment study and a quality dataset effects the fidelity of AI-enabled digital twins greatly. The digital twins provide a collection of data from different sources in real-time [11]. The AI-digital twin technology is very helpful in testing a prototype of a product in the required environment and conditions without creating one. It can help to observe the product in a desired environment which makes it a highly asset. For increased effectiveness of data analysis data mining techniques such as clustering, association rules, classification, regression, prediction, and deviation are very helpful.

In [8], the proposed framework for data-driven smart manufacturing consisted of a total of 4 modules that included manufacturing, a data driver, real-time monitoring, and a problem -processing module. It defined the flow of data from manufacturing to real-time data-driven analysis that is used to solve problems. The data-driven structure has advantageous characteristics like having the ability to adapt according to the needs of the environment, self-organization, precise control, self-regulation, and self-learning Figure 3 [13]. Digital twin technology enables the three main tools of knowledge which are conceptualization, comparison, and collaboration. The processing methods of humans and computers differ largely. The capability of digital twins can help us quickly shorten the process of translating real-world problems into a visual representation of the problem. These tools serve as an excellent foundation for solving real-world problems and innovating next-generation solutions. It offers a range of benefits to smart manufacturing processes like efficiency, productivity, and decision-making [14]. The AI techniques that are used in digital twins offer a foundation for automation, predictive analysis and maintenance processes. These techniques can be broadly classified into four categories: supervised learning, unsupervised learning, reinforcement learning and Artificial Neural Networks.

The digital twin architecture in [15], proposed a five-dimension model of the digital twin. It included physical entity, virtual copy, services, data, and connection. In [16], a data-driven framework of digital twins is proposed that provides solutions to the most common challenges in smart manufacturing. These systems improve reliability, real-time monitoring, and predictive maintenance providing customised services to the keyholders. In [17], the small-scale framework for digital twins identified enabling technologies and digital transformations as a main component of digital twin projects. Also, in [18] a framework to improve the simulation using machine learning techniques is discussed.

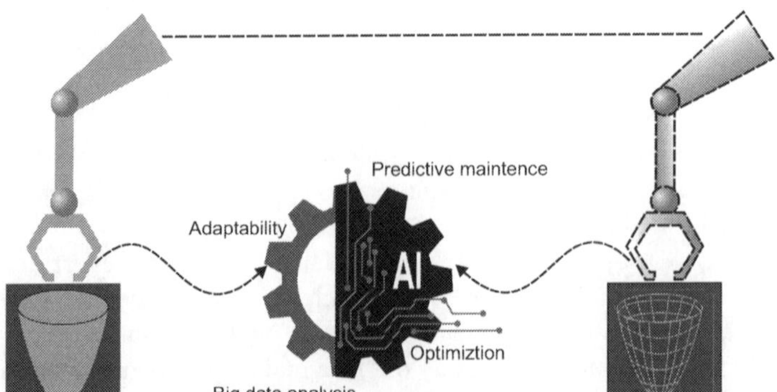

Figure 3: AI-enabled Digital Twin.

Following are some of the aspects of digital twin in smart manufacturing:

i. ***Virtual representation of physical assets***
It helps to visualize the physical assets like cyber-physical systems, equipment, etc. which are easily accessible at any time. It helps to create a detailed view of the physical system that is easy to understand.

ii. ***Prototyping and simulation of products***
It is a great tool to create and analyse the prototype of a product. With added features, we can also analyz e the efficiency of that product virtually.

iii. ***Remote monitoring and control***
As stated earlier, it creates a virtual image of any physical asset which makes it easier to access at any time. This is a great advantage for purposes like remote monitoring and control, and performance analyzation.

iv. ***Predictive maintenance***
Digital twins equipped with technologies like machine learning and AI are useful for prediction. The prediction can be used to analyse the system's efficiency over time and give feedback on its performance which is helpful to predict a problem with the system.

v. ***Performance optimization***
Learning and improving to save resources and time have always been one of the main features of AI-equipped technologies. The learning models regularly analyse the performance in real-time and help in better comparison to assist in performance optimization.

vi. ***Resource optimization***
It is very easy and efficient to visualise the resource that is used by a system which can be done in a digital twin technology. Enhanced with AI, it helps to use the resources effectively for a purpose.

vii. ***Collaborative decision making***
With the digital twin technology, the systems are highly connected and collaborative. The communication between these systems results in collaborative decision-making.

viii. ***Quality Control***
With precise control over the cyber -physical systems and virtual guidance for better prototyping, the quality of the manufacturing can be significantly improved without wasting resources.

2. Data Generation and Collection

In smart manufacturing, a diverse range of data of collected and stored for different manufacturing processes. The data is often generated by interconnected systems that collaborate to optimize processes, improve decision-making, and enhance overall operational efficiency. In AI-enabled digital twin technology for smart manufacturing, data from several sources is collected for different purposes like prediction, analysis, real-time monitoring, etc. The digital twin technology collects and stores the data from actual physical systems. Dealing with data collected from a smart manufacturing factory is associated with several challenges [16]. The data collected is large in volumes obtained from different sources thus we often have to deal with noise and missing values.

The data from the digital twins collectively contribute to the creation and improvement of smart manufacturing processes. It enhances its accuracy, reliability and effectiveness and reduces errors, failures, and faults by enabling smart decision-making and optimization techniques. Here are the primary types of data and their potential sources in AI-enabled digital twin technology.

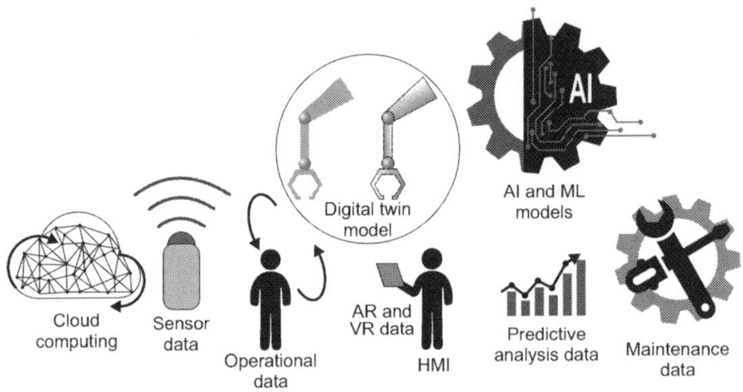

Figure 4: Data Sources in Smart Manufacturing.

i. **Digital Twin Models Data:**
Digital twin model data is the main data that builds up the foundation for all the other data. This data is a virtual replica of the actual physical systems or processes. The twin model data is a collection of all the machinery and technology embedded i n the actual device. The data collected is used for monitoring and predictive maintenance purposes.

ii. **Sensor Data:**
The real-time sensor data is collected for real-time analysis and monitoring. The sensors can be embedded in IoT machinery or various environmental monitoring sensors for temperature, humidity, pressure, etc. The sensory data is the key part of simulation control, it requires accurate data from the sensors to study the external environment and execute commands accordingly. Mostly, real-time data is required for efficient control of actual systems [14]. These data from sensors can help invoke response action if needed, analyse sensory data for predictive precaution or simply record the data for future reference.

iii. **Operational Data:**
The operational data refers to the data related directly to all manufacturing activities. This data is crucial for monitoring and maintaining purposes. This type of data is often real-time or near real-time which is stored to gain insights on any manufacturing organization's activities. The operational data is important in the manufacturing process as it provides information about how processes are running, uptime and downtime of the systems and how efficiently these processes are running.

iv. **Maintenance Data:**

The maintenance data or the diagnostic data is data about the health and maintenance needs of the physical systems. This data is very crucial for an efficient work environment with minimum failure. This type of data is collected from different sources for analysis directly from the machine status indicators, sensors and predictive maintenance algorithms that work on operational data and historical maintenance records.

v. **Predictive Analytics Data:**

The data collected from analysis results from different predictive AI algorithms, machine learning models for equipment failures or quality control models. This data is essential for invoking its response or alert mechanism.

vi. **Communication Network Data:**

The communication data is related to communication between the components within the same or different geographical locations. The interconnections among the devices are necessary for collaboration and communication [16]. Fog Computing and IoT play major roles in the network and communication in a smart manufacturing environment. The data includes a log of data transmission, network performance analysis, information about each system on the network and integration data from CPS.

vii. **Cloud Computing Data:**

Cloud computing data is simply the data stored in cloud environments. It contains historical data for analysis, and stores processing logs and other data needed for systems' overall locations [16]. The advancements in cloud and fog computing result in collection of the comprehensive data about the interconnections of devices and parts of the factory and other necessary information relevant to the modelled digital twin.

viii. **Augmented Reality (AR) and Virtual Reality (VR) Data:**

AR and VR data refers to the data related to user interactions with the simulations and learning models via an AR/VR environment. It contains training and maintenance logs from different AR/VR applications [20] discusses the advantages of combining technologies like Augmented Reality (AR), Virtual Reality (VR) and headsets that provide extended reality. It gives the user a more comprehensive and understandable description of the physical and virtual worlds.

ix. **Human-Machine Interface (HMI) Data:**

The Human-machine interface data includes a history of all the configurations and adjustments made by the user to machines [16]. This data also includes the preferences and personalisation settings of each stakeholder of the manufacturing system. Feedback, responses and other interactions between humans and machines are recorded through HMI.

2.1 The Big Data in Digital Twins

Artificial Intelligence or a machine learning model's efficiency can only be defined by the data it has access to and the information it can retrieve from it. Big data plays an important role in AI digital twins in smart manufacturing. It provides all the necessary raw data for an AI program to learn and extract valuable information from. It helps digital twins to create an accurate virtual representation of the physical assets and processes in a smart manufacturing ecosystem.

Data is the main component in working, learning, and analyz ing control processes. Thus, it is crucial to manage the large volume of data effectively. Data corrupted with noise, errors, missing values, etc. can lead to irrelevancy and incorrect inferences during analysis and other major data-driven procedures [20]. Operational data obtained in smart manufacturing processes tends to be stored as unstructured data which is hard to manage. Any machine learning algorithm is dependent on the data and data with high dimensionality can be difficult to work with. Big Data technologies help to maintain and organize data that can be later used to extract knowledge.

Data in a smart manufacturing environment can come from various sources like feedback, production data, and enterprise requests and are used for different purposes. B ig data technologies like NoSQL, Apache, MongoDB, and Apache Kafka can handle massive amounts of data very efficiently. It improves throughput, scalability and fault tolerance of the system.

The 5 Vs of Big Data in Digital Twins

- **i.** ***Volume***: The simulation of an exact real-world environment and the efficiency of the system in that environment is a complex task. The sensors, IoT devices and other devices produce massive datasets in real-time. Thus, to comprehensively design a virtual replica of a physical system, a large volume of data is required to understand the complexity and nuances of real-world operations. To handle the large amount of data effectively, a big data technology
- **ii.** ***Variety***: There are many sources where the data is collected in digital twins. It can be real-time from sensors, IoT devices, Cyber-physical systems, historically stored data from logs from maintenance records and simulation results or details of individuals in a database. The data can vary from unstructured, semi-structured to structured. It requires a big data platform that can support, store and allow the integration of diversity in the database.
- **iii.** ***Velocity***: Real-time data is extremely necessary in the case of digital twins to simulate the quick movements and processing of the real-world scenario. Thus, the data access should be efficient enough to respond to changes. Big data technologies help in handling the high-velocity data streams ensuring that the digital twin reflect the current state of the physical systems.
- **vi.** ***Veracity***: The real-time or historical data it receives should to accurate to gain actual results and an effective decision-making process. The predictive maintenance, identification of patterns and trends in datasets and simulation of the real space depend on the accuracy of the data. Thus, a big data technology

that facilitates quick and accurate data retrieval is required in a digital twin technology.

v. ***Value*:** Since the emphasis on data is quite large when it comes to AI in digital twins, the data that is to be stored is quite valuable. Handling such large amounts of data securely and efficiently can be easily done with the help of big data technologies. Big data technologies often include robust security features and encryption methods to ensure proper handling of the data.

3. Importance of Data Security and Privacy

As discussed in (2.), a smart manufacturing environment deals with an enormous amount of data every day [21]. Standards and regulations for data security in digital twins are yet to be fully established which makes it easier for cybercriminals to take advantage and threaten the entire manufacturing environment. DTs are intellectual properties of a manufacturing industry company which includes sensitive information representing a digital copy of the entire workspace.

In [21], we see the importance of data security in digital twins. Unauthorized access to any data in digital twins can lead to a havoc situation. It can lead to the manipulation of the behaviour of the twins, control over physical devices, manipulation of parameters, and control over assets and designs.

3.1 Critical Nature of Data Security in Smart Manufacturing

The smart manufacturing environment deals with a lot of data of critical importance due to the high connectivity of systems. The systems exchange various data for different purposes. Low security can bring fatal loss to the organization. The smart manufacturing infrastructure is highly interconnected in nature. It consists of physical devices, IoT devices, cyber-physical systems, applications, digital twins, etc. They are designed to be capable of communicating and collaborating. This network of smart manufacturing systems deals with an enormous amount of data each day. The data transmitted over the network can be both highly confidential property and a small command.

The smallest vulnerability could be seen as an opportunity for attack. It is extremely crucial to prevent vulnerabilities that could lead to data breaches. In extreme cases unauthorised access to sensitive data or control systems can cause operational risks and safety concerns in a manufacturing environment. It may cause unsafe operations to be carried out, equipment damage and even serious accidents. This in turn can lead to severe financial losses that can affect the organization's shareholder value.

Most of the processes in smart manufacturing are dependent on real-time data collected from different sources [14]. In a real scenario, the actual system's movements can be extremely fast. For efficient decision-making of digital twins, a real-time feedback mechanism is extremely necessary. Any disruptions due to any external attacks or unauthorized actions can cause significant loss and impact supply chain commitments [22]. Latency affects the quality of experience for everyone when real-time communication is required. It can also cause disruptions to the supply chain which can result in in production delays. The main purpose of smart manufacturing is to build quality products that make a huge impact on the brand. Therefore, the data fed to the machines is critical and crucial. Any misconduct or tampering with the data input can have a large impact on the production processes. The quality of products may be compromised due to delayed or incorrect data input leading to delays in production, fulfilment challenges and potential harm to suppliers' trust.

Digital twins in smart manufacturing deal with a large quantity of data every day. It involves historical data of manufacturing processes and even sensitive data like design and model training data. Unauthorized access to these data can lead to competitive loss and exposed organizational records and trade secrets. Competitors may gain access to trade secrets, leading to loss of competitive advantage and potential financial harm.

The digital twins are an exact copy of the real-world machine or CPS and leaving it exposed to any vulnerability cause serious damage to the manufacturing infrastructure. Security breaches in smart manufacturing can compromise the control over automation, robotics and IoT devices. It may lead to operational risks and even accidents in the manufacturing environment.

3.1.1 Criticality of Maintaining and Preserving Data

The integration of advanced technologies like AI, ML and digital twins makes it easier in the process of extracting useful insights into the business in the organization, making predictions, and making effective decisions automatically. These technologies require a large repository to store and use training models, feedback data and even historical data. The management of security practices and ensuring that the compliances and regulations are implemented through the entire lifecycle of the data from its collection, storage, access, and disposal is challenging for an organization. Lack of security measures can lead to tampering or manipulation of data can lead to incorrect decision-making, affecting overall operational efficiency. The integrity of data like production data, and processing data is very important to be preserved.

The data needs to be preserved according to the legal compliance. Non-compliance with regulations can result in legal consequences, fines, and damage to the company's reputation. Navigating through all of the compliances and regulations from all around the world while considering the complexities of data storage and processing is a bit challenging. The policies also keep updating to align with rising threats and keeping track of all of these is a big task. Data breaches due to neglecting compliances can lead to financial penalties, legal actions, and the need for comprehensive remediation measures to achieve compliance.

Mishandling critical and data security breaches erode trust, impacting relationships with stakeholders and hindering the widespread adoption of smart manufacturing practices. It leads to negative publicity and loss of trust among customers, partners, and stakeholders. Failure to adapt can result in vulnerabilities being exploited by emerging threats, leading to security incidents.

3.2 Privacy Concerns for Sensitive Manufacturing Data

Privacy concerns for sensitive manufacturing data in AI digital twin-enabled smart manufacturing are significant and must be carefully addressed to ensure the protection of sensitive information. The [23] suggests that the main factors that give rise to privacy concerns are the vast amount of data being generated and the sensitive system data that is to be protected. Considerations for privacy and security of data help in understanding and trusting the digital twin technology [24]. The u navailability of any components over the network leads to financial loss. Balancing the need for data sharing and collaboration with data privacy concerns when dealing with sensitive information is extremely challenging.

Maintaining the balance between the need for manufacturing data to be shared for collaboration and control over proprietary information is the key solution to the challenges faced by data privacy concerns. The implementation of a system that allows authentication and limited access to data according to the role of the individual is essential to reduce over-sharing and unauthorised access to the data.

Furthermore, other than manufacturing data, the protection of PII is also important. PII or Personal Identifiable Information is data that is related to the personal details of individuals. It includes data on employees, operators and even stakeholders that are involved in the manufacturing organization. Exposure of personal data is a serious offense and is against many laws around the world therefore, protection of the personal data should be given more priority. However, the need for effective data analysis of customer feedback or employee throughput seems to be difficult without this data. There should be a clear understanding and consent regarding the collection, processing, and storage of the data between the organization and the individual. Along with transparency, establishing data-sharing protocols and agreements when collaborating with external entities such as suppliers and partners can also be done to improve the privacy and protection of sensitive information.

There are several compliance and regulations for dealing with sensitive information. There should be an agreement and transparency regarding the compliances and regulations followed by the data owner. The organization must ensure that the data will only be used for ethical uses without any biases and discriminatory practices. The concern for designing an effective AI model that follows all the rules and responsible practices that not only protect the privacy of the data but also align with the industry-specific requirements in the smart manufacturing ecosystem is still in question. The employees who handle the data should be aware of the policies and should be trained to handle the data accordingly.

3.3 Challenges in Data Security and Privacy

The increased involvement of advanced technologies also brings security issues with it. Thus, m ore attention to secure sensitive data is required. In the context of smart manufacturing, the manufacturing data, presets, logs, and real-time communication data are some of the examples of sensitive data in smart manufacturing. It is an important yet challenging task to protect data against cyber-attacks and threats due to the dependency of smart manufacturing processes on enabling technologies. It is not a smart move to limit the abilities of the manufacturing processes by eliminating new technologies to protect data. Hence, advanced technologies and compliances along with best security practices are implemented which are discussed later in this chapter.

IoT Devices and Real-time Communication

Digital twins require real-time data to mimic the actual processes of the cyber physical systems. Thus, it requires security measures that can dynamically adapt to the nature of the virtual representation of the physical assets, processes, and systems. Most of the processes in a smart manufacturing industry require real-time data processing. It is quite challenging to provide rapidly changing security measures to match efficient real-time communication while avoiding any latency to the communication.

Another crucial challenge faced is the security of interconnected systems and IoT devices in a manufacturing environment. Communication between the systems is possible due to the IoT devices and systems integration techniques. With the large number of IoT devices like sensors, actuators, cameras, RFIDs, etc. securing each of these devices at each endpoint is a complex and challenging task. Any loss of data or latency can lead to fatal damage. Hence, the security measures should be effective yet not interrupt the smooth flow of communication.

Securing Communication Over the Network

The implementation of end-to-end encryption to maintain integrity and privacy on the network requires high memory and processing capabilities. While handling an enormous amount of data less capable devices may not be able to handle quick encryption-decryption which can lead to latency and even loss of data.

Ensuring the security of cyber-physical systems, where the digital and physical worlds intersect is a challenging and complex task to carry out. Similarly, securing the entire supply chain processes and data including vendors and partners is quite a challenge. Any vulnerability or weakness in third-party security can lead to unauthorized access and loss of privacy of sensitive data revealing the processes and business strategies.

Cloud Security and Scalability

Cloud security is a crucial component of the manufacturing environment for communication among various components located in different geographical locations. Managing the security of cloud infrastructure introduces complexities.

All the security measures and processes should be able to efficiently scale according to the growth of the smart manufacturing environments while considering resource constraints and budget limitations.

Compliance Issues

Keeping up with the local and global compliances and various data regulations and industry-specific compliance standards can be difficult especially when data is stored and processed in different geographical locations with different laws and regulations. Also, the absence of standardized security protocols for smart manufacturing environments leads to inconsistency creating vulnerabilities for others to exploit.

Evolving with Evolving Threats

Evolving and updating security implementations according to the evolving threats is an important yet challenging task. Many manufacturing environments have legacy systems that may lack modern security features. Integrating older systems securely with newer technologies is a complex job. Any overlooked vulnerability leaves an open opportunity for cyberattacks which can compromise data security and privacy.

Training and educating employees on the importance of following security measures can save a lot of effort against unintentional actions done by insiders, including employees or contractors, which can pose significant security risks. Managing insider threats requires a combination of technical and organizational measures which includes handling malicious files and securing systems against a cyberattack.

4. Standards and Regulations

In terms of AI, data security and privacy laws and regulations are yet to be fully defined. Regulations ensure that personal data is protected from any outside threat or data breaches.

Several regulations and standards govern data security and privacy across various industries and regions. Compliance with these regulations is crucial for organizations to ensure the protection of sensitive information. Organizations operating in smart manufacturing and related industries must stay informed about these regulations and standards to ensure compliance and uphold data security and privacy best practices. The landscape is dynamic, with new regulations emerging, and organizations need to adapt their practices accordingly.

The international perspective on data security and privacy compliance involves a complex landscape with diverse regulations and standards across different regions. Here's an overview of the international perspective on key aspects of data security and privacy compliance:

Compliance requirements vary based on factors such as industry, jurisdiction, and the nature of the data being handled. Here are some common compliance requirements related to data security and privacy:

i. **General Data Protection Regulation (GDPR):**

   The GDPR comes under the European Union and European Economic Area. This regulation states the rules related to the protection of personal data. It protects the fundamental right of a person to be able to protect their personal data [25]. It imposes measures to lawfully process any personal data transparently and protect the data from outside threats while processing using encryption algorithms.

ii. **ISO/IEC 27001:**

   The standard ISO/IEC 27001 is designed for Information Security Management Systems (ISMS) [26]. It states the following standards for a company to follow:

   - To establish, implement and maintain the rules for information security and continuously improve the same to eventually improve the ISMS.

- Conduct risk and vulnerability assessments to take preventive measures beforehand and apply security controls at appropriate times.
- Monitor and review the effectiveness of the ISMS.

iii. **Personal Data Protection Act (PDPA):**
PDPA governs the collection, use, and disclosure of personal data in Singapore, emphasizing the need for organizations to obtain consent and protect individuals' personal information. Unlike GDPR PDPA allows implied consent in certain circumstances.

iv. **Health Insurance Portability and Accountability Act (HIPPA):**
HIPPA governs the standards for the secure handling of individual's health information. It safeguards the protected health information (PHI) of an individual. It includes standards and policies to implement physical and technical measures to protect the PHI.

v. **ISO 23247:**
ISO 23247 [18] is a digital twin framework standard for manufacturing which includes four layers namely: observable manufacturing elements, data collection and data control entity, core entity and user entity. This standard is mainly based on the Internet of Things.

vi. **ISA 62443:**
The International Society of Information (ISA) is a standard well-known for Industrial Automation and Control Systems (IACS). ISA-62443 series defines guidelines for establishing a secure and resilient industrial automation environment.

vii. **NIST Cybersecurity Framework:**
The NIST Cybersecurity Framework, originating in the U.S., is widely adopted globally as a best practice guide. It was developed by the National Institute of Standards and Technology (NIST), the framework provides a set of cybersecurity best practices and guidelines for organizations to manage and improve their cybersecurity risk.

5. Security Measures and Best Practices

In smart manufacturing that implements an AI-based digital twin technology, a large amount of data is retrieved from different sources. These data are used in small and big processes that also produce resultant data. S mart manufacturing may hold multiple advanced technologies with different types of input data needed. For example, to implement an efficient machine learning model it needs training data to accurately distinguish between low and high -priority issues; the digital twin needs data about the environment of an actual physical system to replicate conditions and processes. These technologies are interconnected with each other for better coordination which raises the need for a scalable cloud infrastructure and a secured network. Various cryptographic algorithms are also implemented for maintaining data privacy during transmission which requires strong computational units.

Securing all the components of smart manufacturing is extremely important t o prevent any kind of threats to data or manufacturing operations. Certain measures to secure the data exchange over a network between the twins or devices should be taken [22]. The source of data that is uploaded should be authenticated to prevent any unauthorized source from tampering with twins [21]. The cybersecurity issues in digital twins are yet to be explored because of their vulnerable nature. The digital twins are considered critical systems and working with them in automation is difficult. It holds the most important intellectual property of the manufacturing system which is a digital copy of the physical systems.

According to [23], the enabling technologies can help in giving insights about the measures and security practices ensuring privacy and security overcoming trust issues with the digital twin technology.

5.1 The CIA Triad of Security

a) **Confidentiality**

Maintaining confidentiality of data refers to access management and ensuring the privacy of the data in a network. It is the main component of data security that reduces the risk of data breaches. To maintain confidentiality certain security measures can be taken such as:

Access Control—A robust access control system should be designed and implemented to restrict the access of data to authorised entities only.

Data Encryption—Since the digital twins are only a virtual simulation of a physical entity, a real-world entity and its digital twin can be in different locations. It means that the transmission of data like alerts, feedback, commands and configurations would take place over a network. To prevent the interception or tampering, data encryption is important. Many networking protocols implement DES/AES algorithms using public/private keys.

b) **Integrity**

Maintaining the integrity of data is another important aspect of digital twins. The valid and accurate data results in effective analysis and results. According to [22], it is a main requirement in most network protocols to identify the data source. It will also give us control over what data should be allowed and which ones should be blocked. The f ollowing measures can be taken to ensure the integrity of data:

Data Validation—Techniques for input data validation to ensure the reliability of data in digital twins should be implemented. Also, data quality checks and integrity verification systems should be deployed for data to be transmitted.

Digital Signatures—The digital signatures can be used to ensure the originality of data and detect any unauthorized modifications in data.

Version Control—The version control mechanisms help to track changes and restore to a needed state to maintain the integrity of the digital twin models and datasets.

c) **Availability**

The access and retrieval of data in digital twins is an extremely important factor that builds the foundation of real-time processing and analysis. Thus, the availability of data is another main factor to be maintained. This can be done as follows:

Redundancy—Designing a digital twin architecture with redundancy to ensure uninterrupted processes even in case of hardware failures or other disruptions.

Load Balancing—Implementing multiple load servers to distribute load is a very effective way to prevent overload situations in case of high traffic or DoS attacks.

5.2 Data Protection Measures

- **PII Protection**
  Preventing the exposure of personal and identifiable information related to employees, operators, or individuals involved in the manufacturing process. Implementing robust anonymization and encryption techniques to protect PII while still enabling effective data analysis.
- **Data Ownership and Control**
  Determining and maintaining control and ownership of manufacturing data generated by digital twins. Balancing the need for data sharing and collaboration with maintaining control over proprietary information
- **Compliance with Privacy Regulations**
  Adhering to global and regional privacy regulations, such as GDPR, PDPA, and HIPPA, that impact the handling of sensitive manufacturing data. Regularly updating policies and practices to comply with evolving privacy regulations and ensuring alignment with industry-specific requirements.

5.3 Customer Data and Consent

[23] The end users and organizations should discuss the digital twin technology ensuring that end users understand the concept of the technology, data collection and processing methods and security and privacy measures.

- **Informed Consent and Transparency**
  Obtaining informed consent from individuals whose data is used in the manufacturing process and providing transparency about data usage. Developing clear and understandable consent mechanisms and communicating transparently about data collection, processing, and storage practices.
- **Ethical Use of Data**
  Ensuring the ethical use of manufacturing data, avoiding biases, and preventing discriminatory practices. Establishing ethical guidelines for data usage and promoting responsible AI practices within the smart manufacturing ecosystem.
- **Sharing Supplier and Partner Data**
  Sharing manufacturing data with suppliers and partners while ensuring the privacy of sensitive information. Establishing secure data-sharing protocols and agreements to protect sensitive data when collaborating with external entities.

5.4 Precautions and Preparing

- **Employee Training**
  Ensuring that employees are aware of privacy policies and best practices for handling sensitive manufacturing data. Providing ongoing training to employees to keep them informed about evolving privacy issues and the importance of data protection. Thus, they need to be regularly trained to ensure that the data is handled efficiently and securely.
- **Incident Response**
  Establishing and maintaining an effective incident response model is extremely necessary for smart manufacturing processes [24]. Precautions and prediction models should be implemented to recognize the threat in the early stages.
- **Human Factor and Training**
  Employees may inadvertently contribute to security risks through actions such as weak password practices or falling victim to phishing attacks. Regular training is essential but can be challenging to enforce effectively.

6. Advanced Technologies for Enhanced Security

The key aspect of smart manufacturing is to make the system self-sufficient with intelligent capabilities for optimization and precision. Therefore, the focus is to integrate with several advanced technologies to automate different processes and enable automated data-driven decisions. The involvement of advanced technologies also brings the advanced threats posed by them. Cyber threats are evolving rapidly with time and so are the challenges associated with the security of the data. Thus, more advanced measures are needed to be taken for enhanced security and better privacy maintenance [27]. The focus of the industry is to design an intelligent environment using modern technologies like machine learning, big data, blockchain, etc. Although the most critical information of the organization should be provided enhanced security and privacy to keep a competitive edge in any sector.

6.1 Blockchain Technology

Since 2008 when Satoshi Nakamoto first introduced the world to Bitcoin in his paper [28], blockchain technology has rapidly been applied in several sectors other than just cryptocurrency like healthcare, education, finance, agriculture, transportation and supply chains. The decentralized nature of blockchain along with its unmatched fault tolerance and transparency unlike other technologies has defined new applications of blockchain in data security and data transmission. The Fig. 5 lists various possible advantages of smart manufacturing. In [20], it is stated that—"blockchain is a security-focused structure with excellent potential, efficient transparency and decentralization." Blockchain is the most useful tool when it comes to maintaining integrity and storing data records. It can be the most highly valuable integration in the digital twin technology.

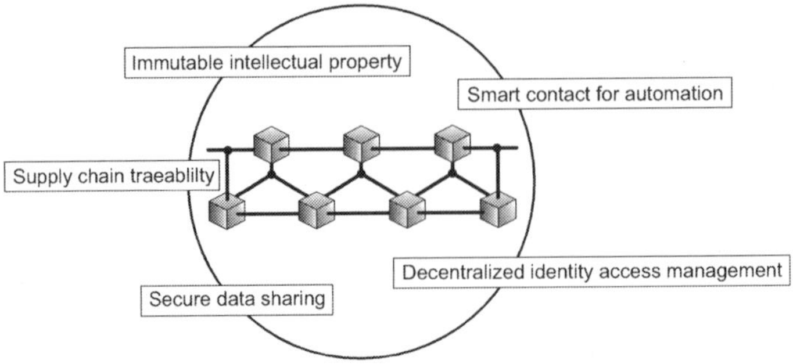

Figure 5: Blockchain Technology for Advanced Security.

In terms of smart manufacturing, the data is critical and prone to attacks. There should be a system that ensures that only the intended data is accessible to each user. For example, the manufacturer has access to read, write, and update product data whereas a simple user has access to only read the product data and update its details. In [20], the authors have proposed such an integrated system of blockchain using Hyperledger.

In [26], the authors have described how authenticity can be maintained in an organization using blockchain and other security measures in digital twins of actual manufacturing processes. It explores the concept and need for process authentication to secure access to business data using blockchain technology.

The paper [29] describes the role of permissioned blockchains in the SE lifecycle. It suggests that securing and limiting access to the data assets of a business is important. It enables connection to the services without giving access to copy data by managing the time-stamp dataset. The blockchain not only efficiently handles the distribution of data but also ensures that the data is immutable. The key feature of this work [29] is that not every participant should be given access to all of the data and certain permissions should be grated selectively based on their role in the manufacturing process.

6.2 AI and Machine Learning—Based Analysis and Anomaly Detection

Artificial Intelligence or AI is technically a software program that is designed to learn, improve, and mimic the intelligence of a human brain. When it comes to machine learning is the most important part of any industrial process. M achine learning is a part of AI that trains itself based on statistical or numerical data values. The three powerful tools of human knowledge [13] conceptualization, collaboration and comparison are much more advanced than any machine learning model. Yet humans due to their limitations are prone to make errors due to negligence or carelessness. M achine learning can be used to train a model to run repetitive tasks accurately every time. It can train itself using different data inputs. It can analyse results and feedback provided to improve itself.

The smart manufacturing system collects data from different sources in real-time for processing. ML and AI present several opportunities for analyz ing and producing insights. It makes decision making faster using patterns and prediction models that help in the automation of different processes as shownin Fig. 6. It also plays a vital role in predictive maintenance and real-time quality control. Training a model requires a lot of effort from the collection of relevant datasets to model stability analysis. The study [30] proposed a to detect anomalies based on time information and periodic packet patterns using sample mean and range for each repeating message. There are many techniques to model a machine learning model such as clustering, monitoring, classification, etc.

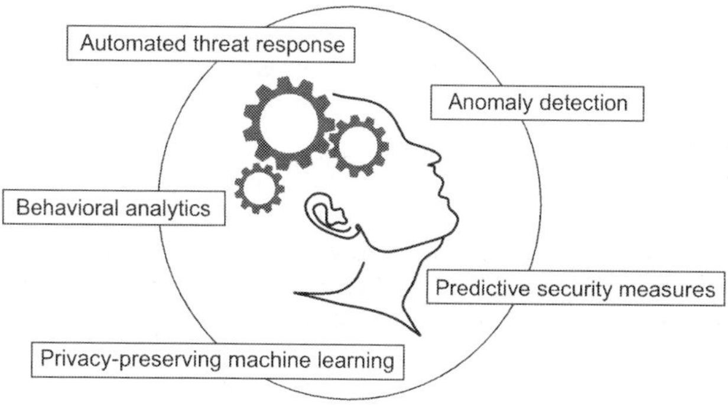

Figure 6: Artificial Intelligence for Automated Security Mechanisms.

7. Conclusion

Digital Twin technology can be an extremely valuable contribution to smart manufacturing and Industry 4.0. Despite the limitations and challenges, there is no denying that Digital Twins has a lot of features that will revolutionize the upcoming industry. It is still in its early stage of development and is yet to reach its full potential. Digital Twins coupled with enabling technologies like IoT, AI/ML, Cloud computing, 5G technology, Blockchain, Advanced Robotics, Sensors and Actuators can define the new era of the smart industry.

The Digital Twins offer many advantages like easy prototyping of the products, simulation of the real-world environment to test the products in, better control over machines, improved collaboration among the devices, AI/ML prediction for analysis and maintenance, robust alert systems due to IoT technology, precision while working with machines, adaptability to its environment and many more.

There are certain limitations and challenges that can be faced in the implementation of digital twins including their complex nature, high cost and high requirement for monitoring and maintenance. Apart from these, the main challenge that we face is the security and privacy of the manufacturing data. It is an emerging technology which is yet to be worked on before fully implementing it in a manufacturing environment.

The standards and regulations discussed in this chapter are not DT-specific which makes it difficult to implement. It brings up the need for the development of DT-specific standards and regulations for designing and maintaining them. A lot of research still needs to be done to introduce the enabling technologies for security.

This book chapter highlighted the major and crucial issues that needs attention in terms of security. It described the security challenges and privacy concerns for the digital twins. The trust issues for the digital twin technologies needs to be addressed to make it an important part of the manufacturing processes.

References

1. Gomez, C., Guardia, A., Mantari, J. L., Coronado, A., Reddy, J. 2021. A contemporary approach to the MSE paradigm powered by Artificial Intelligence from a review focused on Polymer Matrix Composites. *Mechanics of Advanced Materials and Structures, 29,* 1–21. 10.1080/15376494.2021.1886379.
2. Grieves, M. 2014. "Digital Twin: Manufacturing Excellence through Virtual Factory Replication," white Paper, NASA.
3. Y. Chen, Y. 2017. Integrated and Intelligent Manufacturing: Perspectives and Enablers. *Engineering, 3*, 588–595, Oct. 2017.
4. Liu, Z., Meyendorf, N., and Mrad, N. 2018. The role of data fusion in predictive maintenance using digital twin. In Annual Review of Progress in Quantitative Nondestructive Evaluation (Provo, Utah, USA), p. 020023.
5. Zheng, Y., Yang, S., and Cheng, H. 2018. An application framework of digital twin and its case study. *Journal of Ambient Intelligence and Humanized Computing, 10*, 1141–1153, June 2018.
6. Vrabič, R., Erkoyuncu, J. A., Butala, P., and Roy, R. 2018. Digital twins: Understanding the added value of integrated models for through-life engineering services. *Procedia Manufacturing, 16*, 139–146.
7. Madni, A., Madni, C., and Lucero, S. 2019. Leveraging Digital Twin Technology in Model-Based Systems Engineering. *Systems, 7*, p. 7, Jan. 2019.
8. Tao, F., Zhang, M., and Nee, A. Y. C. 2019. Digital twin driven smart manufacturing. Academic P ress.
9. Bergs, T., Gierlings, S., Auerbach, T., Klink, A., Schraknepper, D., and Augspurger, T. 2021.The concept of digital twin and digital shadow in manufacturing, Procedia CIRP 101 (2021) 81–84.
10. Holmes, D., Papathanasaki, M., Maglaras, L., Ferrag, M. A., Nepal, S., and Janicke, H. 2021. Digital Twins and Cyber Security–solution or challenge? In *2021 6th South-East Europe Design Automation, Computer Engineering, Computer Networks and Social Media Conference (SEEDA-CECNSM)*, pp. 1–8. IEEE.
11. Fuller, A., Zhong, F., Day, C. and Barlow, C. 2020. Digital Twin: Enabling Technologies, Challenges and Open Research. IEEE Access. PP. 1–1. 10.1109/ACCESS.2020.2998358
12. Huang, Z., Shen, Y., Li, J., Fey, M., and Brecher, C. 2021. AI-Driven Digital Twins. Sensors 2021, 21, 6340. https://doi.org/10.3390/s21196340.
13. "Digital Twin: Manufacturing Excellence through Virtual Factory Replication"—A White paper by Dr. Michael Grieves. *https://www.3ds.com/fileadmin/PRODUCTS-SERVICES/DELMIA/PDF/Whitepaper/DELMIA-APRISO-Digital-Twin-Whitepaper.pdf*
14. Glaessgen, E. H. and Stargel, D. S. 2012. The digital twin paradigm for future NASA and U.S. Air force vehicles. In Proceedings of the 53rd AIAA/ASME/ASCE/AHS/ASC Structures, Structural Dynamics and Materials Conference 20th AIAA/ASME/AHS Adaptive Structures Conference 14th AIAA, Honolulu, HI, USA, 23–26 April 2012.
15. Jones, D., Snider, C., Nassehi, A., Yon, J., and Hicks, B. 2020. Characterising the Digital Twin: A systematic literature review. CIRPJ. Manuf. Sci.Technol.2020,29,36–52.
16. Friederich, J., Francis, D. P., Lazarova-Molnar, S., and Mohamed, N. 2022. A framework for data-driven digital twins of smart manufacturing systems. *Computers in Industry* 136(2022): 103586.
17. Loaiza, J. H., Cloutier, R. J., and Lippert, K. 2023. Proposing a Small-Scale Digital Twin Implementation Framework for Manufacturing from a Systems Perspective. Systems 2023, 11, 41
18. ISO. Digital Twin Framework for Manufacturing ISO 23247. 2023. Available online: https://www.ap238.org/iso23247/(accessed on 18 July 2023).
19. Wynn, M., and Irizar, J. 2021. Digital Twin Applications in Manufacturing Industry: A Case Study from a German Multi-National. *Future Internet* 2023, 15, 282. https:// doi.org/10.3390/fi15090282.

20. Harrison, R., Vera, D., and Ahmad, B. 2021. A Connective Framework to Support the Lifecycle of Cyber–Physical Production Systems. *Proc. IEEE 2021, 109*, 568–581.
21. Homaei, M. H., Gutiérrez, O. M., Sancho Núñez, J. C., Vegas, M. A., and Lindo, A. C. 2023. A Review of Digital Twins and their Application in Cybersecurity based on Artificial Intelligence. arXiv preprint arXiv:2311.01154.
22. Mashaly, M. 2021. Connecting the twins: A review on digital twin technology & its networking requirements. *Procedia Computer Science 184*(2021): 299–305.
23. Čolaković, A., and Hadzialic, M. 2018. Internet of Things (IoT): A Review of Enabling Technologies, Challenges, and Open Research Issues. *Computer Networks.* 144. 10.1016/j.comnet.2018.07.017.
24. Santhosh, N., Srinivsan, M., and Ragupathy, K. 2020. Internet of Things (IoT) in smart manufacturing. In *IOP Conference Series: Materials Science and Engineering*, *764*(1), p. 012025. IOP Publishing.
25. "GDPR- General Data Protection Regulation" https://gdpr-info.eu
26. "ISO/IEC 27001- ISMS" https://www.iso.org/standard/27001
27. Demetrios, J., Kalawsky, R., Martínez-García, M., Fowler, C., and Fowler, K. 2020. Realizing the role of permissioned blockchains in a system engineering lifecycle. *Systems* 8, no. 4 (2020): 41.
28. Nakamoto, S. 2008. Bitcoin whitepaper. *URL: https://bitcoin. org/bitcoin. pdf-(:17.07. 2019).*
29. Ciampi, M., Romano, D., and Schmid, G. 2022. Process Authentication through Blockchain: Three Case Studies. *Cryptography* 6, no. 4 (2022): 58.
30. Kim, H., and Shon, T. 2022. Industrial network-based behavioral anomaly detection in AI-enabled smart manufacturing. *J Supercomput 78*, 13554–13563.

CHAPTER 12

Application of Cryptography and Deep Learning for Smart Environments

*Raj Pal Singh, Abhishek Rawat** and *Rajat Verma*

1. Introduction

Technology has advanced so quickly in recent years that it has made it possible to create Smart Environments, which have the potential to drastically change several elements of our lives. Smart environments use a variety of networked devices and systems, such as sensors, actuators, Internet of Things gadgets, and intelligent algorithms, to create an environment that is intelligent, adaptable, and sensitive to human needs. But there are a lot of obstacles to overcome before Smart Environments can be widely used, especially when it comes to privacy, data security, and intelligent decision-making [1]. It is critical to guarantee the privacy, veracity, and accuracy of personally identifiable data created and processed by Smart Environments. Additionally, the effective operation of Smart Ecosystems depends on the capacity to derive significant insights from this information and make wise decisions in real-time. This is where Deep Learning and Cryptography, two cutting-edge technologies, are useful. The field of Cryptography, which deals with secure communication, provides a solid basis for protecting personal data in thoughtful contexts. Cryptographic techniques, including authentication, decryption, and encryption, can be used to protect data against manipulation and unwanted access [2]. To guarantee the security and privacy of data transferred and stored inside Smart Environments, cryptography is essential. On the other hand, a subfield of machine learning called Deep Learning is very good at analyzing and interpreting vast volumes of data in clever contexts. Deep Learning algorithms are capable of processing and learning from vast amounts of

Department of Computer Science & Engineering, Pranveer Singh Institute of Technology, Kanpur, Uttar Pradesh, India.
* Corresponding author

sensory input since they are built on multiple-layered neural networks. This enables anomaly identification, pattern recognition, and intelligent prediction-making in intelligent environments. Thanks to Deep Learning, smart settings can now adjust and enhance their operations depending on real-time data analysis. Combining Deep Learning and encryption offers a viable solution to the issues that intelligent settings run against. Secure communication and information security are made possible by Cryptography, and Deep Learning allows for intelligent choices based on the study of enormous volumes of data. By leveraging these technologies, intelligent environments can improve automation, effectiveness, safety, and privacy, leading to a more connected and intelligent future. Let us e xamine how Deep Learning as well as Cryptography can be used in intelligent environments. Deep Learning explores the different cryptographic methods used in Smart Environments to protect data storage and transport [3,59]. Deep Learning algorithms can also help with the quick processing of data and wise decision-making in Smart Environments. The purpose of this investigation is to shed light on the possible advantages, difficulties, and potential paths of using Deep Learning and Cryptography in Smart Environments. We can build strong, safe, and intelligent Environments that improve our daily lives by being aware of the potential and constraints of these technologies. As technology develops further, the idea of smart surroundings is becoming more popular. These networked technologies offer convenience, efficiency, and improved quality of life. Examples of these are smart cities and smart homes. However, these advancements also pose new challenges for preserving the security and privacy of these intelligent environments. Examine how Deep Learning and Cryptography can be used to overcome these security issues in this chapter [4].

2. Understanding Smart Environments

It delves into the intricacies of Smart Environments, discussing their architecture, components, and potential vulnerabilities. These environments rely on various sensors, actuators, and interconnected devices to create seamless experiences for users. However, it also draws attention to the possible dangers that come with these networked devices, including privacy issues, illegal access, and data breaches.

2.1 Introduction to Smart Environments

Smart Environments are dynamic and interconnected systems that use sensors, actuators, and interconnected devices to enhance user experiences. To enable intelligent decision-making, they rely on technology like Deep Learning and the Internet of Things, also known as IoT, for data collection and analysis. These settings do, however, also carry some danger, including the possibility of data breaches and illegal access. Since Cryptography offers methods for safe data storage and transfer, it is essential to the security of Smart Environments. This chapter will explore the architecture, components, vulnerabilities, and the Application of Cryptography and Deep Learning in Smart Environments.

2.1.1 Examples of Smart Environments and their Applications

Smart settings have a broad spectrum of applications in multiple fields. Here are a few instances:

- **Smart Homes:** Smart homes integrate various devices and systems to enhance comfort, convenience, and energy efficiency. These environments utilize interconnected sensors, actuators, and smart devices to automate tasks such as lighting, temperature control, security, and entertainment [5]. They enable remote monitoring and control of home systems through smartphones or voice assistants.
- **Smart Cities:** Smart cities employ technology to optimize resource utilization and raise the quality of life for their residents. They include integrated systems for public safety, disposal of waste, energy management, and traffic control. Smart city initiatives aim to enhance sustainability, efficiency, and liability by utilizing data-driven decision-making and automation.
- **Smart Healthcare:** Smart healthcare environments, also known as Ambient Assisted Living (AAL), utilize technology to support independent living and improve healthcare outcomes [6]. To offer individualized medical care, remote patient monitoring, detection of falls, medication administration, and emergency response systems, these settings incorporate sensors, wearable technology, and monitoring systems.
- **Smart Manufacturing:** Smart Industrial Environments improve production processes, boost efficiency, and enable predictive maintenance using automation and the Internet of Things (IoT). These environments utilize interconnected sensors, robotics, and data analytics to monitor and control manufacturing operations, enhance productivity, and reduce downtime [7].

These examples provide a glimpse into the diverse range of applications where Smart Environments are making a significant impact.

2.2 Overview of the Architecture of Smart Environments

The architecture of Smart Environments refers to the underlying structure and design principles that enable the integration of various technologies and systems to create intelligent and responsive environments. While there are different perspectives and approaches to Smart Environment architecture, the following aspects are commonly considered:

- **Sensors and Actuators:** A network of actuators and sensor arrays is the foundation of Smart Environments, allowing them to gather information and communicate with the outside world. Sensors capture information about temperature, humidity, light, motion, and other environmental parameters, while actuators enable the control of devices and systems based on the collected data [8].
- **Connectivity and Communication:** Smart Environments require robust connectivity to enable seamless communication between devices, systems, and users. Usually, either wireless or wired networks like cellular networks,

Bluetooth, Wi-Fi, Zigbee, or Zigbee are used to do this [9]. Connecting and integrating different systems and devices inside the Smart Environment is made possible in large part using the Internet of Things (IoT).

- **Data Processing and Analytics:** Numerous sensors, gadgets, and human interactions in Smart Environments produce enormous volumes of data [10]. To find patterns, derive valuable insights, and arrive at wise conclusions, data processing & analytics approaches are used. It involves data aggregation, filtering, analysis, and visualization to enable real-time monitoring and intelligent automation.
- **Decision-Making and Control:** Smart Environments utilize intelligent algorithms and decision-making mechanisms to process data and make informed decisions. These decisions can range from simple automated actions, such as adjusting lighting or temperature, to more complex tasks, such as optimizing energy consumption or managing traffic flow in a smart city.
- **User Interfaces and Interactions:** Smart Environments provide user-friendly interfaces and interaction mechanisms to enable users to monitor and control the environment. This can include mobile applications, web interfaces, voice assistants, touchscreens, or gesture-based interfaces. It is intended to be easy, customizable, and accessible to engage with the Smart Environment.
- **Security and Privacy:** Ensuring privacy and security is crucial since Smart Environments gather and handle sensitive data [11]. Robust security procedures, such as control of access, authentication, and encryption, are implemented to protect data and prevent unauthorized access. Privacy considerations involve transparent data collection practices, user consent, and compliance with privacy regulations.

It is worth noting that the Architecture of Smart Environments can vary depending on the specific application domain and requirements. Different research studies and projects have proposed various architectural frameworks and models for Smart Environments, focusing on aspects such as interoperability, scalability, and adaptability. The goal of these architectural frameworks is to offer a thorough grasp of the elements and design ideas required to create settings that are responsive and intelligent. In general, the field of Architectural design of Smart Conditions is multidisciplinary, incorporating aspects from engineering, computer science, architecture, and various other related fields to build intelligent, networked environments that improve modern society's sustainability, efficiency, and quality of life.

3. Cryptography

This section provides a primer to cryptography, the science of secure communication. The basic ideas behind key management, encryption, and decryption. Additionally, t hey explore various cryptographic algorithms commonly used in securing Smart Environments, such as symmetric encryption, asymmetric encryption, and hash functions. It also touches upon the importance of key exchange protocols and secure communication channels. Cryptography is the study and practice of secure communication. It covers the strategies and processes used to safeguard data from unlawful access or alteration, guaranteeing its validity, confidentiality, and integrity. Numerous facts of contemporary life, such as internet operations, secure communication, data security, and privacy preservation, depend heavily on Cryptography. It helps build confidence in digital settings and serves as a foundation for safe systems. This u nit will provide an overview of Cryptography, discussing its fundamental principles and exploring the different cryptographic algorithms commonly used in securing Smart Environments [12]. It also discusses how crucial secure communication routes and key exchange methods are to maintaining data integrity and confidentiality. Gaining knowledge of the fundamentals of Cryptography will help you better grasp how secure communication functions and how sensitive information is protected using Cryptographic Techniques.

3.1 Fundamental Principles of Cryptography

Cryptography is built upon several fundamental principles that form the basis of secure communication. These guidelines guarantee the authenticity, confidentiality, and integrity of data. Let us investigate these ideas in greater depth:

- **Discretion:** Restricting access to information without authorization is known as maintaining confidentiality. It guarantees that the information can only be accessed and understood by authorized people or entities. Plaintext is converted into ciphertext using Cryptographic Techniques like encryption, rendering it unintelligible to anyone lacking the necessary decryption keys [13].
- **Honesty:** Integrity guarantees that data is reliable and unchangeable throughout transmission or storage. Data is hashed using Cryptographic Techniques, including hash functions, to produce unique hash values. By serving as digital fingerprints, these hash values enable the recipient to confirm the accuracy of the material they have received. A distinct hash value will be produced by any alteration to the data, suggesting possible tampering.
- **Verification:** During communication, authentication confirms the identification of the people or things involved. It ensures both the veracity of the data received and the sender's authenticity. Authentication is given through cryptographic methods like digital signatures and certificates [14]. While digital signatures connect the sender's identity to the message using asymmetric encryption, certificates offer a reliable third-party identity verification service.
- **Non-repudiation:** People are unable to deny their participation in a conversation or transaction if they practice non-repudiation. It makes sure that a transaction or message cannot be revoked once it has been delivered or completed. Digital signatures and other cryptographic approaches offer non-repudiation by proving the message's integrity and the sender's identity [15].

These Fundamental Principles of Cryptography form the basis for secure communication and information protection. By applying these principles, cryptographic systems can safeguard sensitive data and ensure the trustworthiness of communication channels.

3.2 Encryption: Protecting Data with Ciphers

Cryptography plays a vital role in safeguarding information by using ciphers to transform data into an incomprehensible format called ciphertext. These ciphers are secure encryption and decryption techniques. Sensitive data is protected using encryption, which is the process of transforming readable plaintext information into encrypted ciphertext. It comprises encrypting the data utilizing Cryptographic Techniques, such as ciphers so that the correct key is required to decrypt it.

- **Ciphers:** In the operations of encryption and decryption, an algorithm is called a cipher [16]. It chooses how the information is converted from unencrypted text to encrypted text and back. Different kinds of ciphers exist, such as public-key and symmetric-key ciphers.
- **Symmetric-Key Encryptions:** One key is used for decryption as well as encryption in symmetric-key ciphers, also known as secret-key ciphers [17]. This secret key must be shared beforehand by the person who sends it and the receiver. Advanced encryption standards, or AES, is one symmetric encryption method that uses the same key over the same data consistently to ensure confidentiality.
- **Ciphers using public keys:** Public-key ciphers, also called asymmetric ciphers, protect and decrypt data using private as well as public keys [18]. Although public keys are made public, private keys are kept private. Due to public-key encryption technologies like RSA, secure encrypted communications can be exchanged without a secret key that is shared.
- **The Value of Cryptography:** Sensitive data must be protected using encryption to prevent illegal access or interception [19]. Information is protected from prying eyes by rendering it unintelligible to those lacking the decryption key. Data integrity is also maintained by encryption, which detects any undesired changes or manipulation. Additionally, encryption ensures the confidentiality of the data received while confirming the sender's identity.
- **Advancements in Encryption:** Over time, encryption algorithms have evolved following security concerns and advancements in technology. For instance, the more robust Triple DES (3DES) method took the role of the standard for encryption (DES), encrypting data three times. Symmetric block ciphers like AES are commonly utilized because of their strong security and effectiveness [20].
- **Applications for Cryptography:** Sensitive data is protected in several fields using encryption. It is frequently used to secure online purchases, messaging services, and email correspondence [21]. Data encryption is also used to protect data while it is at rest. For example, disks or specific files can be encrypted to prevent unwanted access in the event of theft or loss.

3.3 Decryption: Unlocking Encrypted Data

Encryption is the process of converting information to a form that requires a password and decryption key to read. To return encrypted data to its distinctive, readable form, the encryption process must be reversed during the decryption phase 1. Here are some methods and scenarios related to decrypting encrypted data:

- **Decrypting Encrypted Files and Folders:** To decrypt encrypted files and folders, you typically need to use encryption software or specialized tools. Finding the secret file or folder and starting the decryption procedure with the given instructions are the steps in the process [22]. This allows authorized users to regain access to the information.
- **Decrypting USB Drives:** A pen drive or USB flash drive that has been locked can be read-only, write-protected, or encrypted. You can use two different methods for unlocking a USB disk. Requesting authorization to obtain the password is one method. Another method involves editing the Storage Device Policies. If there is any data loss during the write protection removal process, you can use data recovery software to retrieve the lost data [23].
- **Decrypting Encrypted Hard Drives:** It is still possible to retrieve data from an encrypted hard disk even if you have lost the password. The hard drive must first be unlocked using the proper technique or software [24]. Once the hard drive is unlocked, you can proceed with the recommended data recovery software to retrieve your data.
- **Decrypting BitLocker-Encrypted Drives:** Beginning with Windows Vista, BitLocker has been an encryption utility that comes with Microsoft Windows editions. If you have used the application to encrypt a disk and you have lost the recovery code or password, there is no way to unlock the drive. Formatting the BitLocker-encrypted drive will remove the encryption, but the erased data will still be encrypted and require the password or recovery key to decrypt and recover the lost files [25].

3.4 Cryptographic Algorithms

Data is encrypted and decrypted using mathematical operations known as cryptographic algorithms. Information security and data privacy are protected by these algorithms. Key information regarding Cryptography algorithms is as follows:

- **NIST's Special Publication 800-57:** The NIST, offers recommendations for cryptographic algorithms and key management to protect systems. Their work is based on best practices for protecting sensitive, unclassified information.
- **Post-Quantum Cryptography Standardization:** The development of encryption algorithms resistant to quantum computing device attacks is the aim of the NIST Post-Quantum Cryptography Standard project. Cryptography experts from around the world have submitted algorithms for analysis and evaluation [26].
- **Lightweight Cryptography:** The goal of light weight cryptography (LWC) is to develop cryptographic algorithms that perform effectively in restricted contexts, like the Internet of Things (IoT) [27]. Strict requirements for processing speed, security, and power consumption were met in the creation of these algorithms. Examples of LWC algorithms include PRESENT, AES, and SPECK.

- **Cryptographic Techniques for Authentication, Signatures, and Data Encryption:** Cryptographic algorithms find application in data encryption, digital signatures, and authentication, among other areas [28]. To secure the data, these methods rely on cryptographic keys. Cryptographic keys are associated with machine or individual identities using PKI (public key infrastructure) technologies.
- **Public Nature of Cryptographic Algorithms:** The public aspect of encryption algorithms is viewed favorably in today's environment. Due to their considerable testing and study, standard cryptographic methods are now more dependable and safer [29]. It is generally recommended to use well-established algorithms rather than creating private ones.
- **Common Cryptographic Algorithms:** Some commonly used cryptographic algorithms include:
- **The Advanced Encryption Standard, or AES:** is a symmetric encryption technique that is frequently employed by international governments. Many attacks view AES as secure and unbreakable.
- **The acronym RSA stands for Rivest-Shamir-Adleman:** an asymmetric encryption method for internet communication and data encryption. Email encryption frequently uses the RSA algorithm.

3.5 Symmetric Encryption: Securing Data with Shared Keys

One secret key is used for both encryption and decryption when symmetric encryption is used to protect electronic data. The communication partners in this encryption technique must exchange the secret key to use it in the decoding procedure. Here are some key points about symmetric encryption:

- **Shared Secret Key:** Using a single secret key, symmetric encryption techniques encrypt plaintext and decrypt ciphertext. The communication partners two or more share this key as a shared secret [30].
- **Quickness and Effectiveness:** In general, symmetric encryption techniques are quicker and more effective than asymmetric ones. They work well in situations when there is a need for secure data sharing of substantial volumes [31].
- **Bulk Encryption:** When a large amount of information must be encoded or decrypted quickly, symmetric encryption algorithms are usually used. For safeguarding data in transit or at rest, they are especially helpful [32].
- **Key Exchange:** Secure cryptographic key sharing with all authorized parties is one of the primary problems with symmetric encryption. Only symmetric encryption techniques may make it challenging to establish the shared key. To establish a common key between two parties, asymmetric encryption is therefore frequently employed. [33].
- **Algorithms for Symmetric Encryption:** Advanced Encryption Standards (also called AES and Data Encryption Standards (DES), Three Data Decryption Standards (3DES), the European Data Encrypting Algorithm (IDEA), the TLS/SSL protocol [34].

3.6 Asymmetric Encryption: Public and Private Key Pairs

Asymmetric encryption is a type of cryptography that uses matching pairs of private and public keys. It is also known as cryptographic with a public key. These are some essential details of asymmetric encryption:

- **Crucial Pairs:** In asymmetric encryption, each entity possesses a set of keys, which could be either public or private. While the secret key is kept hidden and available to the key owner alone, the general key is freely shared with everybody.
- **Both Encryption and Decryption:** Only the corresponding private key can be utilized for decryption if a common key was previously used for encryption, and vice versa [35]. Consequently, anyone can encrypt a message using the recipient's public key; however, only the recipient, who possesses the corresponding confidential key, can decode the ciphertext and obtain the original message.
- **Secure Communication:** Asymmetric encryption makes it possible for two parties to communicate securely without exchanging secret keys. If public keys as well as private keys are utilized for symmetric encryption, there is no longer a need to safely transmit a shared secret key.
- **Security and Trust:** The security of an asymmetrical encryption system depends on the private key's secrecy. Distribution of the public key is safe and will not compromise security. This makes communication safe even if an enemy intercepts the public key.
- **Applications:** Applications for asymmetric encryption include file encryption, digital signatures, safe messaging apps, and protecting TLS/SSL communication protocols [36].
- **Examples of Asymmetric Encryption Algorithms:** ElGamal, RSA, Diffie-He, and ECC, or elliptic curve cryptography, are a few common asymmetric encryption methods.

3.7 Hash Functions: Preserving the Integrity of Data

Data integrity is ensured in large part by hash algorithms. The f ollowing are some crucial specifics about hash functions and data integrity protection:

- **Accuracy of Data:** The reliability, consistency, and quality of data are referred to as data integrity. It guarantees that during storage, transport, or processing, data stays unaltered and uncorrupted. By producing a distinct hash value for each collection of data, hashing algorithms are used to confirm the validity of data.
- **Hash Function Operation:** A hash function takes input in the form of data and outputs a fixed-size result known as a digest or scrambled value. Frequently, an alphabetical string of characters is the outcome. Even though a tiny alteration in the input will produce a significantly different hash value, a hash value derived from an identical input remains the same.
- **One-Way Function:** Since hash algorithms are one-way functions, one cannot try to determine the initial input to obtain the hash value. By preventing the hash

value from being used to analyze the original input, this feature guarantees that the security of the data cannot be compromised.

- **Data Verification:** The encrypted version of the initial information is determined and stored to guarantee data integrity [37]. The hash value is generated anew to be compared to the previously stored hashed value each time the data is viewed or delivered later. If the hash values match, then the data was not altered or tampered with.
- **Digital Signatures:** Hashing methods are commonly employed with digital signatures. In this instance, a message is hashed first, and the hash value is signed using the sender's private key [38]. The recipient can verify the integrity of the communication by hashing the contents again and comparing the result against the passphrase of the message that was encoded using the sender's public key.
- **Popular Hash Functions:** SHA-256, CRC32, SHA-1, MD5, and other popular hash algorithms are employed for data integrity checks. Each hash function has its special characteristics and security features [39]. It is crucial to select a hash function following the requirements of the program and the required level of security.

4. Deep Learning

Deep Learning computers apply a type of learning technique that uses many layers of artificial neural networks to extract predictions and information from enormous amounts of data. It has significantly improved the state-of-the-art in a few areas, such as natural language processing, computer vision, and speech recognition [40]. The capacity of computers that use Deep Learning algorithms to acquire knowledge on their own by extracting characteristics from raw data, without needing human feature engineering, is one of the primary distinctions between regular machine instruction and deep training. Deep Learning models can recognize complex patterns and correlations in the data since they are trained on large amounts of labeled data. Multiple layers of connected nodes termed neurons make up deep learning, called deep neural networks. Every neuron processes its inputs by a basic calculation and forwards the outcome to the subsequent layer. While deeper layers pick up higher-level representations, those closer to the input oversee extracting low-level information. Deep Learning models are taught by continually adjusting the neuronal weights and biases to reduce the discrepancy between expected and actual outputs. For this, stochastic descent of gradients and other optimization techniques are frequently employed. Deep Learning models are now trained faster thanks to the availability of sophisticated technology, such as graphic processing units (GPUs).

4.1 Types of Deep Learning

Deep Learning is the umbrella term for a variety of neural network architectures that are applied to complicated problem-solving and data-driven prediction. Here are some of the commonly mentioned types of Deep Learning networks:

- **Multi-Layer Perceptron's (MLP):** The most basic kind of large neural network is an MLP. They are made up of several layers of neurons, which are networked nodes. Each neuron in a layer is connected to all the neurons in other layers. Natural language processing, audio recognition, and picture categorization are among the activities that frequently employ MLPs.
- **Convolutional Neural Networks (CNN):** When processing grid-like data, like photos or time series data, CNNs do especially well [41]. To extract hierarchical representations and local characteristics from the input data, they employ convolutional layers. In applications including object detection, picture segmentation, and image classification, CNNs have demonstrated impressive performance.
- **Recurrent Neural Networks (RNN):** Sequential data, where every one of the data points count, is processed by RNNs by design [42]. They can detect dependencies and trends across time because of their connections, which facilitate the flow of information in cycles. Natural language processing, speech recognition, and time series analysis are just a few of the several fields that regularly use RNNs.
- **Networks with Long Short-Term Memory (LSTM):** Vanishing gradient issues can occur when training deep neural networks. One type of RNN that addresses this problem is LSTMs [43]. Language modeling and speech recognition are two applications that benefit greatly from the long-term dependency-tolerant memory cells found in long-sequence training (LSTMs).
- **Recursive Neural Networks (RvNN):** RvNNs are made to handle structured data, like social networks or parse trees [44]. They can capture hierarchical linkages and dependencies in the data by iteratively applying a single neural network feature to different input segments.
- **Functional Neural Network with Radial Basis:** Based on functional neural networks, radial basis functions are employed as activation functions. They are frequently applied to jobs involving pattern recognition and function approximation [45].

4.2 Major Techniques of Deep Learning

Deep Learning refers to the range of methods used to train neural networks to make predictions based on large volumes of data. The following are some key methods frequently employed in Deep Learning:

- **Supervised Learning:** One popular machine learning technique is called "supervised learning," which entails feeding labeled data into a neural network for training. During this procedure, a training set is created by matching the input data with appropriate target labels [46]. By reducing the discrepancy between its expected and actual labels, the network gains the ability to map the information it receives to the appropriate output labels. Backpropagation, an iterative optimization technique, is used to do this, in which the network modifies its internal weights and parameters. The network gradually gets better at making accurate predictions by running training data through it frequently and changing the parameters.

To enable the network to generalize and make predictions on unlabelled data, supervised training aims to educate the network about the fundamental patterns and correlations in the labeled data. For applications like picture classification, audio acknowledgement, speech processing, and many more, supervised learning is therefore appropriate. Supervised Learning uses labeled data to help the network learn from previous examples and then generalize its understanding to predict outcomes on unknown input.

For supervised learning to be effective, the labeled training information must be of a high enough quality and representativeness. A broad variety of scenarios and variables that the network is likely to face in real-world applications should be covered by the training data. The neural network's architecture which includes the quantity of layers and neurons also has a significant impact on the network's capacity to learn from and generalize from labelled data.

- **Unsupervised Learning:** By retraining synthetic brains on unlabelled data, unsupervised artificial intelligence training is a method of machine learning that enables the network to detect correlations, patterns, and patterns among data without the need for present target labels [47]. This method is frequently used for generative modeling, dimensionality reduction, and clustering. When clustering, the network finds organic groups or clusters in the data by comparing related elements. Reducing the overall number of variables that are used as inputs while retaining the required information is the aim of dimensionality reduction. This improves the display of highly dimensional data and increases the efficacy of subsequent machine learning algorithms. Moreover, generative modeling makes use of unsupervised learning, which helps the network figure out the fundamental structure of the data and produce new, comparable data points. Understanding pertinent structures and patterns in the data enables unsupervised learning to yield informative results and opportunities for a variety of uses.
- **Semi-Supervised Learning:** Aspects of unsupervised and supervised learning are combined in hybrid systems like semi-supervised training. To identify structure and pattern in the data, the neural network blends a greater quantity of labeled data with a lesser quantity of unlabelled input [48]. This method efficiently uses the availability of unlabelled data to increase the model's comprehension of the fundamental information distribution by utilizing it to improve the network's performance or generalization. Semi-supervised learning aims to optimize the learning potential from sparsely labeled data by combining labeled and unlabelled data, hence improving the network's capacity for precise prediction and classification.
- **Transfer Learning:** Using artificial neural networks that were previously trained on massive datasets on a certain task and then tweaking them to do a related though unrelated task is known as transfer learning, and it is an efficient technique [49]. By using this technique, even with less labeled data, the network may be able to perform better in the target task and make greater use of the knowledge it acquired during pre-training. Transfer learning efficiently speeds

up the process of acquiring knowledge for the new task by utilizing the insights from the first training, enabling better generalization and efficiency while handling new data.

- **Reinforcement Learning:** Using the dynamic technique of reinforcement learning, an agent is educated to interact with its surroundings and maximize a reward signal [50]. The agent gains the ability to choose actions that maximize its cumulative reward through its relationships with the environment. Through feedback obtained from its actions, an agent adjusts its plan of action to give priority to actions that result in higher rewards. Deep reinforcement learning allows the agent to effectively process and learn from extremely dense input data, empowering it to take on challenging and multifaceted tasks. It does this by merging deep neural networks with reinforcement learning techniques. This combination makes it possible for the agent to understand intricate patterns and make wise choices, which eventually improves its capacity to maneuver through and perform well in difficult situations.
- **GANs, or Generative Adversarial Networks:** The two primary components of a GAN are generative aggressive networks, also known as discriminator networks, and generator networks [51]. The discriminator network's job is to distinguish real data from bogus, while the network of generators creates synthetic data samples. These networks train in an adversarial fashion the generator network seeks to generate ever more realistic samples, while the discriminator network seeks to sharpen its ability to discern between real and fake data. GANs allow the generator to continuously improve its output through this competitive training process, encouraging the production of synthetic data that closely resembles the real data distribution while also putting the discriminator under pressure to get better at telling real samples from synthetic ones.

4.3 Applications of Deep Learning

Deep Learning has demonstrated its versatility and potential through noteworthy advancements across numerous industries and fields. It has proven useful in the medical field for disease detection, customized therapy suggestions, and medical image analysis. The automotive industry has utilized Deep Learning for autonomous driving systems, contributing to advancements in safety and efficiency. Deep Learning-based algorithms have also been applied to risk assessment, algorithmic trading, and fraud detection in the financial industry. In addition, Deep Learning has enabled breakthroughs in natural language processing, including machine translation, chatbot development, and sentiment analysis. These diverse applications underscore the wide-ranging impact of Deep Learning across different sectors, highlighting its role in driving innovation and improvement [52]. Here are some notable applications of Deep Learning:

- **Virtual Assistants:** Deep Learning algorithms are used by virtual assistants like Alexa, Siri, Google Assistant and more to comprehend and efficiently react to user requests [53]. By applying Deep Learning, these artificially intelligent devices can absorb natural language input, enabling them to appreciate the complexities of human requests and offer correct and appropriate responses. Deep Learning also enables these virtual assistants to continuously enhance their performance by learning from extensive datasets, refining their language comprehension, and improving their ability to discern user intent. Moreover, the use of Deep Learning empowers virtual assistants to adapt to individual user preferences and speech patterns, offering personalized interactions. Through ongoing refinement of their language models, virtual assistants can achieve improved natural language understanding, sentiment analysis, and context-aware responses, showcasing the critical role of Deep Learning in facilitating seamless and intuitive human-computer interactions, and ultimately enhancing user experiences across diverse devices and platforms.

- **Automated Driving:** In the automotive sector, Deep Learning is essential, especially for the advancement of autonomous vehicles. Detecting items like stop signs, intersections, and pedestrians is one of its main uses [54]. Through the utilization of Deep Learning algorithms, autonomous vehicles are equipped with the capability to accurately identify and respond to these critical elements of road infrastructure and pedestrian activity. This innovation greatly advances autonomous driving systems by allowing cars to make decisions in real time depending on items they have spotted. The automobile industry is leading the way in improving transportation safety and efficiency by utilizing Deep Learning toward object detection, which will pave the path for the eventual general adoption of autonomous vehicles.
- **Medical Research:** Deep Learning serves as a valuable tool in medical research, particularly in tasks related to cancer cell detection and analysis. By harnessing Deep Learning algorithms, researchers can automate the identification of cancer cells, leading to significant advancements in diagnostic accuracy. With this t echnology, complex systems that can accurately and precisely analyze medical images and detect minute signs of malignant cells can be developed. Through the application of Deep Learning, medical professionals can streamline the process of cancer detection, leading to earlier diagnoses and more effective treatment strategies. Deep Learning in medicine has the potential to transform the area of oncology by improving patient outcomes and enabling more individualized treatment plans, in addition to improving the accuracy of cancer cell detection.
- **Industrial Automation:** By automatically identifying people or items that are too close to heavy machinery, Deep Learning Technology helps to improve worker safety near such machines. Through the application of Deep Learning Algorithms, systems can effectively identify and alert operators when potential hazards are detected, thereby mitigating the risk of accidents and injuries in industrial settings. This proactive strategy for safety reduces the possibility of mishaps at work and creates a safe atmosphere for employees, which eventually improves operational effectiveness and well-being. The integration of Deep Learning for safety enhancement underscores its pivotal role in revolutionizing workplace safety protocols and aligning industrial practices with cutting-edge technological advancements.

- **Visual Recognition:** Applications utilizing Deep Learning as a core technique enable precise item identification and classification within photos. Applications for this capacity are found in many different domains, such as image sorting and facial recognition. Visual detection systems can recognize complex patterns and characteristics in images by using Deep Learning Algorithms [55]. This allows the computers to classify and interpret visual input with extremely high accuracy. The use of Deep Learning to face recognition facilitates the development of trustworthy systems that can identify individuals based on their facial characteristics, hence enhancing security, authentication, and personalized user experiences. Moreover, Deep Learning algorithms are essential for automating the classification and arrangement of visual data in the image -sorting domain. This helps to streamline processes in various industries, including digital asset management, content moderation, and e-commerce. The widespread adoption of Deep Learning for visual recognition underscores its transformative impact on diverse sectors, driving innovation and efficiency in image analysis and interpretation.
- **NLP, or N atural L anguage P rocessing:** Deep Learning is a key component of the processing of natural language or NLP, applications because it allows computers to comprehend and respond to the nuances of human speech [56]. Wide-ranging uses of this revolutionary technology include sentiment analysis, language translation, chatbot building, and many other linguistic tasks. By leveraging Deep Learning Algorithms, NLP systems can effectively interpret the complexities of human language, enabling them to engage in meaningful and contextually relevant interactions with users. In the realm of chatbots, Deep Learning facilitates the creation of intelligent conversational agents capable of understanding and responding to natural language inputs, thereby enhancing user experiences across various domains. Additionally, Deep Learning plays a fundamental role in language translation, empowering systems to accurately translate text between different languages, and fostering global communication and collaboration. Furthermore, in the domain of sentiment analysis, Deep Learning Algorithms enable the automated extraction of sentiment and emotion from textual data, providing valuable insights for businesses and organizations. The pervasive influence of Deep Learning in NLP applications underscores its significance in driving advancements in language understanding, communication, and human-machine interaction.
- **Entertainment Industry:** Deep Learning serves as a cornerstone technology in the entertainment industry, playing a pivotal role in delivering personalized content recommendations on popular platforms such as Netflix and Amazon Prime. Utilizing algorithms that use Deep Learning, these systems are capable of analyzing past data, viewing behaviors, and user preferences to provide personalized content recommendations that improve user satisfaction and engagement. Moreover, Deep Learning facilitates video analysis, enabling the extraction of valuable insights from visual content, such as scene detection, object recognition, and content categorization. This feature improves production workflows and raises the general effectiveness of content development in the

entertainment industry by streamlining content editing procedures and aiding in the automated process of content creation. Deep Learning's incorporation into entertainment highlights how much it has shaped content curation, individualized user experiences, and the growth of media consumption.

- **Manufacturing:** Deep Learning has proven to be extremely beneficial for the manufacturing industry, particularly for applications such as process optimization, identifying problems, maintenance planning, and quality control. Deep Learning Algorithms can be trained to analyze sensor or image data from production lines, enabling the real-time identification of anomalies or flaws in control systems [57]. This automation reduces human error and improves overall product quality. Predictive maintenance is another area where Deep Learning shines, as algorithms can analyze sensor data from machines to predict when maintenance is needed, helping manufacturers avoid unexpected breakdowns and optimize maintenance schedules. Machine learning algorithms can be trained to recognize anomalous patterns in production, sensor, and sales data. This allows for the early diagnosis of irregularities and the avoidance of expensive failures. Anomaly detection is extremely crucial in the manufacturing industry. Lastly, Deep Learning can optimize production processes by analyzing large datasets, including historical production data, sensor data, and environmental factors. By identifying patterns and correlations, Deep Learning models can optimize energy consumption, reduce waste, and improve scheduling, resulting in increased efficiency and cost savings. Overall, Deep Learning Applications in manufacturing contribute to improved product quality, reduced downtime, and enhanced process efficiency.
- **Financial Services:** Deep Learning is extensively used in manufacturing for tasks like quality control, predictive maintenance, and anomaly detection. It helps optimize production processes, improve efficiency, and enhance product reliability. By analyzing enormous amounts of data, machine learning algorithms can identify unusual trends, predict equipment failures, and identify issues in real-time. Manufacturers may limit downtime, cut waste, and improve overall operational efficiency with this proactive strategy.
- **Aerospace and Defence:** Deep Learning is deployed in aerospace and defense for tasks such as object identification in satellite imagery and identifying safe or unsafe zones for troops [58]. It enables efficient analysis of satellite data, aids in surveillance and reconnaissance, and helps make informed decisions about troop movements. Deep Learning's speed and adaptability offer advantages, but careful evaluation and validation are necessary. Overall, Deep Learning enhances situational awareness and improves operational effectiveness in these industries.

5. Conclusion

The integration of Deep Learning and Cryptography in Smart Environments has the potential to significantly change how we interact with the world around us. In smart settings, Cryptography is essential to guaranteeing sensitive data confidentiality and privacy. Smart Environments can improve overall security by preventing unauthorized access and tampering with data by utilizing encryption techniques. Furthermore, Smart Environments can offer more efficient and customized experiences because of the Deep Learning Algorithms' ability to analyze and comprehend human preferences, behaviors, and patterns. Deep Learning Algorithms can adapt and modify many ambient elements, such as lighting, temperature, and settings, based on user interactions and personal preferences. This degree of personalization raises user pleasure while simultaneously improving Smart Environments' general effectiveness and performance. The integration of Deep Learning and encryption in Smart Environments holds the potential to completely change how we interact with the environment around us by giving users more security, privacy, and tailored experiences. The combination of Deep Learning and encryption will be crucial in determining how Smart Environments develop in the future as technology progresses. From smart homes to intelligent cities, the possibilities are vast, offering a glimpse into a world where our surroundings seamlessly cater to our needs while safeguarding our information.

The potential for transforming the way we communicate with the surrounding environment is enormous when encryption and Deep Learning come together in Smart Environments. Smart Environments can guarantee sensitive data privacy and security by utilizing encryption techniques to prevent unauthorized access and modification. These environments can also learn from customer tastes and behaviors and the incorporation of Deep Learning, which creates more efficient and customized experiences. This potent combination not only improves the general security more privacy of smart settings but also makes it possible for them to modify and adapt to specific demands. As a result, it eventually changes how they interact with the world around us and opens the door for an intelligent and secure future.

References

1. Sarker, I. H. 2021. Data science and analytics: an overview from data-driven smart computing, decision-making and applications perspective. *SN Computer Science*, *2*(5), 377.
2. Yang, P., Xiong, N., and Ren, J. 2020. Data security and privacy protection for cloud storage: A survey. *IEEE Access*, *8*, 131723–131740.
3. Saif, S., Biswas, S., and Chattopadhyay, S. 2020. Intelligent, secure big health data management using deep learning and blockchain technology: an overview. *Deep Learning Techniques for Biomedical and Health Informatics*, 187–209.
4. Bae, H., Jang, J., Jung, D., Jang, H., Ha, H., Lee, H., and Yoon, S. 2018. Security and privacy issues in deep learning. *arXiv preprint arXiv:1807.11655*.
5. Jabbar, W. A., Kian, T. K., Ramli, R. M., Zubir, S. N., Zamrizaman, N. S., Balfaqih, M.,... and Alharbi, S. 2019. Design and fabrication of smart home with internet of things enabled automation system. *IEEE access*, *7*, 144059–144074.
6. Syed, L., Jabeen, S., Manimala, S., and Alsaeedi, A. 2019. Smart healthcare framework for ambient assisted living using IoMT and big data analytics techniques. *Future Generation Computer Systems*, *101*, 136–151.
7. Lăzăroiu, G., Andronie, M., Iatagan, M., Geamănu, M., Ştefănescu, R., and Dijmărescu, I. 2022. Deep learning-assisted smart process planning, robotic wireless sensor networks, and geospatial big data management algorithms in the internet of manufacturing things. *ISPRS International Journal of Geo-Information*, *11*(5), 277.
8. Al Mamun, M. A., and Yuce, M. R. 2019. Sensors and systems for wearable environmental monitoring toward IoT-enabled applications: A review. *IEEE Sensors Journal*, *19*(18), 7771–7788.

9. Andronie, M., Lăzăroiu, G., Iatagan, M., Uță, C., Ștefănescu, R., and Cocoșatu, M. 2021. Artificial intelligence-based decision-making algorithms, internet of things sensing networks, and deep learning-assisted smart process management in cyber-physical production systems. *Electronics*, *10*(20), 2497.
10. Alberti, A. M., Santos, M. A., Souza, R., Da Silva, H. D. L., Carneiro, J. R., Figueiredo, V. A. C., and Rodrigues, J. J. 2019. Platforms for smart environments and future Internet design: A survey. *IEEE Access*, *7*, 165748–165778.
11. Sharma, D. K., Singh, N. C., Noola, D. A., Doss, A. N., and Sivakumar, J. 2022. A review on various cryptographic techniques & algorithms. *Materials Today: Proceedings*, *51*, 104–109.
12. Duggineni, S. (2023). Impact of controls on data integrity and information systems. *Science and Technology*, *13*(2), 29–35.
13. Fang, W., Chen, W., Zhang, W., Pei, J., Gao, W., and Wang, G. 2020. Digital signature scheme for information non-repudiation in blockchain: a state of the art review. *EURASIP Journal on Wireless Communications and Networking*, *2020*(1), 1–15.
14. Fang, W., Chen, W., Zhang, W., Pei, J., Gao, W., and Wang, G. 2020. Digital signature scheme for information non-repudiation in blockchain: a state of the art review. *EURASIP Journal on Wireless Communications and Networking*, *2020*(1), 1–15.
15. Limniotis, K. 2021. Cryptography as the Means to Protect Fundamental Human Rights. *Cryptography*, *5*(4), 34.
16. Ahamed, B. B., and Krishnamoorthy, M. 2020. SMS encryption and decryption using modified vigenere cipher algorithm. *Journal of the Operations Research Society of China*, 1–14.
17. Kumar, S., Gaur, M. S., Sharma, P. S., and Munjal, D. 2021, April. A novel approach of symmetric key cryptography. In *2021 2nd International Conference on Intelligent Engineering and Management (ICIEM)* (pp. 593–598). IEEE.
18. Jyothi, V. E., Prasad, B. D. C. N., and Mojjada, R. K. 2020, December. Analysis of cryptography encryption for network security. In *IOP Conference Series: Materials Science and Engineering* (Vol. 981, No. 2, p. 022028). IOP Publishing.
19. National Academies of Sciences, Engineering, and Medicine. 2018. *Decrypting the encryption debate: A framework for decision makers*. National Academies Press.
20. Shetty, V. S., Anusha, R., MJ, D. K., and Hegde, P. 2020, February. A survey on performance analysis of block cipher algorithms. In *2020 International Conference on Inventive Computation Technologies (ICICT)* (pp. 167–174). IEEE.
21. Domingo-Ferrer, J., Farras, O., Ribes-González, J., and Sánchez, D. 2019. Privacy-preserving cloud computing on sensitive data: A survey of methods, products and challenges. *Computer Communications*, *140*, 38–60.
22. Findlay, B. 2024. Techniques and methods for obtaining access to data protected by linux-based encryption–A reference guide for practitioners. *Forensic Science International: Digital Investigation*, *48*, 301662.
23. de Guise, P. 2023. *The Busy IT Manager's Guide to Data Recovery*. CRC Press.
24. Meijer, C., and Van Gastel, B. 2019, May. Self-encrypting deception: weaknesses in the encryption of solid state drives. In *2019 IEEE Symposium on Security and Privacy (SP)* (pp. 72–87). IEEE.
25. Etow, T. R. 2020. Impact Of Anti-Forensics Techniques on Digital Forensics Investigation.
26. Alagic, G., Apon, D., Cooper, D., Dang, Q., Dang, T., Kelsey, J.,... and Liu, Y. K. 2022. Status report on the third round of the NIST post-quantum cryptography standardization process. *US Department of Commerce, NIST*.
27. Kumar, M. 2022. Post-quantum cryptography Algorithm's standardization and performance analysis. *Array*, *15*, 100242.
28. Khalifa, M., Algarni, F., Khan, M. A., Ullah, A., and Aloufi, K. 2021. A lightweight cryptography (LWC) framework to secure memory heap in Internet of Things. *Alexandria Engineering Journal*, *60*(1), 1489–1497.
29. Qin, Y., and Zhang, B. 2023. Privacy-preserving biometrics image encryption and digital signature technique using Arnold and ElGamal. *Applied Sciences*, *13*(14), 8117.
30. Thabit, F., Can, O., Wani, R. U. Z., Qasem, M. A., Thorat, S. B., and Alkhzaimi, H. A. 2023. Data security techniques in cloud computing based on machine learning algorithms and cryptographic algorithms: Lightweight algorithms and genetics algorithms. *Concurrency and Computation: Practice and Experience*, e7691.

31. Mousavi, S. K., Ghaffari, A., Besharat, S., and Afshari, H. 2021. Security of internet of things based on cryptographic algorithms: a survey. *Wireless Networks*, *27*, 1515–1555.
32. Haas, M., & Yarhi-Milo, K. (2020). To disclose or deceive? Sharing secret information between aligned states. *International Security*, *45*(3), 122–161.
33. Seth, B., Dalal, S., Jaglan, V., Le, D. N., Mohan, S., and Srivastava, G. 2022. Integrating encryption techniques for secure data storage in the cloud. *Transactions on Emerging Telecommunications Technologies*, *33*(4), e4108.
34. Sunday, A. E., and Olufunminiyi, O. E. 2023. An Efficient Data Protection for Cloud Storage Through Encryption. *International Journal of Advanced Networking and Applications*, *14*(5), 5609–5618.
35. Halak, B., Yilmaz, Y., and Shiu, D. 2022. Comparative analysis of energy costs of asymmetric vs symmetric encryption-based security applications. *IEEE Access*, *10*, 76707–76719.
36. Kadri, A. F., Arowolo, M. O., Yusuf-Asaju, A. W., Tajudeen, K. O., and Gbolagade, K. A. 2022. An Overview of Enhancing Encryption Standards for Multimedia in Explainable Artificial Intelligence Using Residue Number Systems for Security. *A Roadmap for Enabling Industry 4.0 by Artificial Intelligence*, 247–273.
37. Al-Shabi, M. A. 2019. A survey on symmetric and asymmetric cryptography algorithms in information security. *International Journal of Scientific and Research Publications (IJSRP)*, *9*(3), 576–589.
38. Braeken, A. 2022. Public key versus symmetric key cryptography in client–server authentication protocols. *International Journal of Information Security*, *21*(1), 103–114.
39. Ateniese, G., Francati, D., Nuñez, D., and Venturi, D. 2021. Match me if you can: Matchmaking encryption and its applications. *Journal of Cryptology*, *34*, 1–50.
40. Banoth, R., and Regar, R. 2023. Asymmetric Key Cryptography. In *Classical and Modern Cryptography for Beginners* (pp. 109–165). Cham: Springer Nature Switzerland.
41. Wei, P., Wang, D., Zhao, Y., Tyagi, S. K. S., and Kumar, N. 2020. Blockchain data-based cloud data integrity protection mechanism. *Future Generation Computer Systems*, *102*, 902–911.
42. Cohn-Gordon, K., Cremers, C., Dowling, B., Garratt, L., and Stebila, D. 2020. A formal security analysis of the signal messaging protocol. *Journal of Cryptology*, *33*, 1914–1983.
43. Kairaldeen, A. R., Abdullah, N. F., Abu-Samah, A., and Nordin, R. 2021. Data integrity time optimization of a blockchain IoT smart home network using different consensus and hash algorithms. *Wireless Communications and Mobile Computing*, *2021*, 1–23.
44. Dibia, V., and Demiralp, Ç. 2019. Data2vis: Automatic generation of data visualizations using sequence-to-sequence recurrent neural networks. *IEEE computer graphics and applications*, *39*(5), 33–46.
45. Ribeiro, A. H., Tiels, K., Aguirre, L. A., and Schön, T. 2020, June. Beyond exploding and vanishing gradients: analysing RNN training using attractors and smoothness. In *International conference on artificial intelligence and statistics* (pp. 2370–2380). PMLR.
46. Ma, J., Gao, W., Joty, S., and Wong, K. F. 2020. An attention-based rumor detection model with tree-structured recursive neural networks. *ACM Transactions on Intelligent Systems and Technology (TIST)*, *11*(4), 1–28.
47. Hryniowski, A., and Wong, A. 2019. DeepLABNet: End-to-end learning of deep radial basis networks with fully learnable basis functions. *arXiv preprint arXiv:1911.09257*.
48. Mahadevkar, S. V., Khemani, B., Patil, S., Kotecha, K., Vora, D. R., Abraham, A., and Gabralla, L. A. 2022. A review on machine learning styles in computer vision—Techniques and future directions. *Ieee Access*, *10*, 107293–107329.
49. Taha, K. 2023. Semi-supervised and un-supervised clustering: A review and experimental evaluation. *Information Systems*, 102178.
50. Ahmad, M., Abdullah, M., Moon, H., and Han, D. 2021. Plant disease detection in imbalanced datasets using efficient convolutional neural networks with stepwise transfer learning. *IEEE Access*, *9*, 140565–140580.
51. Shrestha, A., and Mahmood, A. 2019. Review of deep learning algorithms and architectures. *IEEE access*, *7*, 53040–53065.
52. Natale, S. 2020. To believe in Siri: A critical analysis of AI voice assistants.

53. Combs, T. S., Sandt, L. S., Clamann, M. P., and McDonald, N. C. 2019. Automated vehicles and pedestrian safety: exploring the promise and limits of pedestrian detection. *American journal of preventive medicine*, *56*(1), 1–7.
54. Bhinder, B., Gilvary, C., Madhukar, N. S., and Elemento, O. 2021. Artificial intelligence in cancer research and precision medicine. *Cancer discovery*, *11*(4), 900–915.
55. Pishgar, M., Issa, S. F., Sietsema, M., Pratap, P., and Darabi, H. 2021. REDECA: a novel framework to review artificial intelligence and its applications in occupational safety and health. *International journal of environmental research and public health*, *18*(13), 6705.
56. Elgendy, M. 2020. *Deep learning for vision systems*. Simon and Schuster.
57. Hagiwara, M. 2021. *Real-world natural language processing: practical applications with deep learning*. Simon and Schuster.
58. Murthy, C. B., Hashmi, M. F., Bokde, N. D., and Geem, Z. W. 2020 . Investigations of object detection in images/videos using various deep learning techniques and embedded platforms—A comprehensive review. *Applied sciences*, *10*(9), 3280.
59. Rawat, A., Suryavanshi, R. S., Nagar, V. and Yadav, D. (2023, November). Development of Blockchain-Enabled E-Healthcare Appointment System. In International Conference on Trends in Computational and Cognitive Engineering (pp. 359–369). Singapore: Springer Nature Singapore.

Index

S

T

V

Kolibri 360 GmbH
Ferdinand-Jühlke-Straße 7
99095 Erfurt, Deutschland
produktsicherheit@kolibri360.de